Statistical Design and Analysis of Experiments

SIAM's Classics in Applied Mathematics series consists of books that were previously allowed to go out of print. These books are republished by SIAM as a professional service because they continue to be important resources for mathematical scientists.

Classics in Applied Mathematics

C. C. Lin and L. A. Segel, *Mathematics Applied to Deterministic Problems in the Natural Sciences*

Johan G. F. Belinfante and Bernard Kolman, *A Survey of Lie Groups and Lie Algebras with Applications and Computational Methods*

James M. Ortega, *Numerical Analysis: A Second Course*

Anthony V. Fiacco and Garth P. McCormick, *Nonlinear Programming: Sequential Unconstrained Minimization Techniques*

F. H. Clarke, *Optimization and Nonsmooth Analysis*

George F. Carrier and Carl E. Pearson, *Ordinary Differential Equations*

Leo Breiman, *Probability*

R. Bellman and G. M. Wing, *An Introduction to Invariant Imbedding*

Abraham Berman and Robert J. Plemmons, *Nonnegative Matrices in the Mathematical Sciences*

Olvi L. Mangasarian, *Nonlinear Programming*

*Carl Friedrich Gauss, *Theory of the Combination of Observations Least Subject to Errors: Part One, Part Two, Supplement.* Translated by G. W. Stewart

Richard Bellman, *Introduction to Matrix Analysis*

U. M. Ascher, R. M. M. Mattheij, and R. D. Russell, *Numerical Solution of Boundary Value Problems for Ordinary Differential Equations*

K. E. Brenan, S. L. Campbell, and L. R. Petzold, *Numerical Solution of Initial-Value Problems in Differential-Algebraic Equations*

Charles L. Lawson and Richard J. Hanson, *Solving Least Squares Problems*

J. E. Dennis, Jr. and Robert B. Schnabel, *Numerical Methods for Unconstrained Optimization and Nonlinear Equations*

Richard E. Barlow and Frank Proschan, *Mathematical Theory of Reliability*

Cornelius Lanczos, *Linear Differential Operators*

Richard Bellman, *Introduction to Matrix Analysis, Second Edition*

Beresford N. Parlett, *The Symmetric Eigenvalue Problem*

Richard Haberman, *Mathematical Models: Mechanical Vibrations, Population Dynamics, and Traffic Flow*

Peter W. M. John, *Statistical Design and Analysis of Experiments*

*First time in print.

Statistical Design and Analysis of Experiments

Peter W. M. John
University of Texas at Austin
Austin, Texas

Society for Industrial and Applied Mathematics
Philadelphia

This SIAM edition is an unabridged republication of the work first published by The Macmillan Company, New York, 1971.

10 9 8 7 6 5 4 3 2 1

Library of Congress Cataloging-in-Publication Data

John, Peter William Meredith.
Statistical design and analysis of experiments / Peter W.M. John.
p. cm. -- (Classics in applied mathematics ; 22)
"An unabridged republication of the work first published by the Macmillan Company, New York, 1971"--T.p. verso.
Includes bibliographical references (p. -) and index.
ISBN 0-89871-427-3 (softcover)
1. Experimental design. I. Title. II. Series.
QA279.J65 1998 98-29392
001.4'34--dc21

Contents

Preface

This book, about the design of experiments, is for the mathematically oriented reader. It has grown primarily from courses that I have given during the past decade to first- and second-year graduate students in the statistics department at Berkeley and in the mathematics departments at the Davis campus of the University of California and at the University of Texas.

My interest in experimental design began in 1957 when I became a research statistician for what is now Chevron Research Company (Standard Oil of California). I had the particular good fortune to have the benefit of Henry Scheffé as a consultant there, and subsequently as a colleague and friend. That was the time when Scheffé's book on analysis of variance was just about to appear, when Box and Hunter were introducing the chemical engineers to response surface designs, and when engineers were becoming interested in 2^n designs. At Davis I was able to teach a statistical methods course to graduate students in agricultural sciences, and at the same time I became interested in the combinatorial problems of the construction of incomplete block designs.

The book reflects all these phases. Although the treatment is essentially mathematical, I have endeavored throughout to include enough examples from both engineering and agricultural experiments to give the reader the flavor of the practical aspects of the subject.

The main mathematical prerequisite is a working knowledge of matrix algebra and an introductory mathematical statistics course. Unfortunately, courses on matrices nowadays often seem to die before reaching quadratic forms, so I have included an appendix dealing with them. There is no need for a knowledge of measure theory, and the reader who is prepared to accept Cochran's theorem on faith will need no calculus at all.

From a practical standpoint, many readers will find the second chapter more difficult than those that follow it. Unfortunately, this can hardly be avoided, whether one uses Cochran's theorem or Scheffé's geometric approach.

The choice of topics for a one-year course will vary with the interests of the instructor and the mathematical abilities of the students. The first twelve chapters are of interest to researchers and form the basis of a one-year course, with the chapter on Latin squares coming either at the end of the first semester or at the beginning of the second. This allows the first semester to be devoted to analysis of variance and the complete factorial experiment. The second semester then contains 2^n and 3^n designs, fractional factorials, response surfaces, and incomplete block designs.

The last three chapters call for considerably more mathematical maturity than the others, and it is not necessary that the practitioner of experimental design read them. They take the reader to the frontier of research in the construction of incomplete block designs, and a course for graduate students with a theoretical emphasis should include some of these topics. Alternatively, I have sometimes used these chapters as the basis for an advanced seminar.

I am greatly indebted to my wife, Elizabeth, for her continued help and encouragement.

Austin, Texas P. W. M. J.

Preface to the Classics Edition

1. Introduction

In the 27 years since the initial publication of this book, much has changed in both application and theory of the design of experiments. The prime catalysts have been the personal computer (PC) and its increasingly sophisticated software, and the concomitant burgeoning of experimental design in industry. There are key debates over fixed, mixed, and random models in the analysis of variance and over the theory of optimal designs. This second release gives me a welcome opportunity to comment about these topics and how they might have changed this book had it been written today.

2. Computers

When I signed my contract with Macmillan in 1966, some, but by no means all, statisticians had access to computers, which were big mainframes in computer centers. One took one's problem to the center, together with the necessary Fortran cards to suit the software package, and gave it to an employee to have the data keypunched; the only package to which I can now recall having ready access was the Biomed series. The whole thing was handled through the bureaucracy of the computer center, who usually charged for their services. Some statisticians in industry had to send their data away to the computer at company headquarters. So most of us did what we ourselves could do on our desk calculators. Who of my generation does not recall the noise of the Friden desktop machine and such tricks as using a machine with a large register so that one could accumulate x^2 on the left, $2xy$ in the center, and y^2 on the right?

Three decades later, we each have PCs of our own with (relatively) user-friendly software packages that will do, in a few seconds, analyses that used to take hours, and without the problems of making a charge to a budget. Not only can we do many things that we did not have the computer power to do, we can do them while sitting at our own desks and at our own pace. I have not carried a job to the computer center in years. This has had an enormous influence on statistics in several ways. Obviously, we can perform more complex calculations than before and do our analyses more quickly, but there is more.

The "more" is manifested in a change of attitude. Graphics are so good and so speedy that there is no excuse for a scientist not to plot his or her data and to spot obvious outliers or nonnormality. One can vet one's data as a matter of routine. More people can look at their own data and become interested in statistical analysis. They can display their analyses in presentations to colleagues in tables and charts. This can be a mixed blessing, however; someone who does not know what he or she is doing can do a lot of damage. Turning a graduate student loose on a data set armed with a computer package and manual that are new to him or her (and with little knowledge of statistics) is a passport to disaster.

I will not take sides and recommend any particular computer package to the reader, except to say that I use Minitab. I use Minitab for two reasons. The first is that I started my statistical computing with Minitab's father, Omnitab, many years ago, and I am too set in my ways to change, even though I do not question that there are other packages that may do some things better. The second is that I find it a very easy package to use for routine teaching purposes. It is very user friendly, and my students, some of whom are not strong on computer skills, can learn to do many standard things very easily. The package does things at the click of a mouse that I would never even have dreamed of attempting 30 years ago. I have not had to take time to teach the abbreviated Doolittle method of inverting symmetric matrices for so long that I have probably forgotten it.

One impact of the PC revolution is that, if this book were to be rewritten, I would leave some sections out or radically alter them. I particularly have in mind Section 3.14 on analysis of covariance and Section 6.7 on change-over designs with residual effects. The main thrust of these sections was to develop formulae to make the analysis systematic, and, therefore, relatively easy on a desk calculator. Today's reader does not need those formulae and should be spared the pain of reading about them. The missing plot formulae of Section 3.13 and the first exercise in Chapter 6 are obsolescent, if not actually obsolete. Modern computer routines make the calculations for such cases easy.

Readers whose interests in computing go beyond the scope of this book may be interested, for example, in such topics as programs to obtain solutions to the equations that arise when the method of maximum likelihood is used to find estimates of components of variance. I shall mention later the use of computers to find D-optimal experimental designs.

3. The growth of experimental design in industry

In 1971, the vast majority of applications of analysis of variance and design of experiments were still in agriculture. There had been relatively little impact in industry in the United States and Europe outside Imperial Chemical Industries Ltd. in Great Britain, where George Box and his colleagues did pioneering work, and in those oil and chemical companies whose members had learned about 2^n factorials and response surfaces from Box, often at Gordon Research Conferences. Other engineers had little opportunity to learn about that approach to experimentation. After all, Yates's (1937) monograph was published by the Imperial Bureau of Soil Science and dealt with agricultural experiments, and there was little communication between aggies and engineers.

But in the 1980s, the growth of interest in experimental design accelerated as the emerging semiconductor industry became conscious of Japan's industrial success. Much of that success was attributed to the use of statistics that William Edwards Deming had advocated in Japan after World War II and, in particular, to the resulting emphasis on designed experiments. The Japanese engineer, G. Taguchi, who had learned about 2^n designs from Mahalanobis in India, introduced to the semiconductor industry his methods for small complete factorials and orthogonal fractions. He was primarily concerned with main effects and paid little attention to such complications as interactions. Taguchi's method reduces design of experiments to a relatively simple drill, to which some companies require strict adherence. That has some merit, as long as it works, but it does not lead to a very deep understanding. However, Taguchi went further. He made us think not just about experiments to find conditions that optimize yield, but also about experiments to control the variability of a manufacturing process.

At about the same time, the advent of the PC made the requisite calculations much easier. We no longer had to use Yates's algorithm to analyze 2^n factorials. It became easier to analyze them, or their fractions, by regression, especially since the I.C.I. group had persuaded us to think of the levels of the factors as -1, $+1$ instead of 0, 1, as Yates had used. Running nonorthogonal fractions was no longer so daunting because we could actually analyze the data when it came in! The journal *Technometrics*, which had first appeared in 1959 and to which there are numerous references in the 1971 edition of this book, thrived, publishing numerous papers on design of experiments.

Engineers came to realize that they no longer had to associate designed experiments solely with the large complete factorial experiments, planted in the spring and gathered in the fall, that were the tradition of the agronomists. They could think of series of small experiments: running an exploratory fractional experiment and then, after seeing the results, running further points to answer the questions that the analysis of the first experiment raised.

Performing a set of small experiments in sequence quite logically leads one to consider running the points of an individual experiment in a planned sequence. Often they will be run one after another anyway, which itself implies a sequence. Why not arrange the order so that the estimates of the effects will be orthogonal to time trends? Cox (1951) was the first to write about this. Hill (1960), Daniel and Wilcoxon (1966), Steinberg (1988), and John (1990) all wrote about the topic in *Technometrics*. Dickinson (1974) wrote about sequences from another point of view: How can we order the points so as to minimize the cost of all the changes in factor levels that a factorial experiment requires? Another relevant question is how to order the points so that early termination will still leave us with a viable, though incomplete, set of data? (Section 8.14 touches upon the last of these questions.)

4. Fixed, mixed, and random models

In 1971, the question of the expected values of the mean squares in the analysis of factorial experiments seemed to have been settled by Scheffé (1956). In Section 4.7, I gave an account of Scheffé's method of analyzing the two-factor experiment in which one factor, A, is fixed and the other, B, is random. One tests the mean square for A against the mean square for interaction; on the other hand, one tests the mean square for B against the mean square for error. An essential feature of Scheffé's model is that, since A is fixed, we should impose the side condition $\sum_i (\alpha\beta)_{ij} = 0$ for each j. In deriving $E(S_B)$ we can argue that we sum (average) the interactions over all the levels of the fixed factor A; hence, they "cancel out," and no interaction term appears in $E(S_B)$.

Hocking (1973) introduced an alternative model in which the interactions were freed of those side conditions. In Hocking's model the interaction term appears in both $E(S_A)$ and $E(S_B)$, and so M_A and M_B are each tested against M_{AB}. Scheffé's model has come to be called the restricted model and Hocking's the unrestricted model. Practitioners are given a difficult choice. Each model has a logical basis, and the differences are subtle; each model can be defended and each has its advocates. I had no difficulty in 1971. Hocking had not yet written his paper. The only approach of which I was aware was Scheffé's, and so that was the only one that I described.

In Section 5.3 I gave a set of rules for computing the expected values of the mean squares (EMS) in a crossed classification for three factors (using Scheffé's model). One writes down the EMS as if all three factors were random, and then one strikes out certain interaction terms that involve fixed effects, following rule 4. The reader should now call this rule 4r (to correspond to the restricted rule) and add the following rule 4u (if he or she wishes to use the unrestricted rule).

4.1. Rule 4u

In the unrestricted rule, an interaction is called fixed if and only if all its factors are fixed. Otherwise, it is random. In the expected mean square for any effect, strike out any interaction that is fixed and leave the others. Do not strike out an interaction from its own EMS.

In the case of three factors with A random and B and C fixed, the only fixed interaction is BC. We strike it out of the expectations of the mean squares for B and C, but we do not strike it out in the line for BC itself. Because A is random, we do not strike out the ABC interaction from any mean square.

The restricted rule is much more liberal than the unrestricted rule in striking out terms.

These distinctions also apply to the general balanced design in which some factors are crossed and others are nested. We can still begin by following the first four steps of the procedure of Bennett and Franklin that is described in Section 5.6.

Step 5 should be revised as follows.

Write the EMS, assuming that all the factors are random and following the rules that (i) each line of the EMS table contains all the terms that have as subscripts all the letters in the title of the line, and (ii) the coefficient of any component is the product of the levels of all the factors that correspond to absent subscripts. Then strike out terms using step 5r or step 5u.

4.2. Step 5r (the restricted rule)

If some of the factors are fixed, we strike out of the expectation of any mean square all those terms that contain in their *live* subscripts any letters (other than those in the name of the mean square itself) that correspond to fixed effects.

4.3. Step 5u (the unrestricted rule)

We strike out an interaction term *only* if *all* its *live* subscripts belong to fixed effects.

5. Optimal designs

Kiefer (1959) presented to the Royal Statistical Society a paper about his work on the theory of optimal designs. How do we choose between one design for p factors in n points and another? How do we find the best design? This book mentions the criterion of minimizing the average variance of the estimates of the effects, which came to be known as A-optimality or trace-optimality.

Kiefer spoke of D-optimality, choosing the design that maximized the determinant, D, of **X'X**, where **X** is the design matrix. How does one find such a design?

The factors can be discrete or continuous. Users of 2^n factorials have the discrete case with $x_i = \pm 1$. Mitchell (1974) published his DETMAX algorithm, enabling us to find a design that maximizes D when the experimental space has a finite number of points. His algorithm has since been improved and is now available in some standard software packages, so that an engineer has only to type in the values of n and p and the model (with or without interactions), and out comes a design.

It is all very easy, but there are questions in addition to the obvious possibility of local maxima. The criterion is very sensitive to the model. Suppose that the model has q terms. If you drop one of the terms, a design that was optimal for q terms in n runs is not necessarily optimal for $q - 1$ terms in n runs, nor does an optimal design for q factors in n points necessarily contain a subset that is a D-optimal design for $n - 1$ points.

Also, the solution in any case may not be unique. If one permutes the columns of **X**, or if one multiplies any column by minus one, which corresponds to changing the level of that factor at every point, D is unchanged. One can call such solutions isomorphic to one another. But there are non-isomorphic solutions in some cases. In the small problem of fitting main effects and two-factor interactions for a 2^4 factorial in 12 points, there are 4 families of nonisomorphic solutions, all with the same maximum value of D, and one of them is a foldover design! Furthermore, none of them is an optimal plan for main effects only. In this case, an optimal design is given by any 4 columns of the Plackett and Burman design for 11 factors in 12 runs.

The continuous case is beyond the scope of this book. Its mathematical demands go further into measure theory and functional analysis than most of our readers wish. It has become a flourishing field of research among the more mathematically oriented statisticians. A survey of the field written by Atkinson (1982) has more than 100 references, including books by Fedorov (1972) and Silvey (1980). A follow-up survey by the same author in 1988 has many more.

6. Incomplete block designs

I have written of some areas that have grown in interest since 1971. Sadly, I must comment on an area which has quietly disappeared from the radar screen of experimental design: incomplete block designs. In 1971, it was an important topic, and it still provides opportunity for some interesting and elegant mathematics. I have enjoyed working on some of its problems over the years, but I must confess that it is an obsolescent topic, a victim to the advance of computers. I contributed a paper to the 1982 commemorative volume for Yates about running 2^n factorials in incomplete block designs, and published a follow-up paper in 1984, but applications of the results seem to have been few and far between, if, indeed, there have been any!

7. Finis

By contemporary standards, mine has been an unusual career. Nowadays youngsters are expecteded to set a straight path almost as soon as they become upper-division undergraduates and pursue it doggedly to the end. My path has zigged and zagged with the help of some outstanding teachers to whom I am deeply grateful. I began as a pure mathematician and first encountered statistics in 1948 in the postgraduate diploma (now M.Sc.) course at Oxford, where I was taught by David Finney, P.A.P. Moran, and Michael Sampford. Upon coming to the United States, I set statistics aside for a few years while earning a Ph.D. in probability.

Then an extraordinary opportunity occurred in industry. In 1957 I joined Chevron Research Corporation in Richmond, California, as their first statistician. In effect, that gave me a 4-year postdoctoral period working with Henry Scheffé, whom Chevron had retained as its consultant in the previous year. Chevron sent me to Gordon Conferences on Statistics, which were then dominated by George Box and his pioneering work in the application of design of experiments to chemical engineering and, in particular, to response surfaces. The conferences had an electric atmosphere of intellectual ferment. At the same time, Henry set me to read the Box–Wilson (1951) paper and the I.C.I. book on industrial experimentation, edited by my fellow Welshman Owen Davies. I learned a lot from that book. The level of exposition is excellent and, unlike the spring of 1949, when Sampford was lecturing about design to a young mathematician who had never seen a real experiment, I now knew what an experiment was. Real ones were going on all around me.

After 4 invaluable years at Chevron, I chose to return to the academic world, where I have been ever since. I greatly enjoy teaching. Indeed, at Henry's instance, for the last 3 of those years I had taught a course in the statistics department at the University of California, Berkeley, in addition to my work at Chevron. Now I also wanted the freedom to work on any problem that attracted me, regardless of its importance to the company. An ideal opportunity beckoned from the mathematics department at the University of California, Davis, one of the world's foremost agricultural research domains. This involved me in a remarkable new array of experiments, including the large factorials of the agronomists with their fixed, mixed, and random models. Finally, in 1967, I moved to a large mathematics department at the University of Texas, which has no tradition of responsibility for any statistical services to the campus community.

This book, begun in Davis, was largely written in the relative isolation of Austin, where I have remained for personal reasons. Happily, continuing involvement in the Gordon Conferences and consulting kept me in touch with real-world stimuli until another remarkable opportunity emerged from the new semiconductor industry.

In cooperation with the federal government, a consortium of companies in that industry established in Austin a research facility called Sematech Corporation. For several years I have been fortunate to work with members of their statistical methods group. My collaboration with them has led to work in such areas as nonorthogonal fractions, D-optimal designs, and multivariate analysis of variance. I have learned that engineers do indeed carry out large factorial experiments with crossed and nested factors and mixed models; they call them gauge studies!

My path has zigged and zagged but, in retrospect, each turn ultimately afforded stimulating new experiences and opportunity for growth. More than ever, I share Henry Scheffé's (1958) conviction that industry is a valuable source of research problems. The first question that he was asked as a consultant for Chevron led him to invent a new branch of experimental design: experiments with mixtures. He told me repeatedly that there are dozens of interesting problems out there in industry, just below the surface, waiting to be solved, if only one is alert enough to spot them and run with them. That brings me to the role of serendipity and the thesis that a lot of useful research comes from a combination of luck, accident, and being in the right place at the right time.

This is true of two problems, to which I have referred above, from my days at Chevron. The idea of running the points of a factorial experiment in a prescribed sequence so as to protect against early termination came from an accident. A client wanted to do a 2^4 factorial on a pilot plant which he had at his disposal for 2 weeks, enough to do one complete replicate. I knew the "correct" thing to do. We carefully listed the 16 points and arranged them in random order. Fate struck. After 8 points, the pump quit, and could not be repaired within the 2-week period. Have you ever seen an experiment that consists of 8 points chosen at random from all 16? It is a mess. With the benefit of hindsight, we realized that it would have been a lot smarter if, instead of randomizing the 16 points, we had arranged for the first eight points to be either a resolution IV half replicate defined by $I = +ABCD$, or $I = -ABCD$, in random order. Then we would have had a design with good salvage value when the darned pump broke. Clearly, it would have been better to modify the randomization.

The other example led to my interest in nonorthogonal fractions. I had persuaded an engineer to try a 2^3 factorial. The raw material came from a facility in Scotland by slow boat. They sent enough for only 6 runs. I knew what to do with 8 runs; I knew what to do with 4 runs; but 6 was betwixt and between. I had to work out what to do. That led me to 4 factors in 12 runs, and thus to three-quarter replicates.

Both problems arose by accident. They should never have happened. A referee of a paper about one of them objected that the problem was contrived. It certainly was *not.* It was a real-life situation that probably would not have occurred to me had I just spent my time in an ivory-tower office thinking about mathematical theorems in the design of experiments.

Henry Scheffé was right. There are all kinds of practical research problems lurking in industrial applications, just waiting to be found and solved. I advise students to keep their eyes open and spot them. Some can be tremendously interesting and full of theoretical challenge. Above all, being helpful to people is rewarding in itself.

I end with the sentence that concluded my original preface. It is still true. I am greatly indebted to my wife, Elizabeth, for her continued help and encouragement.

Austin, Texas P. W. M. J.

References in the Preface

ATKINSON, A. C. (1982). "Developments in Design of Experiments." *Int. Statist. Rev.*, **50**, 161–177.

ATKINSON, A. C. (1988). "Recent Developments in the Methods of Optimum and Related Experimental Designs." *Int. Statist. Rev.*, **56**, 99–116.

BOX, G. E. P., AND K. J. WILSON (1951). "On the Experimental Attainment of Optimum Conditions." *J. Roy. Statist. Soc. Ser. B.*, **13**, 1–45.

COX, D. R. (1951). "Some Systematic Experimental Designs." *Biometrika*, **38**, 312–323.

DANIEL, C., AND F. WILCOXON (1966). "Factorial 2^{p-q} Designs Robust Against Linear and Quadratic Trends." *Technometrics*, **8**, 259–278.

DAVIES, O. L. (1954). *Design and Analysis of Industrial Experiments.* Oliver and Boyd, London.

DICKINSON, A. W. (1974). "Some Run Orders Requiring a Minimum Number of Factor Level Changes for the 2^4 and 2^5 Main Effect Plans." *Technometrics*, **16**, 31–37.

FEDOROV, V. (1970). *Theory of Optimal Experiments.* Academic Press, New York.

HILL, H. H. (1960). "Experimental Designs to Adjust for Time Trends." *Technometrics*, **2**, 67–82.

HOCKING, R. R. (1973). "A Discussion of the Two-way Mixed Model." *Amer. Statist.*, **27**, 148–152.

JOHN, P. W. M. (1982). "Incomplete Block designs for the 2^3 Factorial." *Utilitas Math.*, **21**, 179–199.

JOHN, P. W. M. (1984). "Incomplete Block Designs for the 2^4 Factorial." *Utilitas Math.*, **26**, 79–88.

JOHN, P. W. M. (1990). "Time Trends and Factorial Experiments." *Technometrics*, **31**, 275–282.

KIEFER, J. (1959). "Optimal Experimental Designs (with discussion)." *J. Roy. Statist. Soc. Ser. B*, **21**, 272–319.

MITCHELL, T. J. (1974). "An Algorithm for the Construction of '*D*-Optimal" Experimental Designs." *Technometrics*, **16**, 203–210.

SCHEFFÉ, H. (1956). "A Mixed Model for the Analysis of Variance." *Ann. Math. Statist.*, **27**, 23–36.

SCHEFFÉ, H. (1956). "Alternative Models for the Analysis of Variance." *Ann. Math. Statist.*, **27**, 251–271.

SCHEFFÉ, H. (1958). "Experiments with Mixtures." *J. Roy. Statist. Soc. Ser. B*, **20**, 344–360.

SCHEFFÉ, H. (1959). *The Analysis of Variance.* John Wiley and Sons, New York.

SILVEY, S. D. (1980). *Optimal Design.* Chapman–Hall, London.

STEINBERG, D. (1988). "Factorial Experiments with Time Trends." *Technometrics*, **30**, 259–270.

YATES, F. (1937). *The Design and Analysis of Factorial Experiments.* Imperial Bureau of Soil Science, Harpenden, England.

CHAPTER 1

Introduction

It is the view of many statisticians, amongst whom the author counts himself, that statistics is a branch of applied mathematics, and that its health and growth depend upon the involvement of the statistician in practical problems. This book deals with one of the interfaces between applied mathematics and scientific investigation: the design of experiments.

To some scientists, the idea that a statistician can contribute substantially to the design of an experiment is, if not patently absurd, at least dubious. A common image of statisticians is one of people who are rather clever with figures and hence useful for reducing data to tables; the statistician might be of some help in the analysis, but only if there are reams of data to feed into computers. One hopes that this false image is fading with the passage of time. Meanwhile, however, this attitude frustrates the statistician and deprives the scientist of valuable help.

It is true that the statistician is not usually an expert in the particular areas of biology or engineering in which the experiments are being made. It is true also that, given a computer, or for that matter enough time, pencils, and paper, one can fit a straight line or a plane to almost any hodgepodge of data whether relevant or not. The bigger the computer, the more variables it can handle in a multiple regression program, and, unfortunately, the more miles of useless numbers it can print out. It is clear, therefore, that some planning before the data is obtained is in order, and that the planning had best be made with the method of analysis of the experiment as well as its objectives in mind.

It follows that the role of the statistician in an experimental program should be not merely to analyze the results of the experiment, but also to help the scientist plan his experiment in such a way as to produce valid results as efficiently as possible. There is little sense in waiting until the end of the experiment and then visiting the statistician for the first time with a pile of data to deal with as best he can. Every applied statistician can tell horror stories of having been asked after the event to analyze data from lengthy and expensive experiments which were so badly planned that no valid conclusions could be drawn from them. There are also numerous experiments which, although valid results have been obtained, would have yielded even more information if skilled knowledge of design had figured in the planning, and others in which the results obtained could, with efficient design, have been obtained with less effort and expense.

This book is written for the mathematically oriented statistician, scientist, and engineer. The basis of all the theory of design and analysis will be the linear model and the analysis of variance, and throughout we shall refer the reader to the book by Henry Scheffé (1959). It is the most authoritative book available on the analysis of variance and is invaluable as a reference book. The examples that we include are drawn from various fields; most of them are taken from the published literature, and some of them come from experiments in which the author has participated as a consultant.

1.1 The Agricultural Heritage

The pioneer in the use of mathematical statistics in the design of experiments was the late Sir Ronald Fisher. He dominated the history of experimental design between the First and Second World Wars, 1918–1939. During the early part of that era he was in charge of statistics at the Rothamsted Agricultural Experiment Station near London. In 1933 he succeeded Karl Pearson as Galton Professor at the University of London and later he moved to a professorship at Cambridge University. It was Fisher who introduced the analysis of variance. He wrote two books about experimental design, each of which has run into several editions, even though neither is by any means easy to read: *Statistical Methods for Research Workers*, which first appeared in 1925, and *The Design of Experiments*. Dr. F. Yates joined Fisher at Rothamsted in 1931, and throughout this book the reader will find references to the major contributions of Yates and his colleagues at the agricultural experiment station.

From this background it has naturally followed that the common terminology of experimental design comes from agriculture, where they sow varieties in or apply treatments to plots and often have the plots arranged in blocks. We shall use this terminology except when we think in terms of engineering applications, and then we shall talk about points or runs. The terms *point* and *run* will be used interchangeably; we make a run on the plant at a predetermined set of conditions, and the result will be an observation or a data point (pedants would say datum point). It sometimes puts engineers off to hear a statistician talk about plots when he discusses their experiments. Some engineers tend to jump to the superficial conclusion that the statistician is locked into a rigid framework of agricultural thinking and hence can contribute little to the solution of engineering problems.

This ignores the essential fact that the statistician really thinks in terms of linear mathematical models. He is distinguished from many experimenters, who are deeply involved in their own specialized sphere, by mathematical training with a certain amount of emphasis on abstract thinking. This makes it easier for him to see the essential mathematical similarity between an experiment comparing varieties of barley and an experiment comparing operating temperatures in a catalytic cracker at an oil refinery.

The process of designing an industrial experiment is not, however, just a matter of taking a standard design from an agricultural handbook and substituting temperature for variety and run for plot. Indeed, differences do exist between agricultural and industrial experimentation, and, with all the incumbent risks of generalizing, we suggest a few. One is that the agronomist usually has to sow his experimental plots in the spring and harvest them in the fall. Plots and plants are relatively cheap, and the emphasis has to be on designing relatively complete experiments; if anything is omitted it will have to wait until next year. On the other hand, much industrial experimentation can be carried out relatively quickly, in a matter of days rather than months, but experimental runs are often very expensive. The emphasis should, therefore, be on small experiments carried out in sequence, with continual feedback of information, so that in designing each stage the experimenter can make use of the results of the previous experiments.

The agronomist is more interested in the tests of significance in the analysis of variance than is the industrial experimenter. One of the reasons is that the agronomist is more often concerned with uniformity trials. Can he produce strains of seed that are essentially equivalent? Can he produce a new variety of seed that will do well, not just on the experimental farm at Davis, California, but all up and down the hot Central Valley and perhaps also in the coastal region, in the other states where the crop is grown, and even in other countries? He wants to be able to accept the null hypothesis of uniformity. The position of the industrial experimenter often differs from that of the agronomist in two ways: he frequently knows before starting the experiment that his treatments are not all the same and is interested in finding out which ones differ and by how much. His emphasis will be on estimation rather than hypothesis testing. He will sometimes argue that failure to reject the null hypothesis is merely the result of taking too small an experiment. Expecting to reject the null hypothesis, he is more interested in confidence intervals.

In the sections that follow we shall introduce the topics of the various chapters by considering a typical experiment.

1.2 An Example of an Experiment

Suppose that a chemical engineer hopes to improve the yield of some petrochemical in an oil refinery plant by comparing several catalysts. Crude oil is fed into the plant, which is charged with the catalyst; some of the crude oil, or feedstock, passes through the plant unchanged; some is converted into the petrochemical, or product. The liquid that comes out of the plant is separated into product and unconverted feedstock, and the yield, or response, is the percentage of feedstock converted into product.

An obvious procedure is to make one or more plant runs using each of the catalysts and to compare the average yields on each catalyst. There are, however, some other considerations that enter the picture. How many catalysts? How many runs? How do we compare these averages after we have obtained them? Let us assume for the moment that the crude supply is plentiful and consider just two catalysts, A and B, which make up the simplest situation. We take r runs on each catalyst, $N = 2r$ runs altogether, and we obtain average yields $\bar{y}_1$ on A and $\bar{y}_2$ on B.

The first of our mathematical models now enters the picture. Let y_{ij} denote the yield on the jth run made with the ith catalyst; y_{ij} is a random variable and we assume $y_{ij} = m_i + e_{ij}$ where m_i is the true value for the ith catalyst and e_{ij} is a random component of error or noise. We further assume that on the average the noise contribution is zero; sometimes it is positive and sometimes negative, but it averages out over the long haul, in the sense that $E(e_{ij}) = 0$, and so $E(y_{ij}) = m_i$. Then $\bar{y}_i$ will be an unbiased estimate of m_i, and $\bar{y}_1 - \bar{y}_2$ will be an unbiased estimate of $m_1 - m_2$.

The word *random* should not be passed over lightly. The use of randomization is the keystone of the application of statistical theory to the design of experiments, and the validity of our deductions rests upon the principle of randomization. The very least that can be said as a recommendation for randomization is that it is only prudent to avoid the introduction of systematic bias into an experiment. An agronomist comparing two varieties of corn would not rationally assign to one variety all the plots that were in the shade and to the other all the plots that were in the sun. If he did, he would not be able to tell after the experiment whether any apparent difference in yields resulted from varietal differences or from the fact that one variety had more sun. Furthermore, Fisher (1947) has shown that the fact that the treatments have been randomly assigned to the runs is itself an adequate basis for obtaining tests of significance and confidence intervals.

The other maxim of experimental design emphasized by Fisher was replication to provide a proper estimate of the error variance, $V(e_{ij})$. If we were to take $r = 1$ and obtain $y_1 = 80$, $y_2 = 79$, we should not be able to make conclusions about the relative potencies of the catalysts with much confidence. We could pick the winner and recommend catalyst A because $y_{11} > y_{21}$, but we should have no idea whether the apparent difference was genuine or a result of noise (error). If we made several runs with each catalyst, and found that the observations were highly repeatable, i.e., that the range of the observations on A, and the range of the observations on B were both very small, we could be happy about concluding that $m_1 > m_2$ when we observed $\bar{y}_1 > \bar{y}_2$. On the other hand, if the observations on both A and B were scattered between 60 and 90, a difference between the means of only one would hardly be convincing.

1.3 Tests of Hypotheses and Confidence Intervals

Deciding whether or not an observed difference between the sample means $\bar{y}_1$ and $\bar{y}_2$ is convincing evidence of a real difference between m_1 and m_2 is equivalent to testing the hypothesis $H_0\colon m_1 = m_2$. There is no need for us to go into the details of the Neyman–Pearson theory of hypothesis testing at this time, or to discuss the merits of one-tailed or two-tailed tests. The reader will be familiar with these topics already. Nor is this the place to rekindle the flames of controversy between the confidence intervals that we shall employ and fiducial intervals; it suffices to say that in practice the intervals are the same and the argument is of little import.

It is at this stage of the proceedings that we have to make some decisions about what we shall assume to be the nature of the random errors e_{ij}. We mention three possibilities:

(i) We could assume only that the treatments were assigned to the runs, or the plots, by some process of randomization. This would lead us to the consideration of randomization models and permutation tests. We shall not take this road because not very much has been done with this approach for designs other than randomized complete blocks and Latin squares. The reader may wish to refer to the discussions of the randomization models and tests given by Kempthorne (1952) and (1955), Scheffé (1959), Wilk and Kempthorne (1955) and (1957). Nor shall we consider the nonparametric rank order tests of the Wilcoxon type.

(ii) We could assume that the errors are (*a*) uncorrelated random variables with zero expectations and the same variance σ^2, or else (*b*) random variables with zero means and a covariance matrix that is of known form.

(iii) We could add to assumption (ii) the additional assumption of normality.

For obtaining our estimates under assumption (ii), we shall rely upon the method of least squares. For testing hypotheses we shall make the normality assumption and use F tests and t tests.

Returning to our earlier model under assumptions (ii*a*) and (iii), we now have $y_{ij} = m_i + e_{ij}$ where the errors e_{ij} are independent normal random variables, each with mean zero and variance σ^2. The least squares estimates $\hat{m}_i$ of the parameters m_i are obtained by minimizing the sum of squares of the residuals,

$$S_e = \sum_i \sum_j (y_{ij} - \hat{m}_i)^2,$$

and they are, as we suspected all the time, $\hat{m}_i = \bar{y}_i$. We can also obtain an estimate of σ^2 by comparing repeated observations on the same treatment. Looking first at A, $\sum_j (y_{1j} - \bar{y}_1)^2$ is an unbiased estimate of $(r - 1)\sigma^2$. If there are t treatments, $E(S_e) = t(r - 1)\sigma^2$.

When two treatments are to be compared, the statistic $z = (\bar{y}_1 - \bar{y}_2)/\sqrt{(2\sigma^2/r)}$ has a normal distribution with mean $m_1 - m_2$ and variance unity, which we shall write as $z \sim N(m_1 - m_2, 1)$. If σ^2 is not known, we shall substitute for σ^2 the estimate $s^2 = S_e/(rt - t)$. The resulting statistic, when $m_1 = m_2$, has Student's t distribution with $t(r - 1)$ degrees of freedom.

No great difficulties are introduced if the numbers of observations are not the same for each treatment. Suppose that there are r_i observations on the ith catalyst. As before, $\hat{m}_i = \bar{y}_i$, and now

$$E(S_e) = E\left\{\sum_i \sum_j (y_{ij} - \bar{y}_i)^2\right\} = \sum_i (r_i - 1)\sigma^2 = (N - t)\sigma^2,$$

where $N = \sum_i r_i$. The estimate of σ^2 is thus $s^2 = S_e/(N - t)$. When the hth and ith treatments are to be compared, the statistic

$$t = \frac{\bar{y}_h - \bar{y}_i}{s(1/r_h + 1/r_i)^{1/2}}$$

has, if $m_h = m_i$, Student's distribution with $(N - t)$ degrees of freedom.

There is, however, a problem if the variance of e_{ij} changes from treatment to treatment. Consider again two treatments and suppose that the variances of the random variables involved are $V(e_{1j}) = \sigma_1^2$ and $V(e_{2j}) = \sigma_2^2$. Then, if there are r_i observations on the ith treatment,

$$V(\bar{y}_1 - \bar{y}_2) = r_1^{-1}\sigma_1^2 + r_2^{-1}\sigma_2^2 = v^2,$$

and $(\bar{y}_1 - \bar{y}_2) \sim N(m_1 - m_2, v^2)$.

If σ_1^2 and σ_2^2 are known, v^2 is easily computed, and the distribution of $(\bar{y}_1 - \bar{y}_2)/v$ is standard normal. If, however, neither σ_1^2 nor σ_2^2 is known, the best estimates of them from the data are $s_i^2 = \sum_j (y_{ij} - \bar{y}_i)^2/(r_i - 1)$.

We let $w_i = r_i^{-1}s_i^2$. Unfortunately, the random variable $t' = (\bar{y}_1 - \bar{y}_2)/(w_1 + w_2)^{1/2}$ does not have the Student's distribution and its distribution has to be approximated. There are several approximations that can be used, and we mention two of them. The disparity between the distributions of t and t' is least when $r_1 = r_2$.

Cochran and Cox (1957, page 101) give a procedure for obtaining an approximation to the critical value of t' for any level α. Let $\phi_i = r_i - 1$. Then the critical value is $t^* = (w_1t_1 + w_2t_2)/(w_1 + w_2)$, where t_i is the critical value, at the desired level α, of Student's t with ϕ_i degrees of freedom.

Welch (1947) approximates the distribution of t' by a Student's t distribution with ϕ d.f., where

$$\phi = \phi_1\phi_2(w_1 + w_2)^2/(\phi_1 w_1^2 + \phi_2 w_2^2).$$

The assumption of equal variances is usually not unreasonable, especially when two similar treatments are being compared. Scheffé (1959) devotes a chapter to the consequences, in the analysis of variance, of departures from assumptions (ii) and (iii). In some cases where the variance is believed to depend upon $E(y)$ in a known manner, the variance can be stabilized by transforming the data. These transformations do not necessarily eliminate the nonnormality, but they seem to alleviate it somewhat.

If $V(y) = \{g(y)\}^2$, and if z is a single-valued function of y, the variance of z is approximated by

$$v^2 = (\partial z/\partial y)^2 \, V(y).$$

If v is to be a constant, we therefore need to have

$$z = c \int [g(y)]^{-1} \, dy,$$

where c is an arbitrary constant. If $g(y)$ is directly proportional to y, the transformation $z = \ln y$ is appropriate; $\ln y$ denotes the natural logarithm of y; $\log_{10} y$ will do just as well. For binomial data the transformation $z = \text{Arcsin } y$ may be used, but this is often not done if the range of y is not large, especially when p is about 0.5. When $p = 0.5$ we have $pq = 0.25$, and when $p = 0.3$ or $p = 0.7$ the product pq only decreases to 0.21.

1.4 Sample Size and Power

If the experimenter has at his disposal a prior estimate of the variance σ^2, he is in a position to make some approximate power calculations. Most readers will already be acquainted with the procedures for computing r in the case of only two treatments, when type I and II errors are given. If the experimenter has decided to use a one-sided normal test with type I error α to compare two means, he will reject the hypothesis $H_0: m_1 = m_2$ in favor of the alternative $H_A: m_1 > m_2$ if $\bar{y}_1 - \bar{y}_2 > c$. The cutoff point c is given by $c = z_\alpha \sigma \sqrt{(2/r)}$ where z_α is the upper α point of the standard normal density, i.e., $P(z > z_\alpha) = \alpha$.

Suppose now that he wants to choose r to be so large that if $m_1 = m_2 + \delta\sigma$ he will have probability $(1 - \beta)$ of deciding that $m_1 > m_2$, i.e., of having $\bar{y}_1 - \bar{y}_2$ fall in the critical region. He must then have

$$P(\bar{y}_1 - \bar{y}_2 < c \mid m_1 - m_2 = \delta\sigma) = \beta.$$

It follows that

$$\delta\sigma - c = z_\beta \sigma/(2/r) \qquad \text{and} \qquad \delta = (z_\alpha + z_\beta)\sqrt{(2/r)}.$$

This gives an approximate value

$$r = 2(z_\alpha + z_\beta)^2/\delta^{-2}.$$

In practice, however, the sample size is often determined not by calculations of power but by real or imagined considerations of the budget. If a prior estimate of the variance is available, a variation of the previous calculation is still worth carrying out. To revert to our earlier example, suppose that previous experience in the plant gives a prior estimate $\sigma^2 = 2$ and that the engineer wishes to detect that the new catalyst A is better than the old catalyst B if $m_1 - m_2 \geq 2 = \sqrt{2}\,\sigma$, but that he is only prepared to make $n = 4$ runs with each of the catalysts. Let us suppose further that he wishes to use $\alpha = 0.05$. We have $z_\alpha = 1.645$, $\delta = \sqrt{2}$.

Then $z_\beta = 0.355$ and $\beta = 0.36$, so that there is about one chance in three that with $n = 4$ he will fail to reveal an actual improvement of 2 units with his experiment. If δ had been one, corresponding to $m_1 - m_2 \geq 1.4$, the chance of detection would have been only about four in ten (0.4). This kind of information, even if it is a rough approximation, is salutary to the engineer. It should at least serve to make his hopes for the outcome of his experiment more realistic. It may even prompt him to abandon or to revamp completely an investigation that was doomed to be unproductive.

When there are several treatments, similar calculations can be made involving the power of the F test. This is discussed by Scheffé (1959), Graybill (1961), and Kempthorne (1952). Scheffé uses the charts by Pearson and Hartley (1951) and by Fox (1956). The other two books contain Tang's tables based upon the incomplete Beta distribution.

1.5 Blocking

There are two ways in which we can reduce the variance of $\hat{m}_1 - \hat{m}_2$. We can take more data points, i.e., increase r and thereby increase the cost of the experiment, or we can somehow improve the design.

Some reduction can be made by refinement in experimental technique and by improved methods of measurement and chemical analysis, but these improvements are limited in their achievement and there still remains what might be called the natural noise of the system. Some engineers have difficulty accepting the fact that a system or a procedure has a built-in variability. This is true for analytical procedures, and the American Society for Testing Materials has tried hard to make its members realize that in standard measurements a certain amount of variability is to be expected because of the nature of the measurement procedure. Hence, failure to have duplicate measurements identical is not necessarily an indication of incompetence on the part of the operator. It is regrettable that in some laboratories duplicate measurements are useless because the atmosphere is such that the operator feels so strong a pressure to have his second reading conform to the first reading that the second reading can by no stretch of the imagination be considered as an independent observation. W. J. Youden has suggested that second readings obtained in this way should be regarded not so much as duplicate readings as duplicity; they certainly give the scientist a false sense of confidence in his results. Genuine duplicates can be obtained by sending the analytic laboratory coded samples, e.g., pouring a gallon of product into eight pint bottles and calling for a single determination on each of the eight "different" samples. It is a nuisance to have to do this, but it is often worth the trouble, and the results can be quite revealing.

One way of improving the precision of estimates of differences, such as $m_1 - m_2$, is by blocking. If there are t catalysts, perhaps we can take a barrel of crude oil, divide it into t parts, and use one part with each catalyst. The agricultural equivalent is to divide a field into blocks and then to subdivide each block into t plots. This topic is discussed at the end of Chapter 3. When we have a completely randomized design in which the runs are made in completely random order, or in which the plots with any given treatment are scattered all over the field, then comparisons between treatments contain additional variability from the fact that some runs are made from one barrel and some from another, or that some are made on good land and some on stony land. The idea behind the blocking is that by balancing the treatments over the blocks we make the differences between the blocks cancel out when we compare the treatments. The individual estimates of m_1 and m_2 are no better than they were before, in the sense that $V(\bar{y})$ has not been reduced, but $V(\bar{y}_1 - \bar{y}_2)$ has hopefully suffered a considerable reduction.

This is the randomized (complete) block experiment, and the reader should note that the word *randomized* appears again. We assign the treatments at random to the runs within each block; we should perhaps be suspicious if we always used the first sample drawn from each barrel for catalyst A and the last sample for catalyst B that there might be some systematic bias introduced. We modify our earlier model and, letting y_{ij} denote the observation on the ith treatment in the jth block, we write $y_{ij} = m_i + b_j + e_{ij}$ where b_j is the effect of the jth block. We shall actually write the model a little differently later when we come to look at it a little more closely.

This procedure works well so long as each barrel of crude oil is big enough to handle t runs, and the randomized complete block design is said to be the most commonly used experimental design. If, however, the blocks are not big enough to accommodate all the treatments, we have to resort to incomplete blocks, with k $(<t)$ treatments being contained in each block. The topic of incomplete block designs is introduced in Chapter 11. If we cannot arrange to have every pair of treatments appear together in every block, perhaps we can arrange the design so that every pair appears in λ blocks where λ is a constant. Such a design is called a balanced incomplete block design. These designs were introduced by Yates in 1936, but long before then such arrangements of symbols had been of interest to pure mathematicians. Number theorists have for years investigated the combinatorial properties of these designs without any care for possible statistical applications. Chapter 13 is devoted to the problems of the construction of balanced incomplete block designs. It is a chapter for mathematicians, and the experimenter who simply wants a particular design for his experiment might be content to look it up in the Fisher and Yates tables.

The restriction that each pair of treatments shall appear in exactly λ blocks is demanding, and designs do not exist for all values of t, k, and λ. In 1939 Bose and Nair introduced partially balanced incomplete block designs. These designs are the topics of Chapters 12 and 14, the latter chapter being devoted to the construction of the designs for different partially balanced schemes. In partially balanced designs each treatment appears with n_1 of the other treatments λ_1 times, with n_2 of them λ_2 times, and so on. There are also some further restrictions, but they can wait until Chapter 12.

Just as one would not normally use an incomplete block design when a suitable complete block design is available, so balanced incomplete block designs are preferable to partially balanced designs because of their higher efficiency. The measure of efficiency is the average of the variances of the estimates $m_h - m_i$ for all pairs of treatments; the measure is made scale free by dividing it by the average that would be obtained for a complete block design with the same number of observations, if such a design could have been used. Kempthorne (1956*a*) conjectured, and J. Roy (1958) proved that for any given set of numbers N, t, k $(k < t)$ the most efficient incomplete block design is the balanced design; that is, if one exists.

1.6 Factorial Experiments

We have just talked about blocking as a technique for improving the precision of the estimates $(\hat{m}_h - \hat{m}_i)$. The block effects b_j canceled out of the comparison between the treatments. We were thinking at the time of the block differences as a nuisance to be eliminated. The idea was to avoid the differences between the crude oils by balancing them out over the treatments. We could, however, purposely choose one barrel of Venezuelan crude oil, one of Sumatran, one of Californian, one of Texan, and so on. This would give us two factors: catalysts and types of oil. The structure is very similar to that of the blocked experiment. Just as each catalyst "sees" each barrel of oil, so each type of oil "sees" each catalyst. The crude oil differences cancel out of comparisons between catalysts; the catalyst differences cancel out of the comparisons between crude oils.

We can rewrite our model to take into account two factors, A and B, the latter standing either for blocks or for another factor:

$$y_{ij} = m + a_i + b_j + e_{ij}.$$

Here m is a grand mean effect, a_i is the main effect of the ith level of A, and b_j is the main effect of the jth level of B. We talk of the several crude oils as being levels of the factor "crude oil." The terminology sounds artificial in this context, but its justification becomes more apparent when we think of a quantitative factor such as temperature at levels of 100, 200, and 300 degrees.

The model that we have just written is an additive model. It may be inadequate. It could happen that the first catalyst is clearly better than the others on all crude oils except the Sumatran, which might perhaps have a higher sulfur content. This failure of the additive model, inasmuch as the expectation of the difference between observations on different crude oils does not remain constant from catalyst to catalyst, is an example of interaction between the factors. Interaction is an added complication, and at this stage the experimenter has to face up to a problem that he should have thought about earlier. From what populations did the levels of the factors that were used come? If the catalysts are some specific types having perhaps an alumina base, a silica base, and so on, we have a fixed factor, meaning that the levels are fixed effects, and the parameters a_i are unknown constants. The objective would then be to investigate comparisons between the parameters a_i.

On the other hand, suppose that the barrels of crude oil were all drawn from day-to-day production in the same California oil field. The effect b_j would be a random variable representing the day-to-day variation in the quality of the oil. Then B would be a random factor, and the objective would usually be to obtain an estimate of the component of variance $V(b_j) = \sigma_b^2$.

If both factors are fixed, we have a fixed effects model. If both are random, we have a random effects model. If there is one of each, we have a mixed model. The analysis, particularly in regard to the choice of denominators for the F statistics, depends upon which of these three types of model is relevant. This is discussed for the case of two factors in Chapter 4.

In Chapter 5 we turn to factorial experiments with several factors. Some of the further development is a routine extension of the results of Chapter 4 to a few more dimensions. There are, however, other complications and variations that will be considered at that time. One of them merits mention now.

Let us return for a moment to the crude oils and the catalysts. Suppose that there is indeed within each oilfield barrel-to-barrel variation in the quality of the crude oil, and that we take three barrels from each of four fields and make one run with each catalyst from each of the twelve barrels. This is an example of what is called a *split-plot experiment*. The catalysts see every barrel, and so barrel differences cancel out in comparing catalysts. However, in comparing crude oils each crude oil average contains the variability only of its own barrels, and this does not cancel out.

The term *split plot* comes from the agricultural background at Rothamsted. A modern example of split plots occurs with the use of crop-dusting aircraft. One possibility is to take several fields; these will be the plots, or the *whole plots*. We now divide the field into strips and plant each strip with a single variety of barley; if there are enough strips we may have several strips to each variety. These strips are the subplots. Suppose now that we also want to test some fertilizers. We can then fertilize each complete field with the same fertilizer by spraying it from an aircraft. Comparisons between fertilizers will be made between whole fields. Comparisons between varieties of barley will be made between strips in the same field. Conversely, one could apply the different fertilizers by hand to the different strips in the fields, and then sow the seeds a field at a time from an aircraft. A more complicated approach would be to sow strips running north-south from the air using different varieties for each strip in each field, and then to spread different fertilizers over the field by having the aircraft fertilize strips running in the east-west direction. There are numerous possibilities.

1.7 Latin Square Designs

Chapter 6 is devoted to Latin square designs. A Latin square is a square of side p in which p letters are written p times in such a way that each letter appears exactly once in each row and each column. Latin square designs figure in two contexts. They were first introduced by Fisher in agricultural experimentation as designs for blocking in two directions (designs for the two-way elimination of heterogeneity), with rows and columns representing the blocking and letters standing for treatments.

Latin squares may also be used for fractional factorial designs. Suppose that we have three factors each at p levels. To carry out a complete factorial would need $p \times p \times p = p^3$ runs. If we use a Latin square we can let the rows represent the levels of factor A, the columns the levels of factor B, and the letters the levels of factor C. Then we shall have a design in which each level of each factor occurs exactly once with each level of every other factor. This balancing enables us to estimate the effects of all three factors (assuming that there is no interaction) with only p^2 out of the p^3 runs, which represents a considerable saving. There are occasions when such a design is forced upon us. This happens when time is one of the factors. If we want to test p gasolines in p cars, and we cannot test more than one gasoline in a car at a time, we must use a design like this. An interesting variation occurs in nutritional experiments on dairy cows, where the different rations are fed in sequence to the cows under test, and where there are residual effects to be considered.

1.8 Fractional Factorials and Confounding

Chapters 7, 8, and 9 continue the investigation of the factorial experiments, but now the emphasis is on several factors at two or three levels each, and the principal area of application is moving away from agriculture to industry. Again, however, the roots are in agriculture. It is to Yates (1937) that we turn for the original work in Chapter 7, and the introduction of fractional factorials by Finney (1945) dates from his days at Rothamsted. In Chapter 7 we introduce 2^n and 3^n designs; these are factorial designs for n factors each at two levels and n factors each at three levels respectively, and because they are so often used they merit special treatment. The main topic of the chapter is confounding.

If the number of points in a factorial design is so great that we cannot accommodate all the points in a single block, then we must run some of the points in one block, and the remainder in one or more other blocks. One of the possible contrasts between the data points will coincide with the difference between the block totals. That degree of freedom is said to be confounded with blocks. Clearly we should not like, for example, to include all the points at high temperature in one block and all the points at low temperature in the other block, because then the temperature effect, if any, would be hopelessly confused or confounded with the block effects. We should like to divide the design into blocks in such a way that the division does as little damage as possible to the design.

In Chapters 8 and 9 we turn to fractionation. We have already mentioned the Latin square design as a 1/3 fraction of a factorial with three factors, each at three levels. There are several reasons for the extensive use of fractional factorials. The relatively high cost of industrial experimentation makes it essential that the designs used be as economical and as efficient as possible. Runs on a refinery unit may cost several thousand dollars a time, and, even though the eventual payoff may be many times more than that, management decision makers are usually loath to spend more money on experimentation than they feel is absolutely necessary.

The fact that it is possible to carry out experiments in relatively quick succession means that it is feasible to consider in the initial stages of an experimental program a large number of factors in the hope of finding a few important ones for further investigation. Such an initial experiment is called a screening experiment, and it is customary to use as small a fraction as possible with hardly any more degrees of freedom than there are factors. We shall consider such designs for $n - 1$ factors in n runs.

An example in which, on the other hand, the scientist hopes to find no important factors is found in the food industry. Suppose that a company has developed a new cake mix. When made in their laboratory ovens it is excellent. Before marketing it, however, the company needs to be sure that it is robust against the vagaries of the outside world, the uncertainties of the equipment in many home kitchens, and the mistakes of some housewives. Will the new mix produce an acceptable cake if the oven is ten degrees too hot or ten degrees too cold, if it is left in too long or taken out a few minutes too soon, if when the cook adds the prescribed two eggs, she uses extra large eggs or the smaller economy size? Here it is appropriate to run a screening experiment to check a whole list of possible mistakes as factors with the hope than none, or at most only a few, will turn out to be damaging to the cake.

Interest in fractional factorials has grown steadily during the last decade. With factors at two levels, we shall present designs that allow us to estimate all the main effects and two factor interactions if the higher order interactions are ignored; smaller designs for estimating main effects but not the two factor interactions; and even smaller designs for estimating all the main effects if all the interactions are ignored. With factors at three or more levels, the number of degrees of freedom taken up by the two factor interactions increases fast with n, and most of the designs that are used are for the estimation of main effects only.

1.9 Response Surfaces

Chapter 10 is industry's own chapter; the ideas were developed in industry for industrial applications. The applications to agriculture are few and far between. We shall be concerned with the basic problem of finding optimum conditions for operating chemical processes. In its simplest form, we can conceive of a chemical process as a black box with two knobs, one marked temperature and the other pressure (with perhaps a third knob at the back for noise). We envision a three-dimensional space with temperature plotted along the x axis, pressure along the y axis, and the yield along the z axis, rising up out of the paper. This response surface is like a mountain with a (hopefully) single peak, and we have to turn the two knobs, i.e., carry out experiments varying the levels of the two factors, until we find that peak. The original work in this area was carried out by members of a group at Imperial Chemical Industries, Great Britain, prominent among whom was G. E. P. Box. One of the first papers on the topic is that of Box and Wilson (1951) on the experimental attainment of optimum conditions. The work has been continued by Box and his students since he left industry to enter the academic world. The chapter concludes with a short section on evolutionary operation, which applies the methods of the earlier part of the chapter to actual working plants in a refinery.

1.10 Matrix Representation

Throughout the book we shall make extensive use of matrices and quadratic forms. This aspect of linear algebra is introduced in Chapter 2, together with a discussion of the distribution of quadratic forms under the normality assumption. We shall conclude this chapter by sketching how the procedure that we shall follow might be applied to the problem of comparing two means.

We shall rewrite the earlier model as

$$y_{ij} = m + t_i + e_{ij},$$

where y_{ij} is the jth observation on the ith treatment ($i = 1, 2; j = 1, 2, \ldots, n$), m is a grand mean, and t_i is the ith treatment effect. We arrange the observations in a vector $\mathbf{Y}$, the first n observations being those on the first treatment. Then

$$\mathbf{Y} = \mathbf{X}\begin{bmatrix} m \\ t_1 \\ t_2 \end{bmatrix} + \mathbf{e},$$

where $\mathbf{e}$ is the vector of errors and $\mathbf{X}$ is a matrix with three columns and $N = 2n$ rows. The first column of $\mathbf{X}$ is $\mathbf{1}_N$, a column of N unit elements which corresponds to the mean. The second column corresponds to t_1 and consists of n ones followed by n zeros. The third column has n zeros followed by n ones. Since the sum of the last two columns is equal to the first column, the rank of $\mathbf{X}$ is two.

We seek a vector $(m, t_1, t_2)'$ of estimates such that

$$S_e = \sum_i \sum_j (y_{ij} - \hat{y}_{ij})^2$$

is a minimum where $\hat{y}_{ij} = \hat{m} + \hat{t}_i$. Differentiating S_e with respect to $\hat{m}$, $\hat{t}_1$, $\hat{t}_2$ in turn gives the three normal equations

$$\sum y_{ij} = N\hat{m} + n\hat{t}_1 + n\hat{t}_2, \qquad n\bar{y}_1 = \sum y_{1j} = n\hat{m} + n\hat{t}_1,$$

$$n\bar{y}_2 = \sum y_{2j} = n\hat{m} + n\hat{t}_2, \qquad \text{or} \qquad \mathbf{X'Y} = \mathbf{X'X}(\hat{m}, \hat{t}_1, \hat{t}_2)'.$$

We cannot invert $\mathbf{X'X}$ because it is singular. Had we stayed with the earlier formulation of the model, we would have avoided this crisis. However, we are really concerned with $(\hat{m} + \hat{t}_1) - (\hat{m} + \hat{t}_2) = \hat{t}_1 - \hat{t}_2$. Subtracting the third normal equation from the second we have $\hat{t}_1 - \hat{t}_2 = \bar{y}_1 - \bar{y}_2$, which comes as no surprise; also, $\hat{y}_{ij} = \bar{y}_i$.

We now consider four quadratic forms. One of them is the raw sum of squares $\sum_i \sum_j y_{ij}^2 = \mathbf{Y'Y} = \mathbf{Y'IY}$. The second is $N\bar{y}^2$, where $\bar{y}$ is the mean of all N observations. This form may be written as $\mathbf{Y'J}_N\mathbf{Y}/N$, where $\mathbf{J}_N$ denotes a square matrix of order N with every element unity.

The identity

$$\sum_i \sum_j y_{ij}^2 = N\bar{y}^2 + \sum_i \sum_j (y_{ij} - \bar{y})^2 = N\bar{y}^2 + n(\bar{y}_1 - \bar{y}_2)^2/2 + \sum_i \sum_j (y_{ij} - \bar{y}_i)^2$$

may be written

$$\mathbf{Y'Y} = \mathbf{Y'A}_1\mathbf{Y} + \mathbf{Y'A}_2\mathbf{Y} + \mathbf{Y'A}_3\mathbf{Y},$$

where

$$\mathbf{A}_1 = N^{-1}\mathbf{J}_N,$$

$$\mathbf{A}_2 = (2n)^{-1}\begin{bmatrix} \mathbf{J}_n & -\mathbf{J}_n \\ -\mathbf{J}_n & \mathbf{J}_n \end{bmatrix},$$

$$\mathbf{A}_3 = \begin{bmatrix} \mathbf{I}_n - n^{-1}\mathbf{J}_n & \mathbf{0} \\ \mathbf{0} & \mathbf{I}_n - n^{-1}\mathbf{J}_n \end{bmatrix}.$$

These three matrices, $\mathbf{A}_1$, $\mathbf{A}_2$, and $\mathbf{A}_3$ are all idempotent, i.e., $\mathbf{A}_i^2 = \mathbf{A}_i$. We shall show later that this implies that the quadratic forms, when divided by σ^2, have χ^2 distributions. Furthermore, $\mathbf{A}_1\mathbf{A}_2 = \mathbf{A}_1\mathbf{A}_3 = \mathbf{A}_2\mathbf{A}_3 = \mathbf{0}$, and this will be seen to imply that the forms are independently distributed. Hence, we may test the hypothesis $m_1 = m_2$, or $t_1 = t_2$, by the statistic

$$\mathscr{F} = (\mathbf{Y'A}_2\mathbf{Y})/(\mathbf{Y'A}_3\mathbf{Y}/(N - 2))$$

(the ranks of the forms are equal to the numbers of degrees of freedom). Under the null hypothesis this statistic has the $F(1, N - 2)$ distribution, and the test is the same as the usual t test.

Indeed, we have

$$\mathscr{F} = \frac{n(\bar{y}_1 - \bar{y}_2)^2}{2} \div \frac{S_e}{N - 2} = \frac{n(\bar{y}_1 - \bar{y}_2)^2}{2s^2} = t^2.$$

CHAPTER 2

Linear Models and Quadratic Forms

2.1 Introduction

In this chapter we derive some of the mathematical results that will be used in the subsequent chapters. The treatment makes considerable use of matrices and quadratic forms; a short discussion of these topics appears in the appendix.

If we arrange the data in an experiment in a vector $\mathbf{Y}$, the analysis of variance procedure involves separating the sum of squares of the observations $\sum y^2$ or $\mathbf{Y}'\mathbf{Y}$ into a set of quadratic forms

$$\mathbf{Y}'\mathbf{Y} = \mathbf{Y}'\mathbf{A}_1\mathbf{Y} + \mathbf{Y}'\mathbf{A}_2\mathbf{Y} + \mathbf{Y}'\mathbf{A}_3\mathbf{Y} + \cdots.$$

If the observations $\mathbf{Y}$ are normally distributed, tests of hypotheses are made by examining some of the ratios of these forms. The distribution of quadratic forms is the topic of one of the later sections. This section is somewhat more demanding mathematically than the rest of the chapter. Some readers may wish to omit it provided they are prepared to accept our assurances that except where we say otherwise:

(i) $\mathbf{Y}'\mathbf{A}_i\mathbf{Y}$ does, with the proper divisors, have a χ^2 distribution with the number of degrees of freedom equal to the rank of $\mathbf{A}_i$,
(ii) The forms are independent, and, hence, (iii) The corresponding ratios have F distributions.

We shall often use the abbreviation d.f. for degrees of freedom.

2.2 Linear Models

We shall be concerned throughout this book with linear models in which an observation y may be represented by

$$y = \beta_1 x_1 + \beta_2 x_2 + \cdots + \beta_p x_p + e,$$

where $x_1, \ldots, x_p$ are coordinates, $\beta_1, \ldots, \beta_p$ are coefficients, and e denotes random error. If there is more than one data point under consideration, we may denote the ith data point by y_i and write

$$y_i = \beta_1 x_{i1} + \beta_2 x_{i2} + \cdots + \beta_p x_{ip} + e_i,$$

with x_{ij} denoting the value of the jth coordinate at the ith data point. We shall customarily use matrix notation and write

$$\mathbf{Y} = \mathbf{X}\boldsymbol{\beta} + \mathbf{e},$$

where $\mathbf{Y}$ is a vector of N observations, $\mathbf{X}$ is a matrix of known constants, $\boldsymbol{\beta}$ is a vector of p parameters, and $\mathbf{e}$ is the vector of random errors.

The matrix $\mathbf{X} = (x_{ij})$ is sometimes called the design matrix; it has N rows and p columns. In the experimental design situations that will be considered, x_{ij} will take only the values zero or one. That restriction is not, however, necessary for the results of this preliminary chapter. The elements of the parameter vector $\boldsymbol{\beta}$ will usually be unknown constants; there will be some occasions in later chapters when we shall consider the special case in which the parameters β_j are random variables. Our main purpose will be to obtain estimates of these parameters β_j, or some functions of the parameters, and to test some hypotheses about them. The hypotheses will usually be that some subset of the parameters $\beta_1, \ldots, \beta_p$ are all zero.

The model is said to be linear because it expresses y as a linear combination of the parameters β_j. There is no restriction that the model has to be a linear function of the x_i or that the latter be independent. We can put the polynomial $\beta_0 + \beta_1 x + \beta_2 x^2 + \beta_3 x^3$ into this form by setting $x_1 = x$, $x_2 = x^2$, $x_3 = x^3$. Some models that are not linear can be made so by transformations. For example, $y = Ae^{Bx}$ becomes $z = \beta_0 + \beta_1 x$ if we take $z = \ln y$; then $\beta_0 = \ln A$ and $\beta_1 = B$. On the other hand, the two models

$$y = \beta_1 + 1/(\beta_2 + \beta_3 x)$$

and

$$y = Ae^{Bx} + Ce^{Dx}$$

are not linear and cannot be made linear by transforming y.

2.3 Distribution Assumptions

In the original linear model, the ith observation consists of two components: $\sum_j x_{ij}\beta_j$ and e_i, the "error" in the ith observation. The following assumptions are made about the errors e_i.

(i) They have zero expectations: $E(e_i) = 0$ for all i;
(ii) They are uncorrelated: $E(e_i e_{i'}) = 0$ if $i \neq i'$;
(iii) They have homogeneous variance: $E(e_i^2) = \sigma^2$ for all i.

These assumptions may be written $E(\mathbf{e}) = \mathbf{0}$, $E(\mathbf{ee}') = \mathbf{I}\sigma^2$, and it follows from assumption (i) that $E(\mathbf{Y}) = \mathbf{X\beta}$.

(In general, if $\mathbf{Z}$ is a vector of random variables, $E(\mathbf{Z})$ is the vector whose ith element is $E(z_i)$. The variance-covariance matrix of the random variables constituting $\mathbf{Z}$ will be denoted by cov ($\mathbf{Z}$). If cov ($\mathbf{Z}$) $= \mathbf{V}\sigma^2$, the covariance of z_i and z_j is $v_{ij}\sigma^2$; we shall sometimes omit the scalar σ^2. If $\mathbf{A}$ is a matrix such that $\mathbf{AZ}$ exists, we have $E(\mathbf{AZ}) = \mathbf{A}E(\mathbf{Z})$, and cov ($\mathbf{AZ}$) $= \mathbf{AVA}'$, where cov ($\mathbf{Z}$) $= \mathbf{V}$.)

In addition to the three assumptions given before, we shall often make a fourth assumption.

(iv) The normality assumption: that the errors e have a multivariate normal distribution with $E(\mathbf{e}) = \mathbf{0}$, cov ($\mathbf{e}$) $= \mathbf{I}\sigma^2$, which will be written $\mathbf{e} \sim N(\mathbf{0}, \mathbf{I}\sigma^2)$.

Equivalently, we may say that $\mathbf{Y}$ has a multivariate normal distribution $\mathbf{Y} \sim N(\mathbf{X\beta}, \mathbf{I}\sigma^2)$. This assumption is important in carrying out tests of significance or in interval estimation, but it is not essential to the subdivision of sums of squares or for obtaining least squares estimates. Indeed, the method of maximum likelihood under the normality assumption gives the same estimates of estimable functions as the method of least squares, which we shall consider in the next section.

2.4 The Method of Least Squares

In estimating the unknown constants, β_j, by the method of least squares, we choose as our estimates a set, $\hat{\beta}_1, \hat{\beta}_2, \ldots, \hat{\beta}_p$, which minimize the sum of the squares of the residuals. The vector of the estimates is written as $\hat{\boldsymbol{\beta}}$, and we shall see later that in some situations it is not unique. The residuals are defined in the following way. Let $\hat{y}_i$ denote the estimate of $E(y_i)$ obtained by substituting $\hat{\boldsymbol{\beta}}$ for the unknown $\boldsymbol{\beta}$, i.e., $\hat{y}_i = \sum x_{ij}\hat{\beta}_j$; then $y_i - \hat{y}_i$ is the ith residual, and we seek to minimize the sum of squares $\sum (y_i - \hat{y}_i)^2$. In matrix notation, we write the sum of the squares of the residuals as $S_e = (\mathbf{Y} - \hat{\mathbf{Y}})'(\mathbf{Y} - \hat{\mathbf{Y}})$. To minimize S_e, we solve a set of p equations obtained by equating to zero the partial derivatives of S_e with respect to each of the estimates $\hat{\beta}_j$. Dropping the common factor -2, the jth of these equations is

$$\sum_{i=1}^{N} x_{ij}y_i - \sum_{s=1}^{p}\sum_{i=1}^{N} x_{is}x_{ij}\hat{\beta}_s = 0.$$

These are called the normal equations. In matrix form they may be written $\mathbf{X'Y} = \mathbf{X'X}\hat{\boldsymbol{\beta}}$.

The matrix $\mathbf{X'X}$ is a symmetric matrix with p rows and p columns. Its rank is the same as the rank of $\mathbf{X}$, which is the number of linearly independent columns of $\mathbf{X}$.

2.5 The Case of Full Rank

This is the usual case that occurs in multiple regression problems. It will, however, be the exception in the experimental designs that will be considered in later chapters. We shall examine it briefly as an introduction to the general case. If the p columns of $\mathbf{X}$ are linearly independent, $r(\mathbf{X}) = r(\mathbf{X'X}) = p$. Thus, $(\mathbf{X'X})^{-1}$ exists, and the normal equations have a unique set of solutions, or solution vector, $\hat{\boldsymbol{\beta}} = (\mathbf{X'X})^{-1}\mathbf{X'Y}$.

The vector $\hat{\boldsymbol{\beta}}$ is an unbiased estimator of $\boldsymbol{\beta}$, for we have

$$E(\hat{\boldsymbol{\beta}}) = (\mathbf{X'X})^{-1}\mathbf{X'}E(\mathbf{Y}) = (\mathbf{X'X})^{-1}\mathbf{X'X}\boldsymbol{\beta} = \boldsymbol{\beta}.$$

Furthermore,

$$\operatorname{cov}(\hat{\boldsymbol{\beta}}) = (\mathbf{X'X})^{-1}\mathbf{X'I}\sigma^2\mathbf{X}(\mathbf{X'X})^{-1} = (\mathbf{X'X})^{-1}\sigma^2.$$

In computing the sum of squares of residuals

$$S_e = \mathbf{Y'Y} - \mathbf{Y'}\hat{\mathbf{Y}} - \hat{\mathbf{Y}}'\mathbf{Y} + \hat{\mathbf{Y}}'\hat{\mathbf{Y}},$$

we substitute $\hat{\mathbf{Y}} = \mathbf{X}\hat{\boldsymbol{\beta}}$ and note that

$$\mathbf{Y'}\hat{\mathbf{Y}} = \hat{\mathbf{Y}}'\mathbf{Y} = \hat{\mathbf{Y}}'\hat{\mathbf{Y}} = \mathbf{Y'X}(\mathbf{X'X})^{-1}\mathbf{X'Y},$$

whence,

$$S_e = \mathbf{Y'}(\mathbf{I} - \mathbf{X}(\mathbf{X'X})^{-1}\mathbf{X'})\mathbf{Y}.$$

We denote the quadratic form $\mathbf{Y'X}(\mathbf{X'X})^{-1}\mathbf{X'Y}$ by S_R and call it the sum of squares for regression. We have a subdivision of the total sum of squares into two components, $\mathbf{Y'Y} = S_R + S_e$.

We shall see later that the expectations of these two component quadratic forms are $E(S_R) = \boldsymbol{\beta}'\mathbf{X}'\mathbf{X}\boldsymbol{\beta} + p\sigma^2$, $E(S_e) = (N - p)\sigma^2$. Since $\mathbf{X}'\mathbf{X}$ is positive definite, $\boldsymbol{\beta}'\mathbf{X}'\mathbf{X}\boldsymbol{\beta}$ is nonnegative and is zero only when $\boldsymbol{\beta} = \mathbf{0}$. Under normality, S_R/σ^2 has a noncentral χ^2 distribution with p d.f., and S_e/σ^2 has a central χ^2 distribution with $(N - p)$ d.f.; furthermore, the two forms are distributed independently. Thus, an appropriate test for the hypothesis $H_0: \boldsymbol{\beta} = \mathbf{0}$, with alternative $H_A: \boldsymbol{\beta} \neq \mathbf{0}$, makes use of the test statistic $\mathscr{F} = p^{-1}S_R/(N - p)^{-1}S_e$. Under the null hypothesis, $\mathscr{F}$ has Snedecor's F distribution with p and $(N - p)$ d.f. The critical region is the upper tail of the F distribution. This is a consequence of the following results about the distribution of quadratic forms in normal variables. Let $\mathbf{Y} \sim N(\boldsymbol{\mu}, \mathbf{I})$. Then $\mathbf{Y}'\mathbf{A}\mathbf{Y} \sim \chi^2(m)$ if, and only if, $\mathbf{A}$ is an idempotent matrix of rank m; $\mathbf{Y}'\mathbf{A}\mathbf{Y}$ and $\mathbf{Y}'\mathbf{B}\mathbf{Y}$ are independent if, and only if, $\mathbf{AB} = \mathbf{0}$.

2.6 The Constant Term and Fitting Planes

The linear model presented in section 2.2 does not appear to contain a constant term. However, such a term is easily included within the scope of the model. We add a coordinate x_0, which always takes the value one, and write

$$E(y) = \beta_0 + \beta_1 x_1 + \cdots + \beta_k x_k = \beta_0 x_0 + \beta_1 x_1 + \cdots + \beta_k x_k.$$

We shall not attempt a comprehensive discussion of fitting planes and hyperplanes to data. The reader who wishes to look further at the practical aspects is referred to the book by Draper and Smith (1966). It has become the custom to refer to the variables $x_1, \ldots, x_k$ as independent variables and the observed response y as the dependent variable. The use of the word *independent* should not be taken seriously. It derives from the habit in beginning algebra courses of labeling y as the dependent variable and x as the independent variable in the formula $y = x + 3$, because one can choose any value of x, substitute in the formula, and obtain the value of y. Nothing, however, is implied about the independence of the x variables. Indeed, as we mentioned before, we can fit a polynomial in x by letting $x^i = x_i$.

In some situations we have no control over the elements of the design matrix $\mathbf{X}$. When we can choose the points at which observations are to be taken, we note that $\operatorname{cov}(\hat{\boldsymbol{\beta}}) = (\mathbf{X}'\mathbf{X})^{-1}\sigma^2$, and choose the experimental points with this in mind. One possibility could be to choose $\mathbf{X}$ in such a way as to minimize the average variance of the $\hat{\beta}_i$. This is equivalent to minimizing the trace of $(\mathbf{X}'\mathbf{X})^{-1}$.

This procedure of fitting hyperplanes is commonly used to obtain equations for predicting future values of the response given $x_1, \ldots, x_k$. If $\beta_i = 0$, we might conclude that x_i is of little use as a predictor variable, although one should be careful about jumping to that conclusion when x_i is highly correlated with other independent variables. One might ask the following question: are the predictor variables $x_1, x_2, \ldots, x_k$ collectively useless as predictors, or, alternatively, does adding the set of x variables give a significant improvement over the simpler prediction equation $E(y) = \beta_0$? More formally, we wish to test the hypothesis $H_0: \boldsymbol{\beta}^* = \mathbf{0}$ where $\boldsymbol{\beta}^* = (\beta_1, \beta_2, \ldots, \beta_k)'$, against the alternative $H_A: \boldsymbol{\beta}^* \neq \mathbf{0}$. We have two nested models:

$$\Omega_0: \quad E(y) = \beta_0; \qquad \Omega_1: \quad E(y) = \beta_0 + \sum_i \beta_i x_i.$$

Under Ω_0, $\hat{\beta}_0 = \bar{y}$ so that $S_R = N\bar{y}^2 = N^{-1}\mathbf{Y}'\mathbf{J}\mathbf{Y}$, and the sum of squares of the residuals is $\sum y_i^2 - N\bar{y}^2 = \sum (y_i - \bar{y})^2$. The matrix $\mathbf{J}$ is a square matrix with every element unity. Under Ω_1 we have, as before, $S_R = \mathbf{Y}'\mathbf{X}(\mathbf{X}'\mathbf{X})^{-1}\mathbf{X}'\mathbf{Y}$. The increase in S_R is attributable to the addition of $x_1, \ldots, x_k$ to the model.

We have

$$\mathbf{Y}'\mathbf{Y} = N^{-1}\mathbf{Y}'\mathbf{J}\mathbf{Y} + \mathbf{Y}'(\mathbf{X}(\mathbf{X}'\mathbf{X})^{-1}\mathbf{X}' - N^{-1}\mathbf{J})\mathbf{Y} + S_e = \mathbf{Y}'\mathbf{A}_1\mathbf{Y} + \mathbf{Y}'\mathbf{A}_2\mathbf{Y} + S_e.$$

Under normality, the three quadratic forms $\mathbf{Y}'\mathbf{A}_1\mathbf{Y}$, $\mathbf{Y}'\mathbf{A}_2\mathbf{Y}$, S_e have, upon division by σ^2, independent χ^2 distributions with $1, k$, and $N - k - 1$ d.f. respectively. The test statistic for H_0 is

$$\mathscr{F} = \mathbf{Y}'\mathbf{A}_2\mathbf{Y}/k \div S_e/(N - k - 1),$$

which has, under the null hypothesis, the $F(k, N - k - 1)$ distribution. Some writers prefer to call $\mathbf{Y}'\mathbf{A}_2\mathbf{Y}$ the sum of squares for regression. We shall not do this, but shall continue instead to let $S_R = \mathbf{Y}'\mathbf{Y} - S_e$. The general question of testing subhypotheses will be considered later.

2.7 The Singular Case

If $\mathbf{X}$ is of less than full rank because the columns are not linearly independent, $\mathbf{X}'\mathbf{X}$ is singular and $(\mathbf{X}'\mathbf{X})^{-1}$ does not exist. Then we must modify the results of the previous sections. This state of affairs can occur either inadvertently or by design. It will, in fact, be true of almost all the experimental design models in the subsequent chapters.

As an example of accidental singularity, suppose that we have a model

$$E(y) = \beta_0 + \beta_1 x_1 + \beta_2 x_2,$$

where $x_1 + x_2 \equiv 100$. This would occur if x_1 denoted the percentage of olefins in a hydrocarbon and x_2 the percentage of nonolefins. The matrix $\mathbf{X}$ is singular because the first column is a linear combination of the last two columns.

Suppose that $\hat{y} = 90 + 3x_1 + 5x_2$ is a predicting plane. Since $x_1 + x_2 \equiv 100$, we may equally well have $\hat{y} = 190 + 2x_1 + 4x_2$ or more generally,

$$\hat{y} = 90 - 100t + (3 + t)x_1 + (5 + t)x_2.$$

Obviously, in this simple example the difficulty can easily be avoided by dropping either x_1 or x_2 (it does not matter which) from the equation. The three parameters β_0, β_1, and β_2 cannot be (uniquely) estimated. However, we note that for all values of t we have

$$\beta_1 - \beta_2 = (3 + t) - (5 + t) = -2, \qquad \beta_0 + 50(\beta_1 + \beta_2) = 490.$$

These two parametric functions are said to be estimable.

An example in which the model was purposely chosen in such a way that $\mathbf{X}$ is singular appears at the end of the previous chapter in the experiment to compare two treatments. The model was $E(\mathbf{Y}) = \mathbf{X}\boldsymbol{\beta}$, where $\boldsymbol{\beta}' = (m, t_1, t_2)$. If y_{ij} denotes the jth observation on the ith treatment, we have $y_{ij} = m + t_i + e_{ij}$. The first column of $\mathbf{X}$ consists entirely of ones. The elements of the other columns are either zero or one, and the first column is the sum of the other two; we have $r(\mathbf{X}) = r(\mathbf{X}'\mathbf{X}) = 2$.

2.8 Pseudo-Inverses and the Normal Equations

We consider now the solution of the normal equations $\mathbf{X}'\mathbf{Y} = \mathbf{X}'\mathbf{X}\hat{\boldsymbol{\beta}}$ in the general case where $\mathbf{X}$ has N rows and p columns and $r(\mathbf{X}) = k \leq p$ and $m = p - k$.

The normal equations are consistent. The augmented matrix $(\mathbf{X}'\mathbf{X}, \mathbf{X}'\mathbf{Y})$ may be written as $\mathbf{X}'(\mathbf{X}, \mathbf{Y})$. Then $r(\mathbf{X}'\mathbf{X}, \mathbf{X}'\mathbf{Y}) \leq r(\mathbf{X}') = r(\mathbf{X}'\mathbf{X})$. However, $r(\mathbf{X}'\mathbf{X}, \mathbf{X}'\mathbf{Y}) \geq r(\mathbf{X}'\mathbf{X})$ since all the columns of $\mathbf{X}'\mathbf{X}$ are contained in the larger matrix, and so $r(\mathbf{X}'\mathbf{X}, \mathbf{X}'\mathbf{Y}) = r(\mathbf{X}'\mathbf{X})$. Let $\tilde{\boldsymbol{\beta}} = \mathbf{P}\mathbf{X}'\mathbf{Y}$ be a set of solutions to the normal equations. We call $\tilde{\boldsymbol{\beta}}$ a solution vector and $\mathbf{P}$ a solution matrix, or a pseudo-inverse of $\mathbf{X}'\mathbf{X}$. $\mathbf{P}$ is not necessarily a symmetric matrix. If $k = p$, there is only one solution matrix, namely $\mathbf{P} = (\mathbf{X}'\mathbf{X})^{-1}$. The use of pseudo-inverses in the solution of the normal equations has been discussed by Rao (1962).

Substituting for $\tilde{\boldsymbol{\beta}}$ in the normal equations, we have $\mathbf{X}'\mathbf{Y} = \mathbf{X}'\mathbf{X}\mathbf{P}\mathbf{X}'\mathbf{Y}$. This is an identity for all $\mathbf{Y}$, and so

$$\mathbf{X}' = \mathbf{X}'\mathbf{X}\mathbf{P}\mathbf{X}'. \tag{2.1}$$

Multiplying both sides of this equation on the right by $\mathbf{XP}$ gives

$$\mathbf{X}'\mathbf{X}\mathbf{P} = \mathbf{X}'\mathbf{X}\mathbf{P}\mathbf{X}'\mathbf{X}\mathbf{P},$$

so that $\mathbf{X}'\mathbf{X}\mathbf{P}$ is idempotent. It may similarly be shown that $\mathbf{X}\mathbf{P}\mathbf{X}'$ and $\mathbf{P}\mathbf{X}'\mathbf{X}$ are also idempotent. Furthermore, multiplying on the right by $\mathbf{X}$ gives $\mathbf{X}'\mathbf{X}(\mathbf{P}\mathbf{X}'\mathbf{X} - \mathbf{I}) = \mathbf{0}$, so that a general solution to the equations $\mathbf{X}'\mathbf{X}\tilde{\boldsymbol{\beta}} = \mathbf{0}$ is $(\mathbf{P}\mathbf{X}'\mathbf{X} - \mathbf{I})\mathbf{Z}$, where $\mathbf{Z}$ is an arbitrary vector. A general solution to the normal equations is thus

$$\tilde{\boldsymbol{\beta}} = \mathbf{P}\mathbf{X}'\mathbf{Y} + (\mathbf{P}\mathbf{X}'\mathbf{X} - \mathbf{I})\mathbf{Z}.$$

If $\mathbf{P} = (\mathbf{X}'\mathbf{X})^{-1}$, the extra term vanishes; if $k < p$, it would appear that we might find ourselves in the awkward position of having numerous solutions. If we seek to estimate a linear function $\psi = \boldsymbol{\lambda}'\boldsymbol{\beta}$ by substituting solution vectors $\tilde{\boldsymbol{\beta}}$ for $\boldsymbol{\beta}$, we obtain a set of estimates

$$\tilde{\psi} = \boldsymbol{\lambda}'\mathbf{P}\mathbf{X}'\mathbf{Y} + \boldsymbol{\lambda}'(\mathbf{P}\mathbf{X}'\mathbf{X} - \mathbf{I})\mathbf{Z}. \tag{2.2}$$

We now define an estimable function [Bose (1944)]. A linear combination $\psi = \boldsymbol{\lambda}'\boldsymbol{\beta}$ of the parameters is said to be estimable if, and only if, there exists a linear combination $\mathbf{c}'\mathbf{Y}$ of the observations such that $E(\mathbf{c}'\mathbf{Y}) = \boldsymbol{\lambda}'\boldsymbol{\beta}$, i.e., such that $\mathbf{c}'\mathbf{X}\boldsymbol{\beta} = \boldsymbol{\lambda}'\boldsymbol{\beta}$ for all vectors $\boldsymbol{\beta}$. The condition is that a vector $\mathbf{c}$ exists such that $\mathbf{X}'\mathbf{c} = \boldsymbol{\lambda}$. A linear function ψ is estimable if, and only if, $\boldsymbol{\lambda}$ is a linear combination of the rows of $\mathbf{X}$. It follows that the set of estimable functions forms a vector space. The dimension of the vector space, i.e., the number of linearly independent estimable functions, is equal to the rank of $\mathbf{X}$.

An equivalent necessary and sufficient condition for a linear combination of the parameters to be estimable is that there exists a solution vector to the system of equations $\mathbf{X}'\mathbf{X}\mathbf{r} = \boldsymbol{\lambda}$. If such a vector $\mathbf{r}$ exists, then we let $\mathbf{X}\mathbf{r} = \mathbf{c}$. Conversely, if there is a vector $\mathbf{c}$ such that $\mathbf{X}'\mathbf{c} = \boldsymbol{\lambda}$, $\mathbf{c}$ is a solution vector to that set of equations, and hence $r(\mathbf{X}') = r(\mathbf{X}', \boldsymbol{\lambda})$. It follows that $r(\mathbf{X}'\mathbf{X})$ and $r(\mathbf{X}'\mathbf{X}, \boldsymbol{\lambda})$ are also equal, and so the set of equations $\mathbf{X}'\mathbf{X}\mathbf{r} = \boldsymbol{\lambda}$ has a solution vector $\mathbf{r}$.

Multiplying Equation 2.1 on the left by $\mathbf{r}'$ and on the right by $\mathbf{X}$ gives $\mathbf{r}'\mathbf{X}'\mathbf{X} = \mathbf{r}'\mathbf{X}'\mathbf{X}\mathbf{P}\mathbf{X}'\mathbf{X}$ or $\boldsymbol{\lambda}' = \boldsymbol{\lambda}'\mathbf{P}\mathbf{X}'\mathbf{X}$. Hence the term $\boldsymbol{\lambda}'(\mathbf{P}\mathbf{X}'\mathbf{X} - \mathbf{I})\mathbf{Z}$ in Equation 2.2 vanishes, and that equation becomes $\tilde{\psi} = \boldsymbol{\lambda}'\mathbf{P}\mathbf{X}'\mathbf{Y}$.

Let $\mathbf{P}^*$ be any other solution matrix. From Equation 2.1 we have $\mathbf{X}'\mathbf{X}\mathbf{P}\mathbf{X}' = \mathbf{X}'\mathbf{X}\mathbf{P}^*\mathbf{X}'$. Multiplying on the left by $\mathbf{r}'$ gives $\boldsymbol{\lambda}'\mathbf{P}\mathbf{X}' = \boldsymbol{\lambda}'\mathbf{P}^*\mathbf{X}'$, whence $\tilde{\psi} = \boldsymbol{\lambda}'\mathbf{P}\mathbf{X}'\mathbf{Y} = \boldsymbol{\lambda}'\mathbf{P}^*\mathbf{X}'\mathbf{Y}$.

Thus, we have shown that the estimates of estimable functions are unique. They are the same, no matter which solution vector $\tilde{\boldsymbol{\beta}}$ to the normal equations we choose. Henceforth, we can denote $\boldsymbol{\lambda}'\mathbf{P}\mathbf{X}'\mathbf{Y}$ by $\hat{\psi}$ and call it the least squares estimate of $\boldsymbol{\lambda}'\boldsymbol{\beta}$.

The estimate $\hat{\psi}$ is unbiased; we have

$$E(\hat{\psi}) = E(\boldsymbol{\lambda}'\mathbf{P}\mathbf{X}'\mathbf{Y}) = \boldsymbol{\lambda}'\mathbf{P}\mathbf{X}'\mathbf{X}\boldsymbol{\beta} = \boldsymbol{\lambda}'\boldsymbol{\beta}.$$

The variance of $\hat{\psi}$ is given by

$$V(\hat{\psi}) = V(\boldsymbol{\lambda}'\mathbf{P}\mathbf{X}'\mathbf{Y}) = \boldsymbol{\lambda}'\mathbf{P}\mathbf{X}'\mathbf{X}\mathbf{P}'\boldsymbol{\lambda}\sigma^2 = \boldsymbol{\lambda}'\mathbf{P}\boldsymbol{\lambda}\sigma^2.$$

This means that we can, in calculating the variances of least squares estimates of estimable functions, act as if $\mathbf{P}\sigma^2$ is the variance-covariance matrix of $\hat{\boldsymbol{\beta}}$. In particular, for any solution matrix $\mathbf{P}$ we have

$$V(\hat{\beta}_i - \hat{\beta}_j) = (p_{ii} + p_{jj} - p_{ij} - p_{ji})\sigma^2.$$

EXAMPLE. Consider the experiment mentioned earlier in which six observations are made, three on each of two treatments. The usual model is $y_{ij} = m + t_i + e_{ij}$ where $i = 1, 2$ and $j = 1, 2, 3$. We have three parameters: $\boldsymbol{\beta}' = (m, t_1, t_2)$, and the transpose of the design matrix is

$$\mathbf{X}' = \begin{bmatrix} 1 & 1 & 1 & 1 & 1 & 1 \\ 1 & 1 & 1 & 0 & 0 & 0 \\ 0 & 0 & 0 & 1 & 1 & 1 \end{bmatrix};$$

$\mathbf{X}'\mathbf{Y} = (G, T_1, T_2)'$ where $T_i = \sum_j y_{ij}$ and $G = \sum_i \sum_j y_{ij}$. The rank of $\mathbf{X}'\mathbf{X}$ is 2, and the normal equations are

$$\mathbf{X}'\mathbf{Y} = \begin{bmatrix} 6 & 3 & 3 \\ 3 & 3 & 0 \\ 3 & 0 & 3 \end{bmatrix}\hat{\boldsymbol{\beta}}.$$

The usual method of solving the equations is to impose the side condition $t_1 + t_2 = 0$ upon the parameters. This makes the first equation $6\hat{m} = G$, whence $6\hat{t}_1 = 2T_1 - G$ and $6\hat{t}_2 = 2T_2 - G$. The corresponding solution matrix $\mathbf{P}_1$ is unsymmetrical.

$$6\mathbf{P}_1 = \begin{bmatrix} 1 & 0 & 0 \\ -1 & 2 & 0 \\ -1 & 0 & 2 \end{bmatrix}.$$

If $\psi = t_1 - t_2$, we have $\boldsymbol{\lambda}' = (0, 1, -1)$ and, if $3\mathbf{r}' = (0, 1, -1)$, $\mathbf{X}'\mathbf{X}\mathbf{r} = \boldsymbol{\lambda}$. Then ψ is an estimable function and

$$\hat{\psi} = (0, 1, -1)\mathbf{P}_1\mathbf{X}'\mathbf{Y} = (T_1 - T_2)/3,$$
$$V(\hat{\psi}) = \boldsymbol{\lambda}'\mathbf{P}_1\boldsymbol{\lambda}\sigma^2 = 2\sigma^2/3.$$

A second solution matrix with $t_1 + t_2 = 0$ is given by

$$12\mathbf{P}_2 = \begin{bmatrix} 1 & 1 & 1 \\ 1 & 1 & -3 \\ 1 & -3 & 1 \end{bmatrix},$$

for which $\hat{m} = (G + T_1 + T_2)/12$ and $\hat{t}_1 = (G + T_1 - 3T_2)/12$. Again we have $\hat{\psi} = (T_1 - T_2)/3$ and $V(\hat{\psi}) = (1 + 3 + 3 + 1)\sigma^2/12 = 2\sigma^2/3$.

Two other solution matrices having $t_2 = 0$ and $t_1 = (T_1 - T_2)/3$ are given by

$$3\mathbf{P}_3 = \begin{bmatrix} 0 & 0 & 1 \\ 0 & 1 & -1 \\ 0 & 0 & 0 \end{bmatrix}, \qquad 3\mathbf{P}_4 = \begin{bmatrix} 0 & 0 & 1 \\ 0 & 1 & -1 \\ 1 & -1 & -1 \end{bmatrix}.$$

These give the same results as before for $\hat{\psi}$ and $V(\hat{\psi})$.

In this example the estimable functions are $\psi_1 = m + t_1$, $\psi_2 = m + t_2$, and any linear combination of them, including in particular $\psi_3 = t_1 - t_2$.

2.9 Solving the Normal Equations

In the example given in the previous section, we solved the normal equations by imposing a side condition upon the parameters. The side condition $t_1 + t_2 = 0$, or $\hat{t}_1 + \hat{t}_2 = 0$, added an extra equation to the normal equations, which then gave us a set of $p + 1$ equations of which p were linearly independent. Any nonestimable linear function of the parameters can be used as a side condition. For example, $m = 0$ or $t_1 = 0$ or $t_2 = 0$ would each suffice. So, for that matter, would a bizarre linear combination such as $72m + 13t_1 + 97t_2 = 0$; the equations would not be convenient to solve and the estimates of m, t_1, and t_2 would look a little peculiar, but it should be emphasized that the estimates $\hat{\psi}$ of the estimable functions ψ would be the same no matter which of the side conditions was imposed. The usual procedure is to choose a side condition which enables us to obtain a solution matrix $\mathbf{P}$ and a solution vector $\hat{\boldsymbol{\beta}}$ as conveniently as possible.

We present now two standard methods of obtaining solution matrices in the case where $r(\mathbf{X}) = k = p - m$ and $m > 0$, with a set of m side conditions. For those experiments in which the normal equations have a simple form, the solution matrix is easily obtained directly; the reader may, therefore, wish to postpone this section until he reaches Chapter 11.

Let $\mathbf{D}$ be a matrix of p rows and m columns such that $r(\mathbf{D}) = m$ and $\mathbf{XD} = \mathbf{0}$. Suppose also that we add m further rows to $\mathbf{X}$ in the form of a matrix $\mathbf{H}$ of m rows and p columns, such that the rank of $(\mathbf{X}', \mathbf{H}')$ is p. Then $\mathbf{HD}$ is nonsingular. If the normal equations are solved, subject to the side conditions $\mathbf{H}\boldsymbol{\beta} = \mathbf{0}$, the solutions $\hat{\beta}_j$ are unique numerically. However, since $\mathbf{D}'\mathbf{X}'\mathbf{Y} = \mathbf{0}$, the solutions may differ in form by some linear combination of the rows of $\mathbf{D}'\mathbf{X}'\mathbf{Y}$.

In the previous example, $\mathbf{D}' = (1, -1, -1)$ and $\mathbf{D}'\mathbf{X}'\mathbf{Y} = G - T_1 - T_2 \equiv 0$. The solutions are obtained with $\mathbf{H} = (0, 1, 1)$. The solution matrix $\mathbf{P}_1$ gives $\hat{\mu} = G/6$; $\mathbf{P}_2$ gives $\hat{\mu} = (G + T_1 + T_2)/12 = (G/6) - (\mathbf{D}'\mathbf{X}'\mathbf{Y}/12)$.

Let $\mathbf{P}$ and $\mathbf{P}^*$ be two solution matrices under the side conditions $\mathbf{H}\boldsymbol{\beta} = \mathbf{0}$. Then $\mathbf{PX}'\mathbf{Y} = \mathbf{P}^*\mathbf{X}'\mathbf{Y}$. This is an identity in $\mathbf{Y}$, and so $(\mathbf{P} - \mathbf{P}^*)\mathbf{X}' = \mathbf{0}$. But $r(\mathbf{X}) = (p - m)$, and $r(\mathbf{D}) = m$. Thus, if $\mathbf{W}$ is a matrix such that $\mathbf{XW} = \mathbf{0}$, there exists a matrix $\mathbf{E}$ such that $\mathbf{W} = \mathbf{DE}'$, and so $\mathbf{P} - \mathbf{P}^* = \mathbf{ED}'$. If $\mathbf{P}$ and $\mathbf{P}^*$

are both symmetric, then $\mathbf{P}^* = \mathbf{P} + \mathbf{DCD}'$, where $\mathbf{C}$ is a symmetric matrix of order m. The two standard methods of solving the normal equations both give symmetric solution matrices.

In the first method [Kempthorne (1952) and Graybill (1961)], the matrix $\mathbf{X}'\mathbf{X}$ is augmented to form the new matrix

$$\mathbf{A}^* = \begin{bmatrix} \mathbf{X}'\mathbf{X} & \mathbf{H}' \\ \mathbf{H} & \mathbf{0} \end{bmatrix}.$$

Then $r(\mathbf{A}^*) = p + m$, and $\mathbf{A}^{*-1}$ exists. Let

$$\mathbf{A}^{*-1} = \begin{bmatrix} \mathbf{B}_{11} & \mathbf{B}_{12} \\ \mathbf{B}_{21} & \mathbf{B}_{22} \end{bmatrix}.$$

The identity $\mathbf{A}^*\mathbf{A}^{*-1} = \mathbf{I}$ gives four equations:

$$\mathbf{X}'\mathbf{X}\mathbf{B}_{11} + \mathbf{H}'\mathbf{B}_{21} = \mathbf{I}, \tag{2.3}$$

$$\mathbf{X}'\mathbf{X}\mathbf{B}_{12} + \mathbf{H}'\mathbf{B}_{22} = \mathbf{0}, \tag{2.4}$$

$$\mathbf{H}\mathbf{B}_{11} = \mathbf{0}, \tag{2.5}$$

$$\mathbf{H}\mathbf{B}_{12} = \mathbf{I}. \tag{2.6}$$

Multiplying Equations 2.3 and 2.4 on the left by $\mathbf{D}'$ and recalling that $\mathbf{D}'\mathbf{X}' = \mathbf{0}$ and that $\mathbf{D}'\mathbf{H}'$ is not singular, we have $\mathbf{D}'\mathbf{H}'\mathbf{B}_{21} = \mathbf{D}'$, whence $\mathbf{B}_{12} = \mathbf{B}_{21}' = \mathbf{D}(\mathbf{HD})^{-1}$ and $\mathbf{D}'\mathbf{H}'\mathbf{B}_{22} = \mathbf{0}$, whence $\mathbf{B}_{22} = \mathbf{0}$.

Multiplying Equation 2.3 on the left by $\mathbf{B}_{11}$ gives $\mathbf{B}_{11}\mathbf{X}'\mathbf{X}\mathbf{B}_{11} = \mathbf{B}_{11}$, and so $\mathbf{X}'\mathbf{X}\mathbf{B}_{11}$ is idempotent.

$$\mathbf{X}'\mathbf{X}\mathbf{B}_{11} = \mathbf{I} - \mathbf{H}'\mathbf{B}_{21} = \mathbf{I} - \mathbf{H}'(\mathbf{D}'\mathbf{H}')^{-1}\mathbf{D}',$$

$$\mathbf{X}'\mathbf{X}\mathbf{B}_{11}\mathbf{X}'\mathbf{Y} = \mathbf{X}'\mathbf{Y} - \mathbf{H}'(\mathbf{D}'\mathbf{H}')^{-1}\mathbf{D}'\mathbf{X}'\mathbf{Y} = \mathbf{X}'\mathbf{Y}.$$

Thus, $\hat{\boldsymbol{\beta}} = \mathbf{B}_{11}\mathbf{X}'\mathbf{Y}$ is a solution to the normal equations, and $\operatorname{cov}(\hat{\boldsymbol{\beta}}) = \mathbf{B}_{11}\mathbf{X}'\mathbf{X}\mathbf{B}_{11}\sigma^2 = \mathbf{B}_{11}\sigma^2$.

In the second method [Scheffé (1959) and Plackett (1960)], the solution matrix is $\mathbf{P}^* = (\mathbf{X}'\mathbf{X} + \mathbf{H}'\mathbf{H})^{-1}$, with $\hat{\boldsymbol{\beta}}^* = \mathbf{P}^*\mathbf{X}'\mathbf{Y}$ as the corresponding solution vector (see particularly Scheffé, p. 17). Numerically, $\mathbf{P}^*\mathbf{X}'\mathbf{Y} = \mathbf{B}_{11}\mathbf{X}'\mathbf{Y}$, but it does not follow that $(\mathbf{X}'\mathbf{X} + \mathbf{H}'\mathbf{H})^{-1} = \mathbf{B}_{11}$.

$$\operatorname{cov}(\hat{\boldsymbol{\beta}}^*) = (\mathbf{X}'\mathbf{X} + \mathbf{H}'\mathbf{H})^{-1}\mathbf{X}'\mathbf{X}(\mathbf{X}'\mathbf{X} + \mathbf{H}'\mathbf{H})^{-1}\sigma^2 = \operatorname{cov}(\hat{\boldsymbol{\beta}}) = \mathbf{B}_{11}\sigma^2.$$

But $(\mathbf{X}'\mathbf{X} + \mathbf{H}'\mathbf{H})(\mathbf{I} - \mathbf{D}(\mathbf{HD})^{-1}\mathbf{H}) = \mathbf{X}'\mathbf{X}$, and so

$$(\mathbf{X}'\mathbf{X} + \mathbf{H}'\mathbf{H})^{-1}\mathbf{X}'\mathbf{X}(\mathbf{X}'\mathbf{X} + \mathbf{H}'\mathbf{H})^{-1} = (\mathbf{I} - \mathbf{D}(\mathbf{HD})^{-1}\mathbf{H})(\mathbf{X}'\mathbf{X} + \mathbf{H}'\mathbf{H})^{-1} = \mathbf{B}_{11}.$$

This gives us a method of obtaining $\mathbf{B}_{11}$ from $(\mathbf{X}'\mathbf{X} + \mathbf{H}'\mathbf{H})^{-1}$. We cannot, however, obtain $(\mathbf{X}'\mathbf{X} + \mathbf{H}'\mathbf{H})^{-1}$ from $\mathbf{B}_{11}$ with this result because

$$(\mathbf{I} - \mathbf{D}(\mathbf{H}\mathbf{D})^{-1}\mathbf{H})$$

is idempotent, and thus is not of full rank.

We now obtain a different relationship between the two solution matrices. We write $(\mathbf{X}'\mathbf{X} + \mathbf{H}'\mathbf{H})^{-1} = \mathbf{B}_{11} + \mathbf{D}\mathbf{C}\mathbf{D}'$. Then

$$\begin{aligned}\mathbf{I} &= (\mathbf{X}'\mathbf{X} + \mathbf{H}'\mathbf{H})(\mathbf{B}_{11} + \mathbf{D}\mathbf{C}\mathbf{D}') = \mathbf{X}'\mathbf{X}\mathbf{B}_{11} + \mathbf{H}'\mathbf{H}\mathbf{D}\mathbf{C}\mathbf{D}'\\ &= \mathbf{I} - \mathbf{H}'(\mathbf{D}'\mathbf{H}')^{-1}\mathbf{D}' + \mathbf{H}'\mathbf{H}\mathbf{D}\mathbf{C}\mathbf{D}',\end{aligned}$$

and so $\mathbf{H}'(\mathbf{D}'\mathbf{H}')^{-1}\mathbf{D}' = \mathbf{H}'\mathbf{H}\mathbf{D}\mathbf{C}\mathbf{D}'$. Multiplying both sides on the left by $(\mathbf{D}'\mathbf{H}')^{-1}\mathbf{D}'$ and on the right by $\mathbf{H}'(\mathbf{D}'\mathbf{H}')^{-1}$ reduces this equation [John (1964)*a*] to

$$(\mathbf{D}'\mathbf{H}')^{-1} = \mathbf{H}\mathbf{D}\mathbf{C},$$

whence

$$\mathbf{C} = (\mathbf{H}\mathbf{D})^{-1}(\mathbf{D}'\mathbf{H}')^{-1} = (\mathbf{D}'\mathbf{H}'\mathbf{H}\mathbf{D})^{-1},$$

so that

$$(\mathbf{X}'\mathbf{X} + \mathbf{H}'\mathbf{H})^{-1} = \mathbf{B}_{11} + \mathbf{D}(\mathbf{D}'\mathbf{H}'\mathbf{H}\mathbf{D})^{-1}\mathbf{D}'.$$

2.10 A Theorem in Linear Algebra

The following theorem will be used later in proving Cochran's theorem. Alternative proofs of the theorem and other relevant results are given by James (1952), Lancaster (1954), Graybill and Marsaglia (1957), and Banerjee (1964).

Theorem 1. *Let* $\mathbf{A}_1, \ldots, \mathbf{A}_k$ *be symmetric matrices of order* n *such that* $\mathbf{I} = \sum_{i=1}^{k} \mathbf{A}_i$. *Each of the following conditions is necessary and sufficient for the other two:*

(i) *All the* $\mathbf{A}_i$ *are idempotent;*
(ii) $\mathbf{A}_i\mathbf{A}_j = \mathbf{0}$ *for all pairs* $i, j, i \neq j$;
(iii) $\sum_i n_i = n$ *where* $n_i = r(\mathbf{A}_i)$.

PROOF. There is an orthogonal matrix $\mathbf{P}$ such that

$$\mathbf{P}'\mathbf{I}\mathbf{P} = \mathbf{I} = \mathbf{P}'\mathbf{A}_1\mathbf{P} + \sum_{j=2}^{k} \mathbf{P}'\mathbf{A}_j\mathbf{P},$$

where

$$\mathbf{P}'\mathbf{A}_1\mathbf{P} = \begin{pmatrix}\mathbf{\Lambda} & \mathbf{0}\\ \mathbf{0} & \mathbf{0}\end{pmatrix} \quad \text{and} \quad \mathbf{P}'\mathbf{A}_j\mathbf{P} = \begin{pmatrix}\mathbf{B}_j & \mathbf{C}_j\\ \mathbf{C}_j' & \mathbf{E}_j\end{pmatrix},$$

where $\mathbf{\Lambda}$ is the diagonal matrix whose diagonal elements are the n_1 nonzero latent roots of $\mathbf{A}_1$, and $\mathbf{B}_j$ is a square matrix of order n_1. Throughout this proof, j takes the values $2 \le j \le k$.

Suppose all the $\mathbf{A}_i$ are idempotent. To establish condition (ii), it is enough to show that $\mathbf{A}_1\mathbf{A}_j = \mathbf{0}$ for all j. We have $\mathbf{\Lambda} = \mathbf{I}$ and $\sum_j \operatorname{tr}(\mathbf{B}_j) = 0$; but $\mathbf{P}'\mathbf{A}_j\mathbf{P}$ is idempotent, and so the diagonal elements of $\mathbf{B}_j$ cannot be negative. Hence $\mathbf{B}_j = \mathbf{0}$ and $\mathbf{C}_j = \mathbf{0}$ for each j. It follows that $\mathbf{A}_1\mathbf{A}_j = \mathbf{P}\mathbf{P}'\mathbf{A}_1\mathbf{P}\mathbf{P}'\mathbf{A}_j\mathbf{P} = \mathbf{0}$.

Since $\mathbf{P}'\mathbf{A}_j\mathbf{P}$ is idempotent, $\mathbf{E}_j$ is idempotent. Then $n = \operatorname{tr}(\mathbf{I}_n) = n_1 + \sum \operatorname{tr}(\mathbf{E}_j) = n_1 + \sum n_j$, and condition (iii) is established.

We now show that conditions (ii) and (iii) each imply condition (i). Suppose $\mathbf{A}_i\mathbf{A}_j = \mathbf{0}$ for all pairs i, j where $i \ne j$. Then for each i,

$$\mathbf{A}_i = \mathbf{A}_i\mathbf{I} = \mathbf{A}_i(\mathbf{A}_1 + \cdots + \mathbf{A}_i + \cdots + \mathbf{A}_k) = \mathbf{A}_i^2.$$

If condition (iii) holds, we can show that $\mathbf{A}_1$ is idempotent. We write $\mathbf{B}$, $\mathbf{C}$, $\mathbf{E}$, $\mathbf{Q}$ for $\sum \mathbf{B}_j$, $\sum \mathbf{C}_j$, $\sum \mathbf{E}_j$, $\sum \mathbf{A}_j$. Then, since $\mathbf{P}'\mathbf{A}_1\mathbf{P} + \mathbf{P}'\mathbf{Q}\mathbf{P} = \mathbf{I}$, we have $\mathbf{E} = \mathbf{I}_{n-n_1}$, $\mathbf{C} = \mathbf{0}$. But $r(\mathbf{P}'\mathbf{Q}\mathbf{P}) = r(\mathbf{Q}) = n - n_1 = r(\mathbf{E})$, and so $\mathbf{B} = \mathbf{0}$ and $\mathbf{\Lambda} = \mathbf{I}$, so that $\mathbf{A}_1$ is idempotent. ∎

2.11 The Distribution of Quadratic Forms

If $\mathbf{Y}$ is a vector of N random variables, the quadratic form $\mathbf{Y}'\mathbf{A}\mathbf{Y}$ is also a random variable. Most of the results in this section about the distribution of quadratic forms hold only when $\mathbf{Y}$ has the multivariate normal distribution. The following lemma is true even when the y_i are not normally distributed.

Lemma. *Let* $E(\mathbf{Y}) = \boldsymbol{\mu}$ *and* $\operatorname{cov}(\mathbf{Y}) = \mathbf{I}\sigma^2$. *Then* $E(\mathbf{Y}'\mathbf{A}\mathbf{Y}) = \boldsymbol{\mu}'\mathbf{A}\boldsymbol{\mu} + \sigma^2 \operatorname{tr}(\mathbf{A})$.

PROOF.

$$\mathbf{Y}'\mathbf{A}\mathbf{Y} = \sum_i a_{ii}\, y_i^2 + 2 \sum_{i<j}\sum a_{ij} y_i y_j,$$

$$E(y_i^2) = \mu_i^2 + \sigma^2, \qquad E(y_i y_j) = \mu_i\mu_j.$$

Hence,

$$E(\mathbf{Y}'\mathbf{A}\mathbf{Y}) = \sum_i a_{ii}\mu_i^2 + \sum_i a_{ii}\sigma^2 + 2\sum_{i<j}\sum a_{ij}\mu_i\mu_j = \boldsymbol{\mu}'\mathbf{A}\boldsymbol{\mu} + \sigma^2 \operatorname{tr}(\mathbf{A}). \quad ∎$$

The next result concerns the normal distribution.

Theorem 1. *Let* $\mathbf{Y} \sim N(\boldsymbol{\mu}, \mathbf{I})$. *The moment generating function of* $\mathbf{Y}'\mathbf{A}\mathbf{Y}$ *is* $M_{\mathbf{Y}'\mathbf{A}\mathbf{Y}}(t) = |\mathbf{I} - 2\mathbf{A}t|^{-1/2} \exp\{-\tfrac{1}{2}[\boldsymbol{\mu}'\boldsymbol{\mu} - \boldsymbol{\mu}'(\mathbf{I} - 2\mathbf{A}t)^{-1}\boldsymbol{\mu}]\}$.

PROOF. Suppose $\mathbf{X} \sim N(\mathbf{U}, \mathbf{V})$. The probability density function of $\mathbf{X}$ is

$$f_{\mathbf{X}}(\mathbf{x}) = \frac{1}{(2\pi)^{N/2}|\mathbf{V}|^{1/2}} \exp\{-\tfrac{1}{2}(\mathbf{X} - \mathbf{U})'\mathbf{V}^{-1}(\mathbf{X} - \mathbf{U})\},$$

and the corresponding multiple integral over the range $-\infty < \mathbf{X} < +\infty$ is unity. The moment-generating function of $\mathbf{Y}'\mathbf{A}\mathbf{Y}$ is

$$E[e^{t\mathbf{Y}'\mathbf{A}\mathbf{Y}}] = (2\pi)^{-N/2} \int_{-\infty}^{\infty} \exp\{-[(\mathbf{Y} - \boldsymbol{\mu})'(\mathbf{Y} - \boldsymbol{\mu}) - 2t\mathbf{Y}'\mathbf{A}\mathbf{Y}]/2\}\, d\mathbf{Y}.$$

Writing the exponent as $-Q/2$, we have

$$Q = \mathbf{Y}'(\mathbf{I} - 2t\mathbf{A})\mathbf{Y} - \mathbf{Y}'\boldsymbol{\mu} - \boldsymbol{\mu}'\mathbf{Y} + \boldsymbol{\mu}'\boldsymbol{\mu}.$$

Completing the square and writing $\mathbf{U}$ for $(\mathbf{I} - 2\mathbf{A}t)^{-1}\boldsymbol{\mu}$,

$$\mathbf{Q} = (\mathbf{Y} - \mathbf{U})'(\mathbf{I} - 2\mathbf{A}t)(\mathbf{Y} - \mathbf{U}) + \boldsymbol{\mu}'(\mathbf{I} - (\mathbf{I} - 2\mathbf{A}t)^{-1})\boldsymbol{\mu},$$

so that

$$\begin{aligned} M(t) &= (2\pi)^{-N/2} \int_{-\infty}^{\infty} \exp\{-\tfrac{1}{2}(\mathbf{Y} - \mathbf{U})'(\mathbf{I} - 2\mathbf{A}t)(\mathbf{Y} - \mathbf{U})\} \\ &\qquad\qquad d\mathbf{Y} \exp[-\boldsymbol{\mu}'(\mathbf{I} - (\mathbf{I} - 2\mathbf{A}t)^{-1})\boldsymbol{\mu}/2] \\ &= |\mathbf{I} - 2\mathbf{A}t|^{-1/2} \exp\{-\tfrac{1}{2}[\boldsymbol{\mu}'\boldsymbol{\mu} - \boldsymbol{\mu}'(\mathbf{I} - 2\mathbf{A}t)^{-1}\boldsymbol{\mu}]\}. \end{aligned}$$

■

If $\mathbf{A} = \mathbf{I}$, we have for the moment-generating function of $\mathbf{Y}'\mathbf{Y}$

$$M_{\mathbf{Y}'\mathbf{Y}}(t) = (1 - 2t)^{-N/2} \exp[-\lambda + \lambda(1 - 2t)^{-1}].$$

This is the moment-generating function of the noncentral χ^2 distribution with N d.f. and noncentrality parameter $\lambda = \boldsymbol{\mu}'\boldsymbol{\mu}/2$. If $\mathbf{A} = \mathbf{I}$ and $\boldsymbol{\mu} = \mathbf{0}$, $M(t) = (1 - 2t)^{-N/2}$, which is the moment-generating function of the central χ^2 distribution with N d.f.

If $\operatorname{cov}(\mathbf{Y}) = \mathbf{V}$, we may prove in similar fashion that

$$M(t) = |\mathbf{I} - 2t\mathbf{A}\mathbf{V}|^{-1/2} \exp\{-\tfrac{1}{2}[\boldsymbol{\mu}'\mathbf{V}^{-1}\boldsymbol{\mu} - \boldsymbol{\mu}'\mathbf{V}^{-1}(\mathbf{I} - 2t\mathbf{V}\mathbf{A})^{-1}\boldsymbol{\mu}]\}.$$

The next theorem points out the importance of idempotent matrices in the theory of the distribution of quadratic forms.

Theorem 2. *Let* $\mathbf{Y} \sim N(\boldsymbol{\mu}, \mathbf{I})$. *A necessary and sufficient condition that* $\mathbf{Y}'\mathbf{A}\mathbf{Y}$ *have a* χ^2 *distribution with* $m \leq N$ d.f. *is that* $\mathbf{A}$ *be an idempotent matrix of rank* m.

PROOF. We give a proof for the case $\boldsymbol{\mu} = \mathbf{0}$ when the distribution is central χ^2. The proof consists of equating the moment-generating function of $\mathbf{Y}'\mathbf{AY}$ with that of a χ^2 distribution. The proof in the noncentral case is similar:

$$M_{\mathbf{Y}'\mathbf{AY}}(t) = |\mathbf{I} - 2\mathbf{A}t|^{-1/2}.$$

$\mathbf{A}$ is a symmetric matrix; hence, there exists an orthogonal matrix $\mathbf{P}$ such that $\mathbf{P}'\mathbf{AP} = \boldsymbol{\Lambda}$. Then

$$|\mathbf{I} - 2\mathbf{A}t| = |\mathbf{P}'|\,|\mathbf{I} - 2\mathbf{A}t|\,|\mathbf{P}| = |\mathbf{I} - 2\boldsymbol{\Lambda}t|.$$

Sufficiency: Let $\mathbf{A}$ be idempotent of rank m. Then $\boldsymbol{\Lambda}$ is a diagonal matrix with m of the diagonal elements unity and the remainder zero; $\mathbf{I} - 2\boldsymbol{\Lambda}t$ is a diagonal matrix with m diagonal elements $(1 - 2t)$ and the others unity. Hence, $|\mathbf{I} - 2\mathbf{A}t| = (1 - 2t)^m$ and $M(t) = (1 - 2t)^{-m/2}$, which is the moment-generating function of a χ^2 distribution with m d.f.

Necessity: In general, $|\mathbf{I} - 2\boldsymbol{\Lambda}t| = \Pi(1 - 2\lambda_i t)$. Suppose that $\mathbf{Y}'\mathbf{AY}$ has a χ^2 distribution with m d.f. Then we have $(1 - 2t)^m = \Pi(1 - 2\lambda_i t)$. This is an identity for all t, and it follows that the latent roots of $\mathbf{A}$ are $\lambda = 0, 1$ with multiplicities $N - m$ and m, so that $\mathbf{A}$ is an idempotent matrix of rank m. ∎

Two extensions of this theorem are of interest: we state them without proof as corollaries.

Corollary 1. *If* $\mathbf{Y} \sim N(\boldsymbol{\mu}, \mathbf{I}\sigma^2)$, *the quadratic form* $\mathbf{Y}'\mathbf{AY}/\sigma^2$ *has a noncentral* χ^2 *distribution with m d.f. and noncentrality parameter* $\lambda = \boldsymbol{\mu}'\mathbf{A}\boldsymbol{\mu}/2\sigma^2$ *if, and only if,* $\mathbf{A}$ *is idempotent of rank m.*

Corollary 2. *If* $\mathbf{Y} \sim N(\boldsymbol{\mu}, \mathbf{V})$, *the form* $\mathbf{Y}'\mathbf{AY}$ *has a noncentral* χ^2 *distribution with m d.f. and noncentrality parameter* $\lambda = \boldsymbol{\mu}'\mathbf{A}\boldsymbol{\mu}/2$ *if, and only if,* $\mathbf{AV}$ *is an idempotent matrix of rank m.*

The following theorem is concerned with the independence of quadratic forms in normal variables. There are at least five proofs in the literature, mostly for the case $\boldsymbol{\mu} = \mathbf{0}$ [Craig (1943), Hotelling (1944*b*), Ogawa (1949), Aitken (1950), and Lancaster (1954)]. The validity of the first two proofs has been questioned. Hotelling questioned Craig's proof, and his proof in turn was questioned by Ogawa. The proof that we shall give is a modification by Lancaster of Aitken's proof.

Theorem 3. *Let* $\mathbf{Y} \sim N(\boldsymbol{\mu}, \mathbf{I})$. *Two quadratic forms* $\mathbf{Y}'\mathbf{AY}$, $\mathbf{Y}'\mathbf{BY}$ *are independent if and only if* $\mathbf{AB} = \mathbf{0}$.

PROOF. [Lancaster (1954)]. This proof is for the case $\boldsymbol{\mu} = \mathbf{0}$. We have shown that the moment-generating functions of $\mathbf{Y}'\mathbf{AY}$ and $\mathbf{Y}'\mathbf{BY}$ are $|\mathbf{I} - 2\mathbf{A}s|^{-1/2}$ and

$|\mathbf{I} - 2\mathbf{B}t|^{-1/2}$. By a similar argument, the joint moment-generating function of the two forms is seen to be $|\mathbf{I} - 2\mathbf{A}s - 2\mathbf{B}t|^{-1/2}$. We have to show that the joint function is the product of the two individual functions, i.e.,

$$|\mathbf{I} - 2\mathbf{A}s|\,|\mathbf{I} - 2\mathbf{B}t| = |\mathbf{I} - 2\mathbf{A}s - 2\mathbf{B}t|, \tag{2.3}$$

for all values of s and t in an interval around the origin if, and only if, $\mathbf{AB} = \mathbf{0}$.

The sufficiency part is trivial since, if $\mathbf{AB} = \mathbf{0}$, we have

$$|\mathbf{I} - 2\mathbf{A}s|\,|\mathbf{I} - 2\mathbf{B}t| = |\mathbf{I} - 2\mathbf{A}s - 2\mathbf{B}t + 4\mathbf{AB}st| = |\mathbf{I} - 2\mathbf{A}s - 2\mathbf{B}t|.$$

Conversely, suppose Equation 2.3 is true. Let $\mathbf{P}$ be an orthogonal matrix such that

$$\mathbf{P}'\mathbf{AP} = \begin{pmatrix} \mathbf{\Lambda} & \mathbf{0} \\ \mathbf{0} & \mathbf{0} \end{pmatrix}, \qquad \mathbf{P}'\mathbf{BP} = \begin{pmatrix} \mathbf{C}_{11} & \mathbf{C}_{12} \\ \mathbf{C}_{21} & \mathbf{C}_{22} \end{pmatrix},$$

where $\mathbf{\Lambda}$ is diagonal and $\mathbf{\Lambda}$ and $\mathbf{C}_{11}$ are square matrices of order $r(\mathbf{A})$. It follows from Equation 2.3 that for any $\lambda \neq 0$ we may replace s by s/λ and t by t/λ to obtain

$$|\lambda\mathbf{I} - 2\mathbf{A}s|\,|\lambda\mathbf{I} - 2\mathbf{B}t| = \lambda^n|\lambda\mathbf{I} - 2\mathbf{A}s - 2\mathbf{B}t|$$

or

$$|\lambda\mathbf{I} - 2\mathbf{P}'\mathbf{AP}s|\,|\lambda\mathbf{I} - 2\mathbf{P}'\mathbf{BP}t| = \lambda^n|\lambda\mathbf{I} - 2\mathbf{P}'\mathbf{AP}s - 2\mathbf{P}'\mathbf{BP}t|.$$

Equating coefficients of $s^{r(\mathbf{A})}$ on each side of the equation gives

$$|\lambda\mathbf{I} - 2\mathbf{P}'\mathbf{BP}t| = \lambda^{r(\mathbf{A})}|\lambda\mathbf{I} - 2\,\mathbf{C}_{22}t|.$$

Thus $\mathbf{P}'\mathbf{BP}$ and $\mathbf{C}_{22}$ have the same nonzero latent roots. This implies that the sum of the squares of elements of $\mathbf{P}'\mathbf{BP}$ is equal to the sum of the squares of the elements of $\mathbf{C}_{22}$, each being the sum of the squares of the latent roots. Since all the elements of $\mathbf{B}$ are real, this implies that $\mathbf{C}_{11}$, $\mathbf{C}_{12}$, and $\mathbf{C}_{21}$ are all zero matrices so that $\mathbf{P}'\mathbf{APP}'\mathbf{BP} = \mathbf{0}$ and $\mathbf{AB} = \mathbf{0}$.

In the case $\boldsymbol{\mu} \neq \mathbf{0}$, we must, in addition, take into account the exponents in the moment-generating functions. Thus we must show that

$$\boldsymbol{\mu}'\{\mathbf{I} - (\mathbf{I} - 2\mathbf{A}s)^{-1}\}\boldsymbol{\mu} + \boldsymbol{\mu}'\{\mathbf{I} - (\mathbf{I} - 2\mathbf{B}t)^{-1}\}\boldsymbol{\mu} = \boldsymbol{\mu}'\{\mathbf{I} - (\mathbf{I} - 2\mathbf{A}s - 2\mathbf{B}t)^{-1}\}\boldsymbol{\mu}$$

for all s and t in some interval containing $s = t = 0$, i.e.,

$$\mathbf{I} - (\mathbf{I} - 2\mathbf{A}s)^{-1} + \mathbf{I} - (\mathbf{I} - 2\mathbf{B}t)^{-1} = \mathbf{I} - (\mathbf{I} - 2\mathbf{A}s - 2\mathbf{B}t)^{-1},$$

if, and only if, $\mathbf{AB} = \mathbf{0}$.

For sufficiently small s we have

$$\mathbf{I} - (\mathbf{I} - 2\mathbf{A}s)^{-1} = \mathbf{I} - (\mathbf{I} + 2\mathbf{A}s + (2\,\mathbf{A}s)^2 + \cdots + (2\mathbf{A}s)^n + \cdots),$$

and the required condition becomes

$$2\mathbf{A}s + (2\mathbf{A}s)^2 + \cdots + 2\mathbf{B}t + (2\mathbf{B}t)^2 + \cdots$$
$$= (2\mathbf{A}s + 2\mathbf{B}t) + (2\mathbf{A}s + 2\mathbf{B}t)^2 + \cdots.$$

Equating coefficients, this is true if, and only if, $\mathbf{A}^m\mathbf{B}^n s^m t^n = \mathbf{0}$ for all m, n, and for all s, t, which in turn is true if, and only if, $\mathbf{AB} = \mathbf{0}$. ∎

In the analysis of variance, $\mathbf{Y'Y}$ is subdivided into several quadratic forms (sums of squares). The next theorem is concerned with the joint distribution of these sums of squares. It is due to Cochran (1934) and is essentially an algebraic result expressed in a statistical context. He originally proved it in the central case. The noncentral case is due to Madow (1940).

Theorem 4. *Let* $\mathbf{Y} \sim N(\boldsymbol{\mu}, \mathbf{I})$ *and let* $\mathbf{Y'Y} = \mathbf{Y'A}_1\mathbf{Y} + \cdots + \mathbf{Y'A}_k\mathbf{Y}$ *where* $\mathbf{Y}$ *has n observations and* $r(\mathbf{A}_i) = n_i$. *A necessary and sufficient condition for the forms* $\mathbf{Y'A}_i\mathbf{Y}$ *to be distributed independently with* $\mathbf{Y'A}_i\mathbf{Y} \sim \chi^2(n_i)$ *is that* $\sum n_i = n$.

PROOF. We have already shown that $\sum n_i = n$ if, and only if, each $\mathbf{A}_i$ is idempotent and if, and only if, $\mathbf{A}_i\mathbf{A}_j = \mathbf{0}$ where $i \neq j$. But $\mathbf{A}_i$ idempotent is equivalent to $\mathbf{Y'A}_i\mathbf{Y} \sim \chi^2(n_i)$, and $\mathbf{A}_i\mathbf{A}_j = \mathbf{0}$ is equivalent to $\mathbf{Y'A}_i\mathbf{Y}$ and $\mathbf{Y'A}_j\mathbf{Y}$ independent. ∎

We may rephrase the statement of the theorem in the following way. Let $\mathbf{Y} \sim N(\boldsymbol{\mu}, \mathbf{I})$, and let $\mathbf{Y'Y} = \sum \mathbf{Y'A}_i\mathbf{Y}$ where $r(\mathbf{A}_i) = n_i$. Then any one of the following five conditions implies the other four:

(i) $\sum n_i = n$;
(ii) Each $\mathbf{A}_i$ is idempotent;
(iii) $\mathbf{A}_i\mathbf{A}_j = \mathbf{0}$ for all pairs i, j, where $i \neq j$;
(iv) $\mathbf{Y'A}_i\mathbf{Y} \sim \chi^2(n_i)$ for each i;
(v) $\mathbf{Y'A}_i\mathbf{Y}$, $\mathbf{Y'A}_j\mathbf{Y}$ are independent for all i, j where $i \neq j$.

Each of the forms $\mathbf{Y'A}_i\mathbf{Y}$ may be written as $\mathbf{Y'B}_i'\mathbf{B}_i\mathbf{Y}$ where $\mathbf{B}_i$ is a matrix of n_i rows and n columns such that $\mathbf{B}_i\mathbf{B}_i' = \mathbf{I}_{n_i}$, i.e., such that the columns of $\mathbf{B}_i'$ form an orthonormal set of n_i vectors, It follows from the fact that $\mathbf{A}_i\mathbf{A}_j = \mathbf{0}$ that $\mathbf{B}_i\mathbf{B}_j' = \mathbf{0}$ also, for we have

$$\mathbf{A}_i\mathbf{A}_j = \mathbf{B}_i'\mathbf{B}_i\mathbf{B}_j'\mathbf{B}_j = \mathbf{0},$$

and, multiplying on the left by $\mathbf{B}_i$ and on the right by $\mathbf{B}_j'$,

$$\mathbf{I}\mathbf{B}_i\mathbf{B}_j'\mathbf{I} = \mathbf{B}_i\mathbf{B}_j' = \mathbf{0}.$$

Let $\mathbf{B}$ be the matrix defined by $\mathbf{B}' = (\mathbf{B}_1', \mathbf{B}_2', \ldots)$. Then $\mathbf{B}$ is an orthogonal matrix, and the transformation $\mathbf{Z} = \mathbf{BY}$ is an orthogonal transformation taking $\mathbf{Y'Y}$ into $\mathbf{Z'Z}$.

2.12 The Gauss–Markoff Theorem

We have shown that the method of least squares provides unique unbiased estimates of the estimable functions. Apart from stating that under normality the least-squares estimates are the same as the maximum likelihood estimates, we have made no claim that these estimates are in any other way particularly desirable. The Gauss–Markoff theorem shows that, in a certain sense, the least-squares estimates are optimal and justifies our choice of this procedure.

Definition. An estimate of a function of the parameters is called a linear estimate if it is a linear combination of the observations, i.e., $\mathbf{a}'\mathbf{Y}$ where $\mathbf{a}$ is a vector.

Theorem. (*Gauss–Markoff*). *Let* $\mathbf{Y}$ *be distributed with* $E(\mathbf{Y}) = \mathbf{X}\boldsymbol{\beta}$ *and* $\text{cov}(\mathbf{Y}) = \mathbf{I}\sigma^2$; *let* $\psi = \boldsymbol{\lambda}'\boldsymbol{\beta}$ *be an estimable function. Then, in the class of all unbiased linear estimates of* ψ, *the least squares estimate* $\hat{\psi} = \boldsymbol{\lambda}'\hat{\boldsymbol{\beta}}$ *has the minimum variance and is unique.*

PROOF. Suppose that $\hat{\boldsymbol{\beta}} = \mathbf{P}\mathbf{X}'\mathbf{Y}$ where $\mathbf{P}$ is any solution matrix. Let ψ^* be an unbiased estimate of ψ, other than $\hat{\psi}$; $\psi^* = \mathbf{a}'\mathbf{Y}$. Let $\mathbf{a}' = \boldsymbol{\lambda}'\mathbf{P}\mathbf{X}' + \mathbf{b}'$. Then

$$E(\psi^*) = E[(\boldsymbol{\lambda}'\mathbf{P}\mathbf{X}' + \mathbf{b}')\mathbf{Y}] = \psi + \mathbf{b}'\mathbf{X}\boldsymbol{\beta}.$$

Since ψ^* is unbiased, $E(\psi^*) = \psi$ and $\mathbf{b}'\mathbf{X}\boldsymbol{\beta} = 0$ for all parameters $\boldsymbol{\beta}$ so that we have $\mathbf{b}'\mathbf{X} = \mathbf{0}$.

Then

$$\begin{aligned} V(\psi^*) &= (\boldsymbol{\lambda}'\mathbf{P}\mathbf{X}' + \mathbf{b}')(\boldsymbol{\lambda}'\mathbf{P}\mathbf{X}' + \mathbf{b}')'\sigma^2 \\ &= (\boldsymbol{\lambda}'\mathbf{P}\mathbf{X}'\mathbf{X}\mathbf{P}\boldsymbol{\lambda} + \mathbf{b}'\mathbf{X}\mathbf{P}'\boldsymbol{\lambda} + \boldsymbol{\lambda}'\mathbf{P}\mathbf{X}'\mathbf{b} + \mathbf{b}'\mathbf{b})\sigma^2 \\ &= V(\hat{\psi}) + \mathbf{b}'\mathbf{b}\sigma^2 = V(\hat{\psi}) + \sum_i \mathbf{b}_i^2\sigma^2. \end{aligned}$$

Thus $V(\psi^*) \geq V(\hat{\psi})$ with equality only if $\mathbf{b} = \mathbf{0}$. We have already shown that $\hat{\psi}$ is unique. Hence, $\hat{\psi}$ is the unique unbiased estimate of ψ with the minimum variance. ∎

We now show that the least squares procedure produces an unbiased estimate of σ^2. The expected values of the observations are estimable functions with estimates $\hat{\mathbf{Y}} = \mathbf{X}\hat{\boldsymbol{\beta}}$, and we have minimized the sum of squares for error $S_e = (\mathbf{Y} - \hat{\mathbf{Y}})'(\mathbf{Y} - \hat{\mathbf{Y}})$. The subdivision of $\mathbf{Y}'\mathbf{Y}$ into sums of squares, $\mathbf{Y}'\mathbf{Y} = S_R + S_e$, was sketched briefly at the beginning of this chapter for the case in which $r(\mathbf{X}'\mathbf{X}) = p$. In the general case, recalling that $\mathbf{X}' = \mathbf{X}'\mathbf{X}\mathbf{P}\mathbf{X}'$, we have

$$\hat{\mathbf{Y}}'\hat{\mathbf{Y}} = \hat{\boldsymbol{\beta}}'\mathbf{X}'\mathbf{X}\hat{\boldsymbol{\beta}} = \mathbf{Y}'\mathbf{X}\mathbf{P}'\mathbf{X}'\mathbf{X}\mathbf{P}\mathbf{X}'\mathbf{Y} = \mathbf{Y}'\mathbf{X}\mathbf{P}\mathbf{X}'\mathbf{Y} = \hat{\mathbf{Y}}'\mathbf{Y} = \mathbf{Y}'\hat{\mathbf{Y}},$$

and

$$S_e = \mathbf{Y}'\mathbf{Y} - \mathbf{Y}'\hat{\mathbf{Y}} - \hat{\mathbf{Y}}'\mathbf{Y} + \hat{\mathbf{Y}}'\hat{\mathbf{Y}} = \mathbf{Y}'(\mathbf{I} - \mathbf{XPX}')\mathbf{Y},$$

$$S_R = \mathbf{Y}'\mathbf{XPX}'\mathbf{Y} = \hat{\boldsymbol{\beta}}'\mathbf{X}'\mathbf{Y}.$$

Since $E(\mathbf{Y}) = \mathbf{X}\boldsymbol{\beta}$ and $\text{cov}\,(\mathbf{Y}) = \mathbf{I}\sigma^2$,

$$E(\mathbf{Y}'\mathbf{Y}) = \boldsymbol{\beta}'\mathbf{X}'\mathbf{IX}\boldsymbol{\beta} + \sigma^2\,\text{tr}\,(\mathbf{I}) = \boldsymbol{\beta}'\mathbf{X}'\mathbf{X}\boldsymbol{\beta} + N\sigma^2,$$

and

$$E(S_R) = E(\mathbf{Y}'\mathbf{XPX}'\mathbf{Y}) = \boldsymbol{\beta}'\mathbf{X}'\mathbf{XPX}'\mathbf{X}\boldsymbol{\beta} + \sigma^2\,\text{tr}\,(\mathbf{XPX}').$$

$\mathbf{XPX}'$ is idempotent, and so $\text{tr}\,(\mathbf{XPX}') = r(\mathbf{XPX}')$. Now $r(\mathbf{XPX}') \le r(\mathbf{X}) = k$; but $\mathbf{X}' = \mathbf{X}'\mathbf{XPX}'$, and so $r(\mathbf{X}) = r(\mathbf{X}') \le r(\mathbf{XPX}')$. Hence, $r(\mathbf{XPX}') = r(\mathbf{X}) = k$, and we have

$$E(S_R) = \boldsymbol{\beta}'\mathbf{X}'\mathbf{X}\boldsymbol{\beta} + k\sigma^2,$$

and, by subtraction,

$$E(S_e) = (N - k)\sigma^2.$$

Thus, we see that the sum of squares for error provides an unbiased estimate $s^2 = S_e/(N - k)$ of the error variance σ^2.

The matrices $\mathbf{XPX}'$ and $(\mathbf{I} - \mathbf{XPX}')$ of the quadratic forms S_R, S_e are both idempotent. Under normality, the conditions of Cochran's theorem are satisfied if we replace $\mathbf{Y}$ by $\sigma^{-1}\mathbf{Y}$, and so S_R/σ^2 and S_e/σ^2 have independent χ^2 distributions with k and $(N - k)$ d.f. respectively.

2.13 Testing Hypotheses

Throughout this section we shall assume that $\mathbf{Y}$ has a multivariate normal distribution. The distribution of S_R/σ^2 is noncentral χ^2 with noncentrality parameter $\boldsymbol{\beta}'\mathbf{X}'\mathbf{X}\boldsymbol{\beta}/(2\sigma^2)$. Thus, as we have already mentioned, an appropriate test statistic for the null hypothesis $H_0\colon \boldsymbol{\beta} = \mathbf{0}$ against the general alternative $H_A\colon \boldsymbol{\beta} \neq \mathbf{0}$ (i.e., at least one of the elements of $\boldsymbol{\beta}$ is not zero) is $\mathscr{F} = k^{-1}S_R/(N - k)^{-1}S_e$, where $k = r(\mathbf{X})$. Under the null hypothesis, $\mathscr{F}$ has Snedecor's F distribution with k and $(N - k)$ d.f. The critical region is the upper tail of $F(k, N - k)$. We shall not discuss here the optimum properties of the F test in this situation. The reader who wishes to pursue the matter further is referred to Scheffé (1959), especially Chapter 2.

When a sum of squares is divided by the number of degrees of freedom in the corresponding χ^2 distribution, the quotient is called a mean square. Thus, s^2 is the mean square for error, and S_R/k is the mean square for regression.

It may be argued that in some practical situations the hypothesis $\boldsymbol{\beta} = \mathbf{0}$ is not especially interesting, and the experimenter might be more interested in testing either the hypothesis that some particular estimable function is zero or the hypothesis that all the parameters in some subset of $\boldsymbol{\beta}$ are zero.

Consider first a particular estimable function $\psi = \boldsymbol{\lambda}'\boldsymbol{\beta}$, which is estimated by $\hat{\psi} = \boldsymbol{\lambda}'\hat{\boldsymbol{\beta}} = \boldsymbol{\lambda}'\mathbf{PX'Y}$. Then $\hat{\psi}^2 = \mathbf{Y'XP'}\boldsymbol{\lambda}\boldsymbol{\lambda}'\mathbf{PX'Y}$, and $\hat{\psi}^2/(\boldsymbol{\lambda}'\mathbf{P}\boldsymbol{\lambda}\sigma^2)$ has a χ^2 distribution with one d.f. Furthermore, $\mathbf{X'(I - XPX')} = \mathbf{0}$, and so $\hat{\psi}^2/(\boldsymbol{\lambda}'\mathbf{P}\boldsymbol{\lambda}\sigma^2)$ and S_e/σ^2 have independent χ^2 distributions. Thus, the hypothesis $H_0: \psi = 0$ may be tested by the F test using the test statistic $\mathscr{F} = \hat{\psi}^2/(\boldsymbol{\lambda}'\mathbf{P}\boldsymbol{\lambda}s^2)$, or, equivalently, by the t test with $t^2 = \mathscr{F}$. Confidence intervals for ψ can be constructed using Student's t in the usual way. For example, a two-sided interval could be obtained as

$$\hat{\psi} - t^*(\boldsymbol{\lambda}'\mathbf{P}\boldsymbol{\lambda}s^2)^{-1/2} \leq \psi \leq \hat{\psi} + t^*(\boldsymbol{\lambda}'\mathbf{P}\boldsymbol{\lambda}s^2)^{-1/2}.$$

More will be said on this topic later.

We next consider testing hypotheses of the type $H_0: \boldsymbol{\beta}_1 = \mathbf{0}$ where $\boldsymbol{\beta}_1$ is a subvector of $\boldsymbol{\beta}$. Suppose that $\boldsymbol{\beta}$ is divided into two subvectors, $\boldsymbol{\beta} = (\boldsymbol{\beta}_1, \boldsymbol{\beta}_2)$, and that $\mathbf{X}$ is similarly partitioned into $(\mathbf{X}_1, \mathbf{X}_2)$. Let $r(\mathbf{X}_1) = k_1$, $r(\mathbf{X}_2) = k_2$. The complete model for which we have just estimated the parameters is

$$\mathbf{Y} = (\mathbf{X}_1, \mathbf{X}_2)(\boldsymbol{\beta}_1', \boldsymbol{\beta}_2') + \mathbf{e}.$$

Under the hypothesis $H_0: \boldsymbol{\beta}_1 = \mathbf{0}$, we have the reduced model

$$\mathbf{Y} = \mathbf{X}_2\boldsymbol{\beta}_2 + \mathbf{e}.$$

We write $S_R(\boldsymbol{\beta}_2)$, $S_R(\boldsymbol{\beta})$ for the sums of squares for regression under the reduced and complete models, respectively, and $S_R(\boldsymbol{\beta}_1|\boldsymbol{\beta}_2)$ for the difference $S_R(\boldsymbol{\beta}) - S_R(\boldsymbol{\beta}_2)$. Let $\mathbf{M}$ be a solution matrix for the reduced normal equations $\mathbf{X}_2'\mathbf{Y} = \mathbf{X}_2'\mathbf{X}_2\boldsymbol{\beta}_2$; $\mathbf{X}_2\mathbf{M}\mathbf{X}_2'$ is idempotent and $S_R(\boldsymbol{\beta}_2) = \mathbf{Y}'\mathbf{X}_2\mathbf{M}\mathbf{X}_2'\mathbf{Y}$.

We now have a subdivision of $\mathbf{Y'Y}$ into three quadratic forms:

$$\mathbf{Y'Y} = S_R(\boldsymbol{\beta}_2) + S_R(\boldsymbol{\beta}_1|\boldsymbol{\beta}_2) + S_e,$$

where $S_R(\boldsymbol{\beta}_1|\boldsymbol{\beta}_2) = \mathbf{Y}'(\mathbf{XPX}' - \mathbf{X}_2\mathbf{M}\mathbf{X}_2')\mathbf{Y}$.

Now $\mathbf{X}' = \mathbf{X'XPX'}$, and so $\mathbf{X}_2' = \mathbf{X}_2'\mathbf{XPX}'$, and $(\mathbf{XPX}' - \mathbf{X}_2\mathbf{M}\mathbf{X}_2')$ is idempotent. Thus, $S_R(\boldsymbol{\beta}_2)/\sigma^2$, $S_R(\boldsymbol{\beta}_1|\boldsymbol{\beta}_2)/\sigma^2$, S_e/σ^2 have independent χ^2 distributions with k_2, $k - k_2$ and $N - k$ d.f. The hypothesis $H_0: \boldsymbol{\beta}_1 = \mathbf{0}$ is tested by comparing the sum of squares for error under the reduced model, $S_e + S_R(\boldsymbol{\beta}_1|\boldsymbol{\beta}_2)$, to the sum of squares for error, S_e, under the complete model, and asking ourselves whether the reduction in error, $S_R(\boldsymbol{\beta}_1|\boldsymbol{\beta}_2)$, made by including $\boldsymbol{\beta}_1$ in the model is large enough to be regarded as significant. The test statistic $F = (k - k_2)^{-1}S_R(\boldsymbol{\beta}_1|\boldsymbol{\beta}_2)/s^2$ has the $F[k - k_2, N - k]$ distribution under the null hypothesis.

2.14 Missing Observations

It occasionally happens that in an experiment one or more of the observations is spoiled or omitted. One might have planned a setup in which there were several cars and several gasolines, each gasoline to be used in each car (a nicely balanced experiment) only to find that something went wrong. Perhaps one of the cars broke down before being used to test all the gasolines and could not be repaired in time; perhaps, because of a failure in communications, not enough of one of the fuels was prepared and the supply ran out before it could be tested in every car; perhaps one of the observations is so clearly incompatible with the others that it should be rejected as containing gross error. What can be done? If possible, the experimenter should try to take another observation duplicating the experimental conditions prescribed for the missing observation. If this cannot be done, we ask the following question: can we obtain a convenient formula or recipe for generating from the actual data an estimate for what the missing value should have been? Because this question first arose in agricultural experimentation [Yates (1933)] where the experimental units are called plots, these formulae are called *missing plot formulae* in the literature. We shall discuss the general case in this section. Specific formulae for some of the standard designs will be given in later chapters, either in the text or as exercises.

Suppose that the experimental plan called for $N + 1$ observations or plots. We have N observations, $\mathbf{Y}$, for which the design matrix is $\mathbf{X}$ $(N \times p)$. The $(N + 1)$th observation is missing; its design matrix is a single row $\mathbf{x}$. We shall choose for our estimate of the missing observation the value y that minimizes the sum of squares for error when we carry out the least squares procedure for all $(N + 1)$ points (the N actual data and y).

Writing $\hat{\boldsymbol{\beta}}$ for the vector of estimates obtained in this way, we minimize

$$S_e = (\mathbf{Y} - \mathbf{X}\hat{\boldsymbol{\beta}})'(\mathbf{Y} - \mathbf{X}\hat{\boldsymbol{\beta}}) + (y - \mathbf{x}\hat{\boldsymbol{\beta}})^2.$$

The first term is minimized if we take $\hat{\boldsymbol{\beta}}$ to be any solution vector to the normal equations for the N actual data points, and the second term is zero if we take $y = \mathbf{x}\hat{\boldsymbol{\beta}}$.

From an abstract point of view we appear to have gained nothing except to say that the best estimate y is obtained by solving the normal equations for the actual data and substituting $\hat{\boldsymbol{\beta}}$ for $\boldsymbol{\beta}$ to obtain $y = \mathbf{x}\hat{\boldsymbol{\beta}}$. In practice, however, the design matrix $(\mathbf{X}', \mathbf{x}')'$ is usually chosen in such a way that the augmented normal equations can be solved very easily, whereas the equations $\mathbf{X}'\mathbf{X}\boldsymbol{\beta} = \mathbf{X}'\mathbf{Y}$ may be more awkward to handle computationally. Since $y - \mathbf{x}\hat{\boldsymbol{\beta}} = 0$, an equivalent procedure, which we shall use later, is to choose for y the value that will fit the model with zero residual when the least squares procedure is carried out on the set of $(N + 1)$ points consisting of the N data points and the missing plot $(\mathbf{x}, y)$.

Exercises

1. Let y_1 and y_2 be independent random variables each $\sim N(\mu, 1)$. Then $Z = (y_1 - y_2)/\sqrt{2} \sim N(0, 1)$ and $Z^2 \sim \chi^2(1)$. Writing $Z^2 = \mathbf{Y}'\mathbf{A}\mathbf{Y}$ where $\mathbf{Y}' = (y_1, y_2)$ show that the expression for the moment-generating function of Z^2: $M(t) = |\mathbf{I} - 2\mathbf{A}t|^{-1/2} \exp -\boldsymbol{\mu}'\{\mathbf{I} - (\mathbf{I} - 2\mathbf{A}t)^{-1}\}\boldsymbol{\mu}/2$ reduces to $(1 - 2t)^{-1/2}$.
2. Let $\mathbf{Y} \sim N(\boldsymbol{\mu}, \mathbf{I})$. We proved the theorem that $\mathbf{Y}'\mathbf{A}\mathbf{Y} \sim \chi^2(n_i)$, if, and only if, $\mathbf{A}$ is an idempotent matrix of rank n_i for the case $\boldsymbol{\mu} = \mathbf{0}$. Prove it for the case $\boldsymbol{\mu} \neq \mathbf{0}$.
3. Prove that if $\mathbf{Y} \sim N(\mathbf{0}, \mathbf{V})$ then $\mathbf{Y}'\mathbf{A}\mathbf{Y} \sim \chi^2(n)$ if, and only if, $\mathbf{A}\mathbf{V}$ is an idempotent matrix of rank n.
4. If $\mathbf{Y} \sim N(\mathbf{0}, \mathbf{V})$ show that the two forms $\mathbf{Y}'\mathbf{A}\mathbf{Y}$, $\mathbf{Y}'\mathbf{B}\mathbf{Y}$ are independent if, and only if, $\mathbf{A}\mathbf{V}\mathbf{B} = \mathbf{0}$.
5. If $\mathbf{A}$ and $\mathbf{B}$ are nonnegative definite matrices and if $\mathbf{Y} \sim N(\mathbf{0}, \mathbf{I})$, show that the forms $\mathbf{Y}'\mathbf{A}\mathbf{Y}$, $\mathbf{Y}'\mathbf{B}\mathbf{Y}$ are independent, if and only if, $\operatorname{tr}(\mathbf{A}\mathbf{B}) = 0$.
6. Let $\mathbf{Y} \sim N(\mathbf{0}, \mathbf{D})$ where $\mathbf{D}$ is diagonal, and let

$$\mathbf{A} = \mathbf{D}^{-1} - (\mathbf{D}^{-1}\mathbf{J}\mathbf{D}^{-1}/\mathbf{1}'\mathbf{D}^{-1}\mathbf{1}).$$

   Show that $\mathbf{Y}'\mathbf{A}\mathbf{Y} \sim \chi^2(n - 1)$.
7. Let $\mathbf{Y}$ be a vector of n independent normal random variables with $E(y_i) = \mu_i$ and $V(y_i) = w_i^{-1}$, and let $\bar{y} = \sum w_i y_i / \sum w_i$. Show that

$$\sum w_i(y_i - \bar{y})^2 \sim \chi^2(n - 1).$$

CHAPTER 3

Experiments with a Single Factor

If an agronomist wishes to compare the yields of five varieties of wheat, an obvious procedure would be to mark off twenty plots of land and to sow variety A on four of them, variety B on four more of them, and so on. The varieties are assigned to the plots by some random procedure, and from each plot comes a single observation, the yield on that plot. This simple experiment is called a completely randomized experiment because of the method of assigning the varieties. We shall begin our discussion of experimental designs by considering this kind of experiment.

An industrial example would be the comparison of four different operating temperatures on the yield of a product from a chemical reactor. Twenty runs could be made, five at each temperature, and the yield of product at each run recorded. Again, the temperatures should be assigned to the runs at random. Suppose that the process involves the use of an acid catalyst which deteriorates with time. If the experimenter were to make the runs at the highest temperature first, then the runs at the next highest temperature, and finally the runs at the lowest temperature, he would have difficulty interpreting his results. If the yields at the high temperatures were greater than the yields at low temperatures, could he be sure whether the apparent drop in yield resulted from the reduction in the operating temperature or from the deterioration of the acid? The random assignment of the temperatures to the runs, or of the varieties to the plots, is essential to our analysis of the experiment. The models for the experiments will be linear models like those of the last chapter. In the model for each observation there will be an error term, e; these errors are assumed to be uncorrelated.

In general, a set of different treatments are applied to N experimental units, one treatment to each unit, and an observation is made on each unit. The theory of experimental design was first developed in agriculture and much of the jargon is taken from that field. We shall refrain from using the expression *experimental unit*. Instead we shall most often use the agricultural term *plot*, or, when talking of industrial experiments, the terms *run* or *point*. An observation will commonly be referred to as a *yield* or a *response*.

Some of our examples will be concerned with octane numbers. Producers of gasoline are interested in the octane number of their product. If a driver uses a gasoline with too low an octane number, the engine of his car will knock. Larger cars with engines that have higher compression ratios need higher octane than do compact cars to prevent knocking. Thus, high octane number is a desirable characteristic of a gasoline. The octane number may be determined in the laboratory or on the road by a procedure that may be described in a simplified fashion as follows.

There is a standard engine on which the compression ratio may be varied. The engine is run with the gasoline to be tested. The compression ratio is increased until knocking occurs, and the reading on the meter is noted. Then the procedure is repeated using reference fuels. These are mixtures of normal heptane and iso-octane. A reference fuel composed of X per cent iso-octane and $(100 - X)$ per cent normal heptane is defined as having octane number X. Two reference fuels are used, one with a slightly higher octane number than the unknown gasoline and one with a slightly lower octane number. The readings on the compression ratio meter at which the engine knocks using each of the reference fuels are noted, and the octane number of the gasoline being tested is obtained by interpolation. An observed octane number is a random variable. A standard deviation of about 0.3 ON is not unusual.

A similar procedure can be followed using automobiles on the road to obtain "road-octane numbers." The manufacturers of both cars and gasolines conduct tests under road conditions to find both the road-octane numbers of commercial gasolines and the octane requirements of commercial automobiles. Because the modern automobile engines differ somewhat from the standard engines, and from each other, and also because the components in commercial gasoline are not iso-octane and normal heptane, it often happens that even apart from random error a test fuel will have different octane ratings in different cars and these ratings will differ from those obtained in the laboratory. The octane number of a gasoline can be increased by the use of special additives such as tetra-ethyl lead or tetra-methyl lead, or by adding some of the aromatic hydrocarbons found in crude oil. The latter, however, are also in demand for petrochemicals, and the managers of an oil company have to decide how much of their supply should be used for each purpose.

An engineer wishing to compare five gasolines might take twenty cars, assign each gasoline to four cars, and observe the octane number for each car–gasoline combination. The choice of cars is important to the analysis of the experiment and to the use of the results. One should not experiment solely on Volkswagens if the results are to be used to make decisions about Cadillacs, nor should one expect crops to yield the same amounts in the Central Valley of California as they do in the coastal area.

3.1 Analysis of the Completely Randomized Experiment

In either of the two experiments just mentioned we can take as a model

$$y_{ij} = \mu_i + e_{ij}, \qquad i = 1, 2, \ldots, t; j = 1, 2, \ldots, r,$$

where y_{ij} denotes the jth observation on the ith gasoline, or the yield on the jth of the plots that were sown with the ith variety of wheat, μ_i denotes the "true" value of the octane number or the yield of the ith gasoline or variety; i.e., $E(y_{ij})$, and e_{ij} is the random error of the (ij)th observation or plot. We assume that the μ_i are unknown constants, and that the e_{ij} are uncorrelated random variables, each distributed with the same mean, zero, and the same variance σ^2. We use the general term *treatments* for the gasolines or the varieties and speak of an experiment with t treatments.

In addition we may make the normality assumption, namely that the e_{ij} are normally distributed, i.e., the vector $\mathbf{e}$ of the e_{ij} has a multivariate normal distribution with mean vector $\mathbf{0}$ and covariance matrix $\mathbf{I}\sigma^2$.

The least squares estimate of μ_i, which is the maximum likelihood estimate under the normality assumption, is the average of the observations on the ith treatment. Writing

$$G = \sum_i \sum_j y_{ij}, \qquad T_i = \sum_j y_{ij}, \qquad y_{i\cdot} = \bar{y}_i = T_i/r,$$

we have

$$\hat{\mu} = y_{i\cdot}, \tag{3.1}$$

$$S_e = t(r-1)s^2 = \sum_i \sum_j (y_{ij} - y_{i\cdot})^2 = \sum_i \sum_j y_{ij}^2 - \sum_i T_i^2/r. \tag{3.2}$$

If the number of observations is not the same for each treatment, we denote the number of observations on the ith treatment by r_i, and let $y_{i\cdot} = T_i/r_i$. Then Equation 3.2 becomes $S_e = \sum_i \sum_j y_{ij}^2 - \sum_i T_i^2/r_i$.

We shall use N to denote the total number of observations, G for the grand total $\sum_i \sum_j y_{ij}$ of the observations, and $y_{\cdot\cdot}$ or $\bar{y}$ for the grand mean G/N. The case where $r_i = r$ for all i will be called the *equireplicate case*. (Some people like to say that the ith treatment is replicated r_i times; we shall do so when we consider incomplete block designs in a later chapter).

Under the normality assumption the hypothesis $\mu_i = \mu_{i'}$ can be tested by Student's t test. In the equireplicate case the test statistic

$$t = (y_i - y'_i)\sqrt{(r/2s^2)}$$

has, under the hypothesis, Student's t distribution with $t(r-1)$ d.f.

An alternative formulation of the model commonly used is

$$y_{ij} = \mu + \tau_i + e_{ij},$$

where y_{ij} and e_{ij} are the same as before, μ is a general mean, and τ_i is an unknown constant parameter associated with the ith treatment. We shall call τ_i the main effect of the ith treatment. Neither μ nor τ_i is estimable; if we were to add any arbitrary constant c to μ and subtract c from each τ_i, the expected value of each observation y_{ij} would remain unchanged.

The model may also be written $E(y_{ij}) = \mu + \sum_h x_h \tau_h$ where $x_i = 1$ if the observation is made on the ith treatment and zero otherwise. The design matrix $\mathbf{X}$ has N rows and $t + 1$ columns; assuming that there is at least one observation on each treatment, the rank of $\mathbf{X}$ is t. The first column, corresponding to μ, is $\mathbf{1}_N$, a vector with every element unity. The remaining columns, corresponding to treatments, consist of zeros and ones, and in any row exactly one of these columns has unity as its element. Thus, the first column is equal to the sum of the last t columns, and the last t columns are linearly independent.

If, therefore, we wish to use this model, we must impose a side condition on the parameters. Three possibilities come to mind.

(i) $\mu = 0$. This sends us back to the original model with τ_i instead of μ_i.
(ii) $\tau_1 = 0$. This is useful in some computer programs. It amounts to designating treatment 1 as a base treatment. If treatment 1 is a control or standard treatment, this has some merit.
(iii) $\sum_i r_i \tau_i = 0$. This is the most commonly used side condition. It reduces to $\sum_i \tau_i = 0$ in the equireplicate case. An attractive consequence is that with this side condition $\hat{\mu} = \sum_i \sum_j y_{ij}/N = \bar{y}$. We shall use this condition.

Although μ and τ_i are not estimable, $(\mu + \tau_i)$ is estimable. Its least squares estimate is $y_{i\cdot}$, which is the same as $\hat{\mu}_i$ was before. Similarly, the hypothesis $\mu_i = \mu_{i'}$ is the same as the hypothesis $\tau_i = \tau_{i'}$, and can also be tested by $y_{i\cdot} - y_{i'\cdot}$. One might ask what, if anything, is gained by introducing the additional complication of singular design matrices. It will actually turn out that when we consider several factors, or experiments with blocking, we shall be faced with singular design matrices anyway. The advantage in this simple experiment lies in the fact that the new model gives a sum of squares for treatments.

The simplest model that could have been chosen for the data would have been

$$y_{ij} = \mu + e_{ij},$$

which is equivalent to assuming that all the μ_i are equal, or that with the side condition $\sum \tau_i = 0$ all the τ_i are zero. Under this model $\hat{\mu} = G/N$, and the error sum of squares is

$$S'_e = \sum_i \sum_j y_{ij}^2 - G\hat{\mu} = \sum_i \sum_j y_{ij}^2 - G^2/N.$$

When the treatment parameters τ_i are added to the model the estimate of μ is still $\hat{\mu} = \bar{y}$; the estimate of τ_i is $\hat{\tau}_i = y_{i.} - \bar{y}$, and the new sum of squares for regression is

$$S_R = G\hat{\mu} + \sum_i T_i\hat{\tau}_i = G^2/N + \sum_i T_i(y_{i.} - \bar{y}) = \sum_i T_i^2/r_i.$$

The reduction in the sum of squares for error or the increase in the sum of squares for regression because of the addition of the treatment parameters to the model is thus $\sum T_i^2/r_i - G^2/N$. This is called the sum of squares for treatments and is denoted by S_t. The sum of squares for error, S_e, is the same as it was with the first model.

We now have the subdivision of the sum of the squares of the observations

$$\mathbf{Y'Y} = G^2/N + \left[\sum_i T_i^2/r_i - G^2/N\right] + \sum_i\sum_j (y_{ij} - y_{i.})^2$$

or

$$\mathbf{Y'Y} = C + S_t + S_e.$$

The term G^2/N is called the correction for the mean and is often written as C. $\mathbf{Y'Y}$ is called the raw sum of squares; the corrected sum $\mathbf{Y'Y} - C$ is called the total sum of squares, and we shall occasionally denote it by S_T.

The quadratic forms C, S_t, and S_e have ranks $1, t - 1$ and $N - t$ respectively. Under normality they have, upon division by σ^2, independent chi-square distributions. The expected value of C is $\sigma^2 + N^{-1}(N\mu + \sum r_i\tau_i)^2$. If we write $\sum r_i\tau_i = N\bar{\tau}$, this becomes $\sigma^2 + N(\mu + \bar{\tau})^2$, and if we take the side condition $\sum r_i\tau_i = 0$, $E(C)$ reduces to $\sigma^2 + N\mu^2$. To compute $E(S_t)$ we note that

$$E(T_i^2/r_i) = r_i^{-1}E\left(\sum_j e_{ij} + r_i\mu + r_i\tau_i\right)^2 = \sigma^2 + r_i(\mu + \tau_i)^2.$$

Then

$$\begin{aligned} E(S_t) &= \left(t\sigma^2 + N\mu^2 + 2\mu\sum r_i\tau_i + \sum r_i\tau_i^2\right) - E(C) \\ &= (t - 1)\sigma^2 + \sum r_i(\tau_i - \bar{\tau})^2. \end{aligned}$$

This information is summarized in Table 3.1.

TABLE 3.1
ANALYSIS OF VARIANCE

Source	SS	d.f.	EMS
Total	$\mathbf{Y'Y} - C$	$N - 1$	
Treatments	$\sum T_i^2/r_i - C$	$t - 1$	$\sigma^2 + (t - 1)^{-1}\sum r_i(\tau_i - \bar{\tau})^2$
Error	$\mathbf{Y'Y} - \sum T_i^2/r_i$	$N - t$	σ^2

There has been a tendency in the last few years to replace the words analysis of variance by the single word, Anova. In the column headings SS stands for sum of squares and EMS for the expected value of the mean square, the latter being the sum of squares divided by the number of degrees of freedom. The mean squares for error and treatments may be denoted by M_e and M_t respectively, although we shall usually use s^2 rather than M_e.

The hypothesis $H_0: \tau_1 = \tau_2 = \cdots = \tau_t$, which is the hypothesis that there are no treatment differences, is tested by the statistic $\mathscr{F} = M_t/s^2$. Under the normality assumption $\mathscr{F}$ has, when the null hypothesis is true, a central $F(t-1, N-t)$ distribution; when the null hypothesis is false, the distribution of $\mathscr{F}$ is a noncentral F with a noncentrality parameter $\lambda = \sum r_i\tau_i^2/2\sigma^2$. The test has for its critical region the upper tail of the F distribution, and the null hypothesis is rejected if $\mathscr{F}$ exceeds the upper α value of F. Some authors add a fifth column to the analysis of variance table, in which they enter values of $\mathscr{F}$; these entries are indicated by one asterisk if they exceed the 5 per cent value of F and with two asterisks if they exceed the one percent value. We mean by the upper α value of $F(\phi_1, \phi_2)$ the value F^* defined by $\int_{F^*}^{\infty} f(u)\,du = \alpha$ where $f(u)$ is the probability density function of a random variable, u, having the $F(\phi_1, \phi_2)$ distribution.

This hypothesis, which may also be expressed as $\boldsymbol{\tau} = \mathbf{0}$ where $\boldsymbol{\tau}$ is the vector of the τ_i, is of interest in uniformity trials where, for example, an agronomist wants to ascertain whether several strains of wheat are essentially equivalent or whether the yield of a particular variety is reasonably constant under varying conditions. (Power calculations for the F test involve integrals for the noncentral distribution. Tang (1938) has prepared a set of tables for use in power calculations. Subsequently, charts were made by Pearson and Hartley (1951) and by Fox (1956). These charts are somewhat easier to use than the Tang tables.)

In many industrial situations, however, the hypothesis that there are no treatment differences is of little interest; the scientist expects to find differences and is more concerned with finding which treatments differ or in investigating contrasts among the treatments.

We shall confine ourselves now to the equireplicate case with the normality assumption and consider contrasts in the treatment effects, $\psi = \sum c_i\tau_i$ where $\sum c_i = 0$ (or, equivalently $\psi = \mathbf{c}'\boldsymbol{\tau}$ where $\mathbf{c}'\mathbf{1} = 0$). The difference between two effects, $\tau_i - \tau_{i'}$, will sometimes be called a comparison. The least squares estimate of ψ is $\hat{\psi} = \sum c_i\hat{\tau}_i = \mathbf{c}'\hat{\boldsymbol{\tau}} = \sum c_i y_{i.}$; $\hat{\psi}$ is normally distributed with variance $\mathbf{c}'\mathbf{c}\sigma^2/r$, and an obvious test for the hypothesis $\psi = 0$ is Student's t test.

To carry out a t test for a hypothesis such as $\tau_i = \tau_{i'}$ we can use a least significant difference (LSD). We have $\hat{\tau}_i - \hat{\tau}_{i'} = y_{i.} - y_{i'.}$. Under the null hypothesis, $(y_{i.} - y_{i'.})/(2s^2/r)^{1/2}$ has Student's distribution with $(N-t)$ d.f. The hypothesis is rejected in the two-tailed test if $|y_{i.} - y_{i'.}| > t^*(2s^2/r)^{1/2}$ where t^* is the critical value of t with the desired significance level α.

The quantity $t^*(2s^2/r)^{1/2}$ is the LSD. The multiple t test is carried out by comparing the treatment means and by declaring any pair of treatments whose means differ by more than the LSD to be different.

This is a simple and convenient procedure, but the criticism has been made that when this LSD is applied to all possible pairs of treatments there is an increased chance of type-one errors. The argument of the critics is essentially that whereas it is true that the probability is 0.95 that any two observations made at random from a normal population will differ by less than 1.96σ, if we take a sample of $n > 2$ observations, the probability that the largest and the smallest of them differ by more than 1.96σ is greater than 0.05 and increases as n increases. This has led to the development of tests with the property that the probability of a type-one error per experiment (rather than per comparison) is α. Three such tests in common use are due to Newman (1939), Duncan (1955), and Tukey (1953); Newman's test was revived by Keuls (1952) and is more commonly known as the Newman–Keuls test. We shall derive Tukey's test. Accounts of the other two tests are found in the references cited and in the books by Federer (1955) and Miller (1966).

3.2 Tukey's Test for Treatment Differences

The Studentized range is defined as follows. Suppose that $x_1, x_2, \ldots, x_n$ is a sample of n observations from a normal population with mean μ and variance σ^2 and that s^2 is a quadratic estimate of σ^2 with ϕ d.f., i.e., s^2/σ^2 is distributed as chi-square with ϕ d.f. The range $R(x)$ is the difference between the largest and smallest members of the sample, and we suppose further that $R(x)$ and s^2 are independent. Then the random variable $R(x)/s$ is said to be distributed as the Studentized range with ϕ d.f. With probability $(1 - \alpha)$ we have $R(x) \leqq q^{\alpha}_{n:\phi}s$, where $q^{\alpha}_{n:\phi}$ is the upper α point of the Studentized range.

In Tukey's test we replace the x_i of the previous paragraph by the treatment means $y_{i\cdot}$ and argue as follows. Under the null hypothesis, the probability is $(1 - \alpha)$ that $R(y_{i\cdot}) \leq q^{\alpha}_{t:\phi}s/\sqrt{r}$ where $\phi = N - t$. Thus the probability is $(1 - \alpha)$ that $\hat{\tau}_i - \hat{\tau}_{i'} \leq q^{\alpha}_{t:\phi}s/\sqrt{r}$ for all pairs i, i'. We may therefore use $q^{\alpha}_{t:\phi}s/\sqrt{r}$ as the least significant difference and declare that $\tau_i \neq \tau_{i'}$ if, and only if, $|y_{i\cdot} - y_{i'\cdot}| > sq^{\alpha}_{t:N-t}/\sqrt{r}$. Tukey's test is somewhat too conservative to suit some experimenters. The tests of Newman and Keuls and of Duncan provide compromises between Tukey's test and the t test.

3.3 Scheffé's S Statistic

The three tests mentioned before are useful for testing comparisons. The experimenter usually has a few specific contrasts in mind when he plans his experiment, and it is appropriate to use the t test for these. However, if he is going to wait until after the data have been collected and then test those contrasts that strike his eye during the analysis, the criticisms of blanket use of the t test for the whole set of $t(t - 1)/2$ comparisons are no less valid for the set of all contrasts. Scheffé's statistic provides a general hunting license for the experimenter who wishes to test any estimable function that takes his fancy after he has taken a look at the data. The derivation of this statistic may be found in Scheffé's original paper (1953) or in his book (1959). The procedure is the following.

Let $\hat{\psi} = \mathbf{c}'\hat{\boldsymbol{\tau}}$ be an estimate of an estimable function $\psi = \mathbf{c}'\boldsymbol{\tau}$, and let s^2 be an independent quadratic estimate of the variance based on a χ^2 distribution with ϕ d.f. We write $\hat{\sigma}_{\psi}^2 = \mathbf{c}'\mathbf{c}s^2$. Scheffé's result is as follows: The probability is $(1 - \alpha)$ that, for all possible estimable functions,

$$\hat{\psi} - S\hat{\sigma}_{\psi} \leq \psi \leq \hat{\psi} + S\hat{\sigma}_{\psi},$$

where $S^2 = (t - 1)F_{\alpha}(t - 1, \phi)$, $\phi = N - t$ and $F_{\alpha}(t - 1, \phi)$ is the upper α value of F. This test gives longer intervals than the t test when there are more than two treatments, but that is the price that has to be paid to have an error rate of α per experiment. For simple comparisons, Tukey's method gives shorter confidence intervals than Scheffé's.

We have mentioned only a few of the numerous alternatives to the t test in the analysis of variance. The alternatives to which we have referred all assume normality. There are also nonparametric procedures available, such as those of Kruskal and Wallis (1952) and Steel (1960). The subject of multiple comparisons in the analysis of variance is a topic that fills a book on its own [Miller (1966)], and we shall pursue it no further here. From now on we shall have in mind the t test when we think of tests of significance for contrasts. A great deal has also been written about the robustness of the t and F tests in the face of departures from normality, notably by Box (1953) and by Scheffé (1959, Chap. 10); less is known about the robustness of the other procedures that are based upon the assumption of normality.

3.4 A Worked Example

Consider the following data from an experiment with five gasolines A, B, C, D, E and four observations on each gasoline. Note that $G = 1834.2$ and $\bar{y} = 91.71$.

Gas	Observations				Total	Mean
A	91.7	91.2	90.9	90.6	364.4	91.10
B	91.7	91.9	90.9	90.9	365.4	91.35
C	92.4	91.2	91.6	91.0	366.2	91.55
D	91.8	92.2	92.0	91.4	367.4	91.85
E	93.1	92.9	92.4	92.4	370.8	92.70

In computing the sums of squares in Table 3.2 the arithmetic is simplified if we subtract 90.0 from each observation. If this is done we have $\mathbf{Y}'\mathbf{Y} = 67.96$, $C = 58.482$, and $\sum T_i^2/4 = 64.59$. The analysis of variance table follows.

TABLE 3.2
ANALYSIS OF VARIANCE

Source	SS.	d.f.	MS.	F
Total	$67.96 - 58.482 = 9.478$	19		
S_t	$64.59 - 58.482 = 6.108$	4	$1.527 = M_t$	6.78
S_e	$67.96 - 64.59 = 3.370$	15	$0.225 = s^2$	

The upper 5 per cent and 1 per cent points of $F(4, 15)$ are 3.06 and 4.89. The ratio $M_t/M_e = 6.78$ exceeds both these values, and so the hypothesis $H_0: \boldsymbol{\tau} = \mathbf{0}$ is rejected. We could put two stars after 6.78 in the F column to show that 6.78 exceeds the upper one per cent value.

It should be noted that the fact that $S_t + S_e$ adds up to the total sum of squares does not in itself provide much of a check on the calculations, because it provides no check that the quantities $\mathbf{Y}'\mathbf{Y}$, C and $\mathbf{T}'\mathbf{T}$ have been computed correctly. If we had erroneously calculated $\mathbf{Y}'\mathbf{Y}$ as 77.96, we should still have obtained an analysis of variance table which, superficially at least, would look valid; on the other hand, if we had mistakenly calculated $\mathbf{T}'\mathbf{T}/4$ as 54.59, that would have given a negative value of S_t, which would have indicated that either $\mathbf{T}'\mathbf{T}/4$ or C was wrong. A check can, however, be made by obtaining the sum of the squares of the deviations $\sum_j (y_{ij} - y_{i\cdot})^2$ for each treatment and then checking that $S_e = \sum_i \sum_j (y_{ij} - y_{i\cdot})^2$.

Having rejected the hypothesis of no treatment effects, we now look at the treatment comparisons. The estimated variance of a mean is $s^2/4$.

Using the t test we have, at $\alpha = 0.05$, $t_{\alpha,15} = 2.131$, and the LSD is $2.131\sqrt{(2s^2/4)} = 2.131\sqrt{0.1125} = 0.714$. Comparing differences between means to 0.71 we should conclude that gasoline E has a higher octane number than any of the other gasolines and that D exceeds A.

For the more conservative Tukey test, the upper 5 per cent value of the Studentized range for five samples with 15 d.f. for error is 4.37 [see, for example, Pearson and Hartley (1954)] and the LSD is $4.37\sqrt{(s^2/4)} = 1.036$. Using this test, we should only conclude that $E > A, B, C$.

For an example of the use of Scheffé's S statistic, consider the estimable contrast $\psi = \tau_1 + \tau_2 - \tau_4 - \tau_5$. We have $\hat{\psi} = 1.10 + 1.35 - 1.85 - 2.70 = -2.10$; $V(\hat{\psi}) = 4\sigma^2/4 = \sigma^2$. With $\alpha = 0.05$, $S^2 = 4F_\alpha(4, 15) = 12.24$, and $S = 3.5$. Then

$$-(3.5)(0.474) - 2.10 < \psi < (3.5)(0.474) - 2.10,$$

$$-3.76 < \psi < -0.44,$$

and the hypothesis $\psi = 0$ would be rejected.

3.5 Qualitative and Quantitative Factors and Orthogonal Contrasts

We have spoken so far of an experiment with t treatments. We may also talk about an experiment with a factor at t levels. Thus, the gasoline experiment could have been described either as an experiment with five gasolines or as an experiment with one factor, gasoline, at five levels. The latter description may be more appropriate if, for example, the gasolines differ only by the percentage of olefins present in their compositions. It has already been remarked that if an experimenter is concerned with a specific set of contrasts which have been decided upon beforehand, it is reasonable for him to use the t test or, equivalently, the F test. It is convenient if these contrasts are orthogonal, but it is not essential.

In the previous chapter we mentioned that when $\mathbf{Y}'\mathbf{Y}$ is split into sums of squares $\mathbf{Y}'\mathbf{A}_i\mathbf{Y}$ these quadratic forms can in turn be expressed as sums of the squares of orthogonal linear combinations, z_i, of the observations, corresponding to an orthogonal transformation $\mathbf{Z} = \mathbf{BY}$. If one of these linear combinations is taken to be $\bar{y}/\sqrt{N}$, corresponding to the mean, the others must be contrasts in the observations. In the equireplicate case a contrast in the treatment totals is also a contrast in the observations. Suppose that $\mathbf{a}'\mathbf{Y}$ and $\mathbf{b}'\mathbf{Y}$ are two linear forms. We say that they are orthogonal if $\mathbf{a}'\mathbf{b} = 0$. Similarly, two linear forms $\mathbf{c}'\mathbf{T}$ and $\mathbf{d}'\mathbf{T}$ in the treatment totals ($\mathbf{T}$ being the vector of the treatment totals) are orthogonal if, and only if, $\sum c_i d_i r_i = 0$. If $r_i = r$ for all i this becomes $\mathbf{c}'\mathbf{d} = 0$. We shall again confine ourselves to the equireplicate case.

The correction for the mean G^2/N may be written $\mathbf{T}'\mathbf{JT}/N$ where $\mathbf{J}$ is a $(t \times t)$ matrix with every element unity. Then $S_t = r^{-1}\mathbf{T}'\mathbf{T} - G^2/N$ is actually a quadratic form in the treatment totals, and the $(t - 1)$ linear combinations z_i in the transformation $S_t \rightarrow \mathbf{Z}'\mathbf{Z}$ are orthogonal linear combinations of the treatment totals. Any set of orthogonal contrasts in the treatment totals will suffice, and the experimenter can choose a set that satisfies his particular purposes. If there are four treatments, we may take the set of orthogonal contrasts:

$$z_1 = (T_1 - T_2)/\sqrt{(2r)}; \qquad z_2 = (T_1 + T_2 - 2T_3)/\sqrt{(6r)};$$
$$z_3 = (T_1 + T_2 + T_3 - 3T_4)/\sqrt{(12r)},$$

or, as we shall see a little later in this section, we could also take, with the proper divisors, the set

$$z_1 = (-3T_1 - T_2 + T_3 + 3T_4), \qquad z_2 = (+T_1 - T_2 - T_3 + T_4),$$
$$z_3 = (-T_1 + 3T_2 - 3T_3 + T_4).$$

In either case, $S_t = z_1^2 + z_2^2 + z_3^2$.

We do not need to restrict ourselves to contrasts $z = \mathbf{c}'\mathbf{T}$ with $\mathbf{c}'\mathbf{c} = 1$, i.e., to vectors $\mathbf{c}$ which have unit length. If $z = \mathbf{c}'\mathbf{T}$ is any contrast we can always divide each coefficient by $\sqrt{\mathbf{c}'\mathbf{c}}$ to obtain a coefficient vector with unit norm. The sum of squares for any contrast is thus $z'z/r\mathbf{c}'\mathbf{c}$. Upon division by σ^2 it has a χ^2 distribution (central, if $E(z) = 0$) with one degree of freedom.

Consider an experiment involving six cars, Plymouth, Ford, Chevrolet, Dodge, Mercury, and Buick, and denote the corresponding treatment totals by $T_1, T_2, \ldots, T_6$, respectively. One possible set of orthogonal contrasts follows.

T_1	T_2	T_3	T_4	T_5	T_6	
+1	+1	+1	−1	−1	−1	Low priced cars vs. medium priced cars
+1	0	−1	0	0	0	Plymouth vs. Chevrolet
0	0	0	+1	0	−1	Dodge vs. Buick
+1	−2	+1	0	0	0	Ford vs. other low priced cars
0	0	0	+1	−2	+1	Mercury vs. other medium priced cars

The engineer can also test the Ford vs. the Plymouth if he wishes, even though that contrast is not orthogonal to the others.

The factor used in this experiment is a qualitative factor. The experimenter is interested in establishing and estimating the differences, if any, between the various makes of cars. The prediction of the performance in some subsequent run at some intermediate level of the factor, such as a hybrid of Plymouth and Mercury, is not relevant to the experiment. However, with a quantitative factor such as temperature, which may in a particular experiment have such levels as 100, 200, 300, or 400 degrees, we are usually interested in the whole range covered. We might wish to use the data to predict what the response would be if a subsequent run were to be made at 150°, or even to predict where in the range is the "best" temperature in the sense of the temperature at which the response is either a local maximum or a local minimum.

This leads to the fitting of linear, quadratic, and higher order trend lines, $y = f(t)$, for predicting the response as a polynomial function of the temperature. Consider, for example, the cubic polynomial

$$y = a + bt + ct^2 + dt^3.$$

One approach is to fit this model to the data by least squares, test the hypothesis $d = 0$, and then, if that hypothesis is accepted, to fit the reduced model

$$y = a + bt + ct^2.$$

In the reduced model we now test the hypothesis $c = 0$. If that is accepted, we fit the straight line $y = a + bt$, and so on.

The difficulty with this procedure is that unless the least squares estimates of a, b, c, d are orthogonal contrasts, the equation has to be refitted anew at each stage. We can test in the analysis of the cubic model whether the cubic term is necessary, but unless the contrasts are orthogonal we cannot test simultaneously $c = 0$ and $b = 0$.

An alternative approach is to make use of orthogonal polynomials. This is conveniently done if the levels of the factor are equally spaced. We illustrate the procedure by an example. Suppose that one observation is made at each of the five temperatures, 100, 200, 300, 400 and 500 degrees, and that we wish to fit the cubic function

$$E(y) = \beta_0 + \beta_1 x + \beta_2 x^2 + \beta_3 x^3,$$

where x denotes the temperature and y is the response.

Instead we fit the polynomial

$$E(y) = \gamma_0 u_0 + \gamma_1 u_1 + \gamma_2 u_2 + \gamma_3 u_3,$$

where $u_0 \equiv 1$, and the u_i $(i = 1, 2, 3)$ are ith degree polynomials in x such that $\sum u_i u_j = 0$ for each pair $i, j\,(i \neq j)$, the summation being made over all the values taken by x. In practice it is convenient to take the values of u_1 to be small integers and to derive u_2 and u_3 as polynomials in u_1 rather than x.

Considering first u_1, let $u_1 = a_1 + b_1 x$. The condition $\sum u_0 u_1 = 0$ gives

$$5a_1 + 1500b_1 = 0.$$

This is a single equation in two unknowns, and the solution which gives the smallest integer values of u_1 is $a_1 = -3$, $b_1 = 0.01$, so that u_1 takes the values $(x - 300)/100$ or $-2, -1, 0, +1, +2$.

For $u_2 = a_2 + b_2 u_1 + c_2 u_1^2$, the conditions $\sum u_0 u_2 = 0$, $\sum u_1 u_2 = 0$ give

$$5a_2 + 0 \cdot b_2 + 10c_2 = 0,$$

$$10b_2 \qquad\qquad = 0,$$

whence, $b_2 = 0$, $a_2 = -2c_2$. The choice $a_2 = -2$, $c_2 = 1$, or $u_2 = u_1^2 - 2$, gives

$$u_2 = +2, -1, -2, -1, +2.$$

For $u_3 = a_3 + b_3 u_1 + c_3 u_1^2 + d_3 u_1^3$, we have three conditions: $\sum u_0 u_3 = 0$, $\sum u_1 u_3 = 0$, $\sum u_2 u_3 = 0$. The last of these may be written as $\sum a_2 u_3 + \sum b_2 u_1 u_3 + \sum c_2 u_1^2 u_3 = 0$. By virtue of the first two, this reduces to $\sum u_1^2 u_3 = 0$, giving the three equations:

$$5a_3 + 10c_3 = 0, \qquad 10b_3 + 34d_3 = 0, \qquad 10a_3 + 34c_3 = 0,$$

whence $a_3 = c_3 = 0$ and $b_3 = -17d_3/5$. The choice $d_3 = 5$, $b_3 = -17$, or $u_3 = 5u_1^3 - 17u_1$, gives values $u_3 = -6, +12, 0, -12, +6$. The usual choice is, therefore, $(5u_1^3 - 17u_1)/6$, giving

$$u_3 = -1, +2, 0, -2, +1.$$

The estimates of γ_i are now easily obtained as $\hat{\gamma}_i = \sum u_i y / \sum u_i^2$ with cov $(\hat{\gamma}_i, \hat{\gamma}_{i'}) = 0$ $(i \neq i')$ and $V(\hat{\gamma}_i) = \sigma^2 / \sum u_i^2$. Thus, writing the responses as $y_1, y_2, \ldots, y_5$ in ascending order of temperature and letting the vector $\mathbf{Y}$ be $(y_1, y_2, \ldots, y_5)'$, we have

$$\hat{\gamma}_0 = \sum y/5 = \bar{y}, \qquad \hat{\gamma}_1 = (-2, -1, 0, +1, +2)\mathbf{Y}/10,$$

$$\hat{\gamma}_2 = (+2, -1, -2, -1, +2)\mathbf{Y}/14 \qquad \hat{\gamma}_3 = (-1, +2, 0, -2, +1)\mathbf{Y}/10.$$

The contrasts

$$(-2, -1, 0, +1, +2)\mathbf{Y}, \qquad (+2, -1, -2, -1, +2)\mathbf{Y},$$

and

$$(-1, +2, 0, -2, +1)\mathbf{Y}$$

are called the linear, quadratic, and cubic contrasts.

Once the estimates $\hat{\gamma}_i$ have been obtained, we can, if we wish, compute estimates of the coefficients β_i. For purposes of prediction it is easier to work in terms of $\hat{\gamma}_i$ and u_i. Thus, if it is desired to extrapolate and to predict the response at a temperature of 600 degrees, we have $u_1 = 3$, $u_2 = 7$, $u_3 = 14$, and $\hat{y} = \hat{\gamma}_0 + 3\hat{\gamma}_1 + 7\hat{\gamma}_2 + 14\hat{\gamma}_3$. The sizes of the cubic and quadratic coefficients give an indication of the dangers involved in extrapolation.

The following example illustrates this point. The earnings per share of stock of American Telephone and Telegraph Company for the years 1963–1967 were reported to be \$3.03, \$3.24, \$3.41, \$3.69, and \$3.79. Suppose that we try to predict from this data the earnings per share in 1968. We have $\hat{\gamma}_0 = \bar{y} = 3.432$, $\hat{\gamma}_1 = 0.197$, $\hat{\gamma}_2 = -0.008$, and $\hat{\gamma}_3 = -0.014$. With these coefficients we can obtain linear, quadratic, and cubic estimates of the earnings for 1963–1967 and projections for 1968.

Year	1963	1964	1965	1966	1967	1968 (Estimated only)
Actual earnings	\$3.03	\$3.24	\$3.41	\$3.69	\$3.79	
Linear estimate	\$3.04	\$3.24	\$3.43	\$3.63	\$3.83	\$4.02
Quadratic estimate	\$3.02	\$3.24	\$3.45	\$3.64	\$3.81	\$3.97
Cubic estimate	\$3.04	\$3.22	\$3.45	\$3.66	\$3.80	\$3.77

Notice that throughout the range of actual reported data there are only slight differences among the linear, quadratic, and cubic estimates; the range of the three estimates is never more than three cents. At the extrapolated point, however, the question of whether the higher order terms are included becomes more important. Adding the quadratic term to the linear estimate reduces the estimate by five cents; the cubic term makes a difference of twenty cents (about 5 per cent) in the extrapolated value.

3.6 The Gasoline Experiment (Continued)

Suppose now that the gasolines used in the experiment discussed earlier in this chapter differed only in the amount of some quantitative factor; in particular, suppose that gasolines A, B, C, D, and E were identical except that they contained $x = 0, 1, 2, 3, 4$ cubic centimeters, respectively, of some additive per gallon. We can subdivide S_t into linear, quadratic, cubic, and quartic components. The sum of squares for the linear contrast is

$$S_1 = (-2T_1 - T_2 + T_4 + 2T_5)^2/(10 \times 4) = (14.8)^2/40 = 5.476;$$

the sums for the quadratic and cubic contrasts are

$$S_2 = (2T_1 - T_2 - 2T_3 - T_4 + 2T_5)^2/(14 \times 4) = (5.2)^2/56 = 0.483,$$

$$S_3 = (-T_1 + 2T_2 - 2T_4 + T_5)^2/(10 \times 4) = (2.4)^2/40 = 0.144.$$

The quartic contrast is $(1, -4, 6, -4, 1)\mathbf{Y}$, and the corresponding sum of squares is $(1.2)^2/280 = 0.005$.

These four sums of squares add up to S_t, which provides a check on our calculations. If the corresponding coefficient in the polynomial is zero, S_i/s^2 has an F distribution with 1 and 15 d.f. The upper 5 per cent value of $F(1, 15)$ is 4.54. The only component that exceeds $4.54s^2 = 4.54(0.225) = 1.022$ is S_1. We conclude that the higher order coefficients can be omitted from the polynomial. The estimated line is

$$\hat{y} = \hat{\gamma}_0 + \hat{\gamma}_1 u_1,$$

where $\hat{\gamma}_0 = \bar{y} = 91.71$, $\hat{\gamma}_1 = 14.8/40 = 0.37$, and $u_1 = x - 2$, or $\hat{y} = 90.97 + 0.37x$. The linear estimates and their deviations from the treatment means are

x	0	1	2	3	4
$\hat{y}_i$	90.97	91.34	91.71	92.08	92.45
$y_{i\cdot} - \hat{y}_i$	+0.13	+0.01	−0.16	−0.23	+0.25

We note that $\sum_i (y_{i\cdot} - \hat{y}_i) = 0$ and $4 \sum_i (y_{i\cdot} - \hat{y}_i)^2 = 4(0.1580) = S_t - S_1$.

3.7 The Random Effects Model

In the previous model the levels of the factor were associated with unknown constants τ_i. In the random effects model we have

$$y_{ij} = \mu + a_i + e_{ij},$$

where a_i is a random variable drawn from a (normal) population with mean zero and variance σ_A^2. The random variables a_i are uncorrelated with each other and with the e_{ij}. It is convenient in this model to denote the variance of the errors e_{ij} by σ_e^2.

Thus, one might envision making n_1 batches of a chemical product by the same process and drawing n_2 samples from each batch, in which case y_{ij} would denote the response of the jth sample from the ith batch. We assume that the process gives a mean response μ; σ_A^2 denotes the batch-to-batch variation; σ_e^2 measures the variability of samples in the same batch. Similarly, a geneticist might pick n_1 dams at random from a population under study and then choose n_2 offspring from each dam. This model is also called the components of variance model, σ_e^2 and σ_A^2 being the components of variance.

The N observations are no longer all independent. Observations made on the same batch are now correlated. We take the same subdivision into sums of squares that we used in the fixed effects case. The expectation of the sum of squares between batches, S_t, is $(\sigma_e^2 + n_2\sigma_A^2)(n_1 - 1)$, and the hypothesis $\sigma_A^2 = 0$ is tested in the usual way by the F statistic.

It is convenient to write ϕ_1 for $(n_1 - 1)$ and ϕ_2 for $n_1n_2 - n_1$. The estimates of the components of variance are given by $\hat{\sigma}_e^2 = \phi_2^{-1}S_e = s^2$ and $\hat{\sigma}_e^2 + n_2\hat{\sigma}_A^2 = \phi_1^{-1}S_t = M_t$. Under normality, the two sums of squares S_e/σ_e^2 and $S_t/(\sigma_e^2 + n_2\sigma_A^2)$ have independent chi-square distributions with ϕ_1 and ϕ_2 degrees of freedom, respectively.

In this relatively simple experiment we can obtain the maximum likelihood estimates of σ_A^2 and σ_e^2 without too much difficulty. However, in more complex experiments involving random effects the solutions to the maximum likelihood equations are not so easily obtained. We shall adhere in the future to the procedure that we used this time. We shall take the same breakdown of $\mathbf{Y'Y}$ into sums of squares that we get in the fixed effects case and obtain our estimates of the components of variance from them.

3.8 The Power of the F Test with Random Effects

The hypothesis $\sigma_A^2 = 0$ is tested by the statistic $\mathscr{F} = M_t/s^2$. Under normality, $S_t/(\sigma_e^2 + n_2\sigma_A^2)$ has a χ^2 distribution with $n_1 - 1$ d.f., and S_e/σ_e^2 has an independent χ^2 distribution with $n_1(n_2 - 1)$ d.f., so that

$$\mathscr{F} = (1 + n_2\sigma_A^2/\sigma_e^2)F(n_1 - 1, n_1(n_2 - 1)).$$

We reject the hypothesis $\sigma_A^2 = 0$ if $\mathscr{F} > F_\alpha(n_1 - 1, n_1(n_2 - 1))$, i.e., if

$$F(n_1 - 1, n_1(n_2 - 1)) > \frac{F_\alpha(n_1 - 1, n_1(n_2 - 1))}{1 + n_2\theta}$$

where $\theta = \sigma_A^2/\sigma_e^2$. Thus, in the random effects case the power of the F test depends only upon the central F distribution.

EXAMPLE. Suppose that $n_1 = 4$, $n_2 = 6$, $\theta = 2.5$, and that α is taken to be 0.05. The upper 5 per cent value of $F(3, 20)$ is 3.10. Then the power of the test is $P[F(3, 20) > 3.10(1 + 6(2.5))^{-1}] = P[F(3, 20) > 0.19]$, which is approximately 0.90.

3.9 Confidence Intervals for the Components of Variance

A confidence interval for σ_e^2 can be obtained in the usual way using the χ^2 distribution

$$[S_e/\chi_U^2(\phi_2)] < \sigma_e^2 < [S_e/\chi_L^2(\phi_2)],$$

where for confidence, $(1 - \alpha)$, χ_U^2, and χ_L^2 are the upper and lower $\alpha/2$ points of the chi-square distribution with ϕ_2 d.f.

An approximate confidence interval for σ_A^2 can be obtained in the same way using S_t and assuming that $s^2 = \sigma_e^2$. We should then have

$$[S_t/\chi_U^2(\phi_1)] - s^2 < n_2\sigma_A^2 < [S_t/\chi_L^2(\phi_1)] - s^2.$$

Unfortunately, this method can give negative lower limits even when the F test rejects the hypothesis $\sigma_A^2 = 0$. A preferable approach is to argue, as in the previous section, that $\mathcal{F}/(1 + n_2\theta)$ has a central $F(\phi_1, \phi_2)$ distribution. Then we can obtain a confidence interval with confidence $(1 - \alpha)$ by writing

$$(1 + n_2\theta)F_L < \mathcal{F} < (1 + n_2\theta)F_U,$$

and manipulating it to obtain

$$n_2^{-1}(\mathcal{F}/F_U - 1)s^2 < \sigma_A^2 < n_2^{-1}(\mathcal{F}/F_L - 1)s^2,$$

where F_U and F_L are the upper and lower $\alpha/2$ points of $F(\phi_1, \phi_2)$. F_L can be obtained from the usual tables by the identity $F_{1-\alpha/2}(\phi_1, \phi_2) = 1/F_{\alpha/2}(\phi_2, \phi_1)$.

The general problem of confidence intervals for the components of variance has been investigated by Bulmer (1957) and by Scheffé (1959). There are also interesting problems involving the allocation of observations. For a fixed number $N = n_1n_2$ of observations, what is the best ratio n_1/n_2 for the experimenter to take in designing his experiment? Should he perhaps choose an unbalanced design having unequal numbers of observations on the batches? For the case in which the same number of observations is made on each batch, the problem was solved by Hammersley (1949) [see also Scheffé (1959)].

3.10 The Gasoline Experiment (Continued)

Suppose that instead of being fixed effects the gasoline effects were random. Suppose that the gasolines were drawn from five different tanks, all filled with the product from the same plant, and that apart from random variation the contents of the tanks were essentially the same. We have

$$y_{ij} = \mu + t_i + e_{ij},$$

where $t_i \sim N(0, \sigma_A^2)$. The hypothesis $\sigma_A^2 = 0$ is rejected by the F test. The point estimates of σ_e^2 and σ_A^2 are $\hat{\sigma}_e^2 = M_e = 0.225$ and $\hat{\sigma}_A^2 = (M_t - M_e)/4 = 0.326$.

A two-sided 95 per cent confidence interval for σ_e^2 is obtained in the customary way using χ^2 with 15 d.f. as follows:

$$3.370/27.49 < \sigma_e^2 < 3.370/6.262, \qquad 0.123 < \sigma_e^2 < 0.538.$$

When we try the first method of obtaining an approximate confidence interval for σ_A^2 by putting s^2 for σ_e^2 and using χ^2, we run into trouble. The method gives for a 95 per cent interval with 4 d.f. for χ^2,

$$(1.527/11.14) - s^2 < 4\sigma_A^2 < (1.527/0.4844) - s^2,$$

$$-0.022 < \sigma_A^2 < 0.732,$$

and even though $H_0\colon \sigma_A^2 = 0$ was emphatically rejected, the lower limit turns out to be negative.

The second procedure, which was the one we recommended, has for 95 per cent confidence $F_U = F_{0.025}(4, 15) = 3.80$ and $F_L = 1/F_{0.025}(15, 4) = 0.1155$.

Then the desired confidence interval is

$$0.04 < \sigma_A^2 < 3.25.$$

3.11 The Randomized Complete Block Experiment

At the beginning of this chapter we considered an experiment with gasolines and cars. There were five gasolines and we wanted four observations on the octane number of each gasoline. The experimental design suggested was to choose twenty cars (presumably at random from whatever automobile population was germane to the experiment), assign four cars at random to each gasoline, and then to take one observation in each car. The model then suggested can be rewritten

$$y_{ij} = \mu + \tau_i + f_{ij},$$

where y_{ij} is the observation on the ith gasoline in the jth of the cars assigned to it, τ_i is the effect of the ith gasoline, and f_{ij} is the random error in the observations. The f_{ij} are uncorrelated random variables with zero means and variance σ_f^2. Let us examine these f_{ij} terms a little more closely.

Our assumptions imply that if we were to take several cars and observe the octane number of gasoline A in each of them, the observations would be scattered about the "true" value of the octane number of gasoline A in a random manner with variance σ_f^2. The scatter in the observations comes from two sources: first, different cars rate gasolines differently; second, even in the same car, repeated observations on the same gasoline will differ because of what might be called the "noise" in the system or the general uncooperativeness of nature. We can rewrite the model to take account of these two sources of variation by splitting f_{ij} into two terms:

$$y_{ij} = \mu + \tau_i + c_{j(i)} + e_{ij}.$$

Here $c_{j(i)}$ is a "car" component which corresponds to the jth of the cars that were assigned to the ith gasoline.

Suppose that c_{ij} is a random variable with mean zero and variance σ_c^2, while e_{ij} as usual has mean zero and variance σ^2. We have $\sigma_f^2 = \sigma_c^2 + \sigma^2$. We assume also that the population from which the cars were chosen is so large that for sampling purposes it can be regarded as infinite. Then if we were to take an observation on a gasoline in each of r cars and average them, the variance of the mean of the observations would be $(\sigma_c^2 + \sigma^2)/r$. The situation is analogous to that in which d dams are assigned to a sire and a single offspring is observed from each litter.

What would happen if, instead of taking twenty cars, we were to take four cars and to use each gasoline once on each car? The experiment might take a little longer because each car would now be used for five gasolines, but on the other hand there would be fewer cars to worry about. Are there more pressing arguments in favor of this new design? Does it give data that are in some sense better?

As before, we estimate μ by the grand mean $G/20$ and $(\mu + \tau_i)$ by the average $y_{i\cdot} = \sum_j y_{ij}/4$ (to be derived shortly). Neither μ nor τ_i is estimable, but, as before, we can estimate $\tau_i - \tau_{i'}$ by $y_{i\cdot} - y_{i'\cdot}$. In the original experimental design we had $V(y_{i\cdot}) = \sigma_f^2/4$, and $V(y_{i\cdot} - y_{i'\cdot}) = 2\sigma_f^2/4$. In the new design, $V(y_{i\cdot})$ is still $\sigma_f^2/4$ so that as far as estimating $\mu + \tau_i$ is concerned there has been no improvement in precision in the sense that there has been no reduction in the variance. However, $V(y_{i\cdot} - y_{i'\cdot})$ is reduced to $2\sigma^2/4$. This reduction becomes apparent when we rewrite the model

$$y_{ij} = \mu + \tau_i + c_j + e_{ij},$$

because car j is now the jth car for each of the gasolines, and we no longer need to write $c_{j(i)}$. Then

$$\begin{aligned} T_i - T_{i'} &= \sum_j y_{ij} - \sum_j y_{i'j} = \sum_j (y_{ij} - y_{i'j}) \\ &= \sum_j [\mu + \tau_i + c_j + e_{ij} - \mu - \tau_{i'} - c_j - e_{i'j}] \\ &= 4(\tau_i - \tau_{i'}) + \sum_j (e_{ij} - e_{i'j}) = 4(\tau_i - \tau_{i'}) + \sum_j e_{ij} - \sum_j e_{i'j}, \end{aligned}$$

and $V(\hat{\tau}_i - \hat{\tau}_{i'}) = 8\sigma^2/16 = \sigma^2/2$.

What has happened is that the car effects have canceled out and the precision of the estimates of treatment comparisons has been thereby increased. The same procedure is commonly used in agronomy. Instead of dividing a field into $N = rt$ plots and assigning the treatments to the plots at random, the field is first split up into blocks that are as homogeneous as possible, and each variety is sown on one plot in each block. Thus, each variety gets a plot on the good land, and each gets a plot on the bad land, and so on. The term block has been retained in standard use in all areas of application. In our experiment, each car constituted a block. This kind of experiment is called a randomized complete block experiment or design: complete because each block contains each treatment, and randomized because the plots are assigned to the treatments at random in each block.

We now derive the analysis. The model for t treatments, each appearing exactly once in each of b blocks, is

$$y_{ij} = \mu + \tau_i + \beta_j + e_{ij},$$

where y_{ij} is the observation on the ith treatment in the jth block. We shall assume that the block effects, β_j, are random normal variables, independently distributed with zero means and variance σ_b^2. It does not matter, so far as our derivation of estimates is concerned, whether the β_j are fixed or random. The distinction is of interest when we consider applying the results of the analysis. If the blocks can be considered as chosen at random from some population, it is not unreasonable to suppose that the conclusions about treatment contrasts will hold throughout the population of blocks of which those used are regarded as "typical." However, if the cars chosen come from a small group or if they are to be regarded as fixed effects, there are problems when we wish to apply our results elsewhere. Of what population are the cars used typical? Is it the population of low priced 1969 cars? Or of all 1969 cars? Or the population of cars driven to work by engineers in the research laboratory during the third week in July?

Just as in the case of the random samples from random batches, we have two choices, under normality, when the blocks are random effects. We can solve the maximum likelihood equations for the estimates of the parameters, noting that the observations are no longer uncorrelated because $\text{cov}\,(y_{ij}, y_{ij'}) = \sigma_b^2 \; (j \neq j')$. That approach will be taken up in the exercises at the end of this chapter. The second approach is to act as if the β_j are fixed effects to obtain least squares estimates. We shall follow the latter course. (Indeed, if we did not know the distribution of the β_j it would be the only course open to us.)

In vector notation the model can be written

$$\mathbf{Y} = \mathbf{X\theta} + \mathbf{e},$$

where $\boldsymbol{\theta}' = (\mu, \boldsymbol{\tau}', \boldsymbol{\beta}')$. The design matrix $\mathbf{X}$ has $1 + t + b$ columns. The first column is $\mathbf{1}_N$, a column of ones, corresponding to the mean; the next t columns are associated with treatments, and the others with blocks. The row for y_{ij} has one in the first column, the $(i + 1)$th column and the $(1 + t + j)$th column; the other entries are zero. Columns 2 through $t + 1$, and columns $t + 2$ through $1 + t + b$ both add up to $\mathbf{1}_N$, and so $\mathbf{X}$ is of rank $t + b - 1$.

The normal equations are

$$G = N\hat{\mu} + b\sum_i \hat{\tau}_i + t\sum_j \hat{\beta}_j,$$

$$T_i = b\hat{\mu} + b\hat{\tau}_i + \sum_j \hat{\beta}_j,$$

$$B_j = t\hat{\mu} + \sum_i \hat{\tau}_i + t\hat{\beta}_j,$$

where $B_j = \sum_i y_{ij}$ is the sum of the observations in the jth block. Imposing the side conditions $\sum \tau_i = 0$ and $\sum \beta_j = 0$, we immediately have

$$\hat{\mu} = G/N, \qquad \hat{\tau}_i = T_i/b - \hat{\mu}, \qquad \hat{\beta}_j = B_j/t - \hat{\mu}.$$

We note that the estimates $\hat{\tau}_i$ are the same as they would have been without the β_j in the model, and the $\hat{\beta}_j$ are the same as they would have been without the τ_i. We therefore say that treatments are orthogonal to blocks. This property is a consequence of the balance in the experiment inasmuch as every treatment appears exactly once in every block. We shall have more to say about this when we discuss two-factor experiments. The sum of squares for regression $S_R = C + \sum \hat{\tau}_i T_i + \sum \hat{\beta}_j B_j$ may be split into three forms, C, S_t, which we have seen before, and a similar sum of squares $S_b = \sum_j B_j^2 - C$, the sum of squares for blocks. The analysis of variance table becomes:

TABLE 3.3
ANALYSIS OF VARIANCE

Source	SS.	d.f.	EMS
Total	$\mathbf{Y'Y} - C$	$N - 1$	
Treatments	$\sum_i T_i^2/b - C$	$t - 1$	$\sigma^2 + b\sum_i \tau_i^2/(t-1)$
Blocks	$\sum_j B_j^2/t - C$	$b - 1$	$\sigma^2 + t\sigma_b^2$
Error	$\mathbf{Y'Y} - C - S_t - S_b$	$(t-1)(b-1)$	σ^2

The result of blocking has thus been to remove from the error sum of squares, S_e, a portion with $b - 1$ d.f.; hopefully the amount of residual removed has been appreciable, at least enough to compensate for reduction of d.f. for error. With fewer d.f., the critical values of F and t will be increased.

3.12 The Gasoline Experiment (Continued)

We return now to the gasoline data, assuming that the effects are fixed and that the experiment was a randomized complete block experiment. There were four blocks. The jth block consists of the jth observation on each treatment;

thus block 1 consisted of observations 91.7 on A, 91.7 on B, 92.4 on C, 91.8 on D and 93.1 on E. The block totals (after 90.0 has been subtracted from each observation) are 10.7, 9.4, 7.8, 6.3.

The analysis of variance table is as follows:

TABLE 3.4
ANALYSIS OF VARIANCE

Source	SS.	d.f.	M.S.	F
Total	9.478	19		
Treatments	6.108	4	1.527	15.6
Blocks	$60.676 - C = 2.194$	3	0.731	7.46
Error	1.176	12	0.098	

The new estimate of the error variance is $s^2 = 0.098$ and $s = 0.313$.

Using the t test, the new LSD is $2.179\sqrt{0.049} = 0.482$, and we conclude that $E > A, B, C, D$ and $D > A, B$.

For Tukey's test, $q_{0.05}$ with 12 d.f. is 4.51. The new LSD is 0.71, and we conclude that $E > A, B, C, D$ and $D > A$.

The S test for ψ still accepts $\psi = 0$, but barely. $S^2 = 4F(4, 12) = 13.04$ and $Ss = 1.13$, so that the confidence interval becomes

$$-2.23 < \psi < 0.03.$$

3.13 Missing Plots in a Randomized Complete Block Design

The analysis of the randomized complete block design depends upon having each treatment appear once in each block. What happens if something goes wrong and one of the plots is missing? Perhaps somebody makes a gross mistake and plants the wrong variety, or harvests it too early. In some industrial situations, it is possible to obtain a repeat observation fairly easily. If one of the cars breaks down before all the gasolines have been tested, it might be feasible to wait until the car is repaired and then finish the experiment, or to find another car and, having tested all the gasolines on it, ignore the data on the old car. However, the agronomist has to wait another growing season to obtain a repeat observation, and we can easily think of industrial situations in which a repeat observation would be impossible. An industrial example which is not uncommon occurs when an experiment is conducted in a pilot plant which is available to the project only for a limited period. If the observations require rather lengthy chemical analyses, it is quite likely that by the time the scientist discovers that an observation has been spoiled or lost, the test equipment will already have been dismantled and the pilot plant will be in use on some other project.

The statistician may calculate from the data what the missing observation would have been if only he had it; then he will have a complete set of data and he can perform the usual analysis (with some modification). There is the alternative of analyzing the data as an incomplete block experiment with unequal block sizes and unequal numbers of replications of the treatments. That, as will be seen in a subsequent chapter, calls for messy calculations, and is considerably less convenient than the procedure suggested here.

We discussed the general problem of missing plot formulae in the previous chapter. Suppose that y_{ij} is missing. Let T' denote the sum of the $b - 1$ actual observations on the ith treatment, B' the sum of the $t - 1$ observations on the jth block, and G' the sum of all $N - 1$ observations. We take as our estimate of the missing plot or observation the value u which minimizes S_e when the ordinary analysis of variance procedure is carried out on the N points (the $N - 1$ actual points and u). We showed in the last chapter that this was equivalent to choosing u so as to give an exact fit. In that case

$$u = \hat{\mu} + \hat{\tau}_i + \hat{\beta}_j = \frac{G' + u}{N} + \frac{T' + u}{b} - \frac{G' + u}{N} + \frac{B' + u}{t} - \frac{G' + u}{N},$$

and, solving,

$$u = \frac{tT' + bB' - G}{(b - 1)(t - 1)}.$$

We are mainly concerned with comparisons $y_{i\cdot} - y_{i'\cdot}$. The missing plot value is a linear combination of the observations, and so T_i is correlated with the other treatment totals. It can be shown that, when there is a missing observation in the ith treatment,

$$V(y_{i\cdot} - y_{i'\cdot}) = b^{-2}V(T_i - T_{i'}) = \sigma^2\left[\frac{2}{b} + \frac{t}{b(b - 1)(t - 1)}\right].$$

If there are two missing plots, we have two choices. Either we can repeat the procedure just used with two simultaneous equations to be solved, or we can adopt an iterative procedure of guessing one value and fitting the other, then going back and fitting the first value, and so on.

The formula that we have derived was first obtained by Allan and Wishart (1930). The F test will be slightly biased, and the reader is referred to the paper by Yates (1933) for a discussion of this point.

3.14 The Analysis of Covariance

We have just considered the use of randomized complete blocks to improve the precision of treatment comparisons. Another method that may be used sometimes is the analysis of covariance. This procedure involves adjusting the observed responses for the linear effect of another factor (a concomitant variable). In a chemical experiment, where we have for the treatments different temperatures, we might wish to adjust the yields for the percentage of some impurity in the feed stock at each run. In an experiment to investigate the effect on the growth of young pigs of some change in the standard diet, we may take weight at six months as the response, y, and adjust it for the initial weight, x. The adjustment is linear, and we consider here only the case of a single concomitant variable. The model that we shall take is

$$\Omega_1: \quad y_{ij} = \mu + \tau_i + \beta x_{ij} + e_{ij},$$

where x_{ij} is the measurement made on the concomitant variable on the ijth run, or for the ijth datum point.

We introduce the following notation

$$E_{xx} = \sum_i \sum_j (x_{ij} - x_{i.})^2; \qquad E_{xy} = \sum_i \sum_j (x_{ij} - x_{i.})(y_{ij} - y_{i.});$$

$$E_{yy} = \sum_i \sum_j (y_{ij} - y_{i.})^2;$$

$$T_{xx} = \sum_i \sum_j (x_{i.} - x_{..})^2; \qquad T_{xy} = \sum_i \sum_j (x_{i.} - x_{..})(y_{i.} - y_{..});$$

$$T_{yy} = \sum_i \sum_j (y_{i.} - y_{..})^2;$$

$$S_{xx} = \sum_i \sum_j (x_{ij} - x_{..})^2; \qquad S_{xy} = \sum_i \sum_j (x_{ij} - x_{..})(y_{ij} - y_{..});$$

$$S_{yy} = \sum_i \sum_j (y_{ij} - y_{..})^2,$$

where $S = T + E$. We do not assume that there are necessarily the same number of observations at each level of the factor being considered; $x_{i.}$ is the average of the x measurements for the ith treatment, and $x_{..}$ is the average of all N measurements of the concomitant variable.

The most convenient computational procedure for calculating these quantities is to carry out a one-way analysis of variance for x and for y and a parallel calculation for xy. We then obtain, writing $X_{..}$ for $Nx_{..}$ and $Y_{..}$ for $Ny_{..}$, the following table:

TABLE 3.5
ANALYSIS OF COVARIANCE

Source	d.f.	x	xy	y
Raw total	N	$\sum_i \sum_j x_{ij}^2$	$\sum_i \sum_j x_{ij} y_{ij}$	$\sum_i \sum_j y_{ij}^2$
Mean	1	$X_{..}^2/N$	$X_{..} Y_{..}/N$	$Y_{..}^2/N$
Treatments	$t - 1$	T_{xx}	T_{xy}	T_{yy}
Error	$N - t$	E_{xx}	E_{xy}	E_{yy}
$T + E$	$N - 1$	$S_{xx} = T_{xx} + E_{xx}$	S_{xy}	S_{yy}

Note: The entries in the x and y columns cannot be negative; entries in the xy column may, however, be negative.

We can now proceed in two ways, and it is constructive to consider both. When we introduce blocks, we shall set up and solve the normal equations. At this time we shall use the other approach.

Suppose that we have no treatment effects. Our model would be

$$\Omega_0: \quad y_{ij} = \mu + \beta x_{ij} + e_{ij}.$$

Under this model β is estimated by $\hat{\beta} = S_{xy}/S_{xx}$, and the sum of squares for error is $S_e' = S_{yy} - (S_{xy})^2/S_{xx}$.

Suppose now that we have treatment effects and that the model is Ω_1. This model implies that there is a family of parallel lines, one for each treatment. Some of the lines may coincide; Ω_0 corresponds to the case in which all of the lines coincide. The data for the ith treatment alone give an estimate b_i of β, where

$$b_i = \frac{\sum_j (x_{ij} - x_{i\cdot})(y_{ij} - y_{i\cdot})}{\sum_j (x_{ij} - x_{i\cdot})^2} \quad \text{and} \quad V(b_i) = \frac{\sigma^2}{\sum_j (x_{ij} - x_{i\cdot})^2}.$$

The several estimates, b_i, are clearly uncorrelated because they involve different data. The minimum variance unbiased combined estimate from them is $\hat{\beta} = \sum_i w_i b_i$, where the weight, w_i, given to any b_i is inversely proportional to $V(b_i)$, and $\sum_i w_i = 1$. This gives $w_i = \sum_j (x_{ij} - x_{i\cdot})^2/E_{xx}$ and

$$\hat{\beta} = E_{xy}/E_{xx}.$$

The ith line is thus $y_{ij} = y_{i\cdot} + \hat{\beta}(x_{ij} - x_{i\cdot})$.

The sum of the squares of the deviations about the several lines is

$$S_e = \sum_i \sum_j (y_{ij} - y_{i\cdot} - \hat{\beta}(x_{ij} - x_{i\cdot}))^2 = E_{yy} - (E_{xy})^2/E_{xx}.$$

We can now write down the analysis of variance table. The sum of squares for treatments is the reduction in error made by fitting individual lines rather than a common line, i.e., $S_e' - S_e$.

TABLE 3.6
ANALYSIS OF VARIANCE

Source	d.f.	SS
Total	$N - 1$	S_{yy}
Fitting a single line	1	$(S_{xy})^2/S_{xx}$
Treatments	$t - 1$	$(S_{yy} - (S_{xy})^2/S_{xx}) - (E_{yy} - (E_{xy})^2/E_{xx})$
Error	$N - t - 1$	$E_{yy} - (E_{xy})^2/E_{xx}$

The error variance, σ^2, is estimated by $s^2 = (N - t - 1)^{-1}(E_{yy} - (E_{xy})^2/E_{xx})$.

If the concomitant variable had not been added to the model, the error sum of squares would have been E_{yy}. The reduction resulting from the inclusion of the concomitant variable is, therefore, $(E_{xy})^2/E_{xx}$, and the hypothesis $H_0: \beta = 0$ is tested by the statistic $(E_{xy})^2/(s^2E_{xx})$, which has, under the null hypothesis, the central F distribution with one and $N - t - 1$ degrees of freedom.

It should be noted that if the same x array is used for each of the treatments, we are really back to a complete block design with the blocks corresponding to the values of x.

3.15 Analysis of Covariance for the Randomized Complete Block Design

We confine our discussion to the case in which there is a single observation on each treatment in each block; y_{ij} denotes the observation on the ith treatment in the jth block, and x_{ij} is the corresponding value for the concomitant variable. The complete model is

$$y_{ij} = \mu + \tau_i + \beta x_{ij} + \gamma_j + e_{ij},$$

where γ_j is the jth block effect and $i = 1, \ldots, t; j = 1, \ldots, r$. We shall use capital letters X, Y to denote totals. Thus $Y_{i.} = \sum_j y_{ij}$, and so on. The normal equations are, with the side conditions $\sum_i \tau_i = 0$, $\sum_j \gamma_j = 0$,

$$Y_{..} = tr\,\hat{\mu} + \hat{\beta}X_{..},$$

$$Y_{i.} = r\hat{\mu} + r\hat{\tau}_i + X_{i.}, \qquad Y_{.j} = t\hat{\mu} + t\hat{\gamma}_j + X_{.j},$$

$$\sum_i \sum_j x_{ij}\, y_{ij} = X_{..}\hat{\mu} + \sum_i X_{i.}\hat{\tau}_i + \sum_j X_{.j}\hat{\gamma}_j + \sum_i \sum_j x_{ij}^2\hat{\beta}.$$

Eliminating $\hat{\mu}$, $\hat{\tau}_i$, and $\hat{\gamma}_j$ from the last equation, we have

$$\sum_i \sum_j x_{ij}y_{ij} - t^{-1}\sum_j X_{.j}Y_{.j} - r^{-1}\sum_i X_{i.}Y_{i.} + N^{-1}X_{..}Y_{..}$$
$$= \hat{\beta}\left\{\sum_i \sum_j x_{ij}^2 - r^{-1}\sum_j X_{.j}^2 - t^{-1}\sum_i X_{i.}^2 + N^{-1}X_{..}^2\right\}.$$

The coefficient of $\hat{\beta}$ is the sum of squares for error obtained when an analysis of variance, including blocks, is carried out on the x_{ij}. The left side of the equation is the corresponding sum obtained from the xy terms. We write E'_{xx}, E'_{xy} for these quantities and define similarly E'_{yy}, S'_{xx}, S'_{xy}, S'_{yy}. We then obtain

$$\hat{\beta} = E'_{xy}/E'_{xx}.$$

The other parameters are estimated by

$$\hat{\mu} = (Y_{..} - \hat{\beta}X_{..})/N, \qquad \hat{\tau}_i = \{tY_{i.} - Y_{..} - \hat{\beta}(tX_{i.} - X_{..})\}/N,$$

$$\hat{\gamma}_j = \{rY_{.j} - Y_{..} - \hat{\beta}(rX_{.j} - X_{..})\}/N.$$

The sum of squares for regression (including the mean) is

$$Y_{..}\hat{\mu} + \sum_i Y_{i.}\hat{\tau}_i + \sum_j Y_{.j}\hat{\gamma}_j + \hat{\beta}\sum_i \sum_j x_{ij}y_{ij},$$

whence $S_e = E'_{yy} - \hat{\beta}E'_{xy} = E'_{yy} - (E'_{xy})^2/E'_{xx}$.

If $\beta = 0$, we have a straightforward randomized complete block situation with $S'_e = S'_{yy} - (S'_{xy})^2/S'_{xx}$.

The "recipe" for the calculations is thus to proceed in the same way as for the unblocked case, using S'_{yy}, etc., which are calculated, in the same manner used before by analysis of variance procedures for x, y and xy, including the terms for blocks in the computations.

Exercises

1. Six octane number determinations were made on each of five gasolines. The data are given as follows. The columns correspond to six blocks. Analyze the data as a randomized complete block experiment. Which gasolines differ?

	Blocks					
	I	II	III	IV	V	VI
A	57	61	62	59	58	59
B	65	65	67	64	65	63
C	59	60	65	60	60	59
D	55	54	53	51	50	53
E	54	50	57	55	54	57

2. This data comes from a randomized complete block experiment with three blocks and twelve treatments. The responses are the yields, in ounces, of cured tobacco leaves. Compare the control to the average of the other treatments. [Petersen (1952)].

	Treatments											
	1	2	3	4	5	6	7	8	9	10	11	12 (Control)
Block I	76	82	76	70	76	70	82	88	81	74	67	79
II	70	70	73	74	73	83	74	65	67	67	67	78
III	80	73	77	62	86	84	80	80	81	76	79	63
Means	75	75	75	69	78	79	79	78	76	72	71	73

3. An experiment is to be designed to compare t new treatments with a control treatment. A one-way layout is to be used; there are to be r_0 observations on the control and r observations on each of the new treatments. Show that for any given value of N the variance of the comparisons between the control treatment and the new treatments, $V(\hat{\tau}_0 - \hat{\tau}_i)$, is minimized if $r_0 = r\sqrt{t}$.
4. Let x take the values 1, 2, 3, . . ., n, and let $y = x^2$. Show that the correlation coefficient of x and y is given by

$$r^2 = \frac{15(n + 1)^2}{16n^2 + 30n + 11},$$

which is a monotone decreasing function of n with limit 15/16.
5. Two observations are made at $x = 0$ and one each at $x = 1, 2, 3$, and 4. Find a set of orthogonal linear, quadratic polynomials.
6. If x takes the values 1, 2, . . ., n once each, show that the polynomials $u_1 = x - \bar{x}$, $u_2 = 12u_1^2 - (n^2 - 1)$ are orthogonal.
7. Find a pair of orthogonal linear and quadratic polynomials when x takes the values 0, 1, 2, 4, 8.
8. Find the expected value of S_t for a randomized complete block design when one of the plots is missing and its value has been calculated by the formula.
9. Derive the formula for $V(\bar{y}_h - \bar{y}_i)$ if one of the observations on the hth treatment has been calculated by the missing plot formula.
10. Find the expected value of S_t in the random effects case when there are r_i observations on the ith variety and the r_i are not all equal.
11. In the random effects model for the one-way layout with r observations on each variety, use the method of maximum liklihood under the normality assumption to obtain estimates of σ_e^2 and σ_A^2, and to test the hypothesis H_0: $\sigma_A^2 = 0$.
12. Assume that in a randomized complete block experiment the errors e_{ij} and the block effects β_j are normal random variables. Use the method of maximum likelihood to derive estimates of τ and σ_e^2.
13. Prove that S_t and S_e are independent in the random effects model under normality in the balanced case. Is it true even when there are unequal numbers of observations on the varieties?
14. Suppose that in a randomized complete block design the observation on the first treatment in the first block and the observation on the second treatment in the second block are missing. Obtain formulae for calculating the two missing values.
15. Repeat problem 14 for the case in which the two missing plots are in the same block.
16. Scheffé (1959, p. 70) points out that the F test will reject H_0 if, and only if, there is at least one estimable ψ which has a significant estimate $\hat{\psi}$ by his S test. Is it true that if, in the fixed effects case, the F ratio is significant, then there is at least one contrast that will be declared significantly different from zero by the t test? Is the converse true? If not, give an example.

CHAPTER 4

Experiments with Two Factors

In the previous chapter we considered as an example of a randomized block experiment an experiment in which t treatments (gasolines) were tested in r cars. The cars were chosen at random from some appropriate population and they constituted the blocks of the design. The purpose of the blocking was to improve the precision of the comparisons between gasolines. The differences between cars were eliminated in estimating the gasoline contrasts. In some presentations the analysis of variance tables contain under the heading, "Expected mean square for the sum of squares between blocks," the single word *discard*. The variation between blocks is considered as something to be "backed out" of the experimental analysis.

Suppose, however, that instead of having r nondescript vehicles, we had been at pains to choose a Chevrolet, a Plymouth, a Ford, and so on, with the result that cars could be considered as a second factor occurring at r levels. The two factors are said to be crossed. Each factor appears at least once with each level of the other factor. Indeed, each gasoline sees each car exactly once, and each car sees each gasoline exactly once. How, if at all, does this experiment differ from the randomized block experiment of the previous chapter? We consider first the simple experiment in which each gasoline is tested only once in each car. This experimental setup is called the two-way layout (or two-way crossed classification) with one observation per cell.

4.1 The Two-Way Layout with One Observation per Cell

We shall find it convenient now to think about a factorial experiment involving two factors, A and B, with a and b levels respectively. In our previous example, the factors were gasolines and cars. The factors can be qualitative or quantitative.

The levels of the car factor could be Plymouth, Ford, etc.; the levels of the gasoline factor might be one ml/gal, 2 ml/gal, etc., of some additive like lead. Our model is the simple additive model $\mathbf{Y} = \mathbf{X}\boldsymbol{\theta} + \mathbf{e}$, or

$$y_{ij} = \mu + \alpha_i + \beta_j + e_{ij},$$

where y_{ij} is the observation made on (or the response at) the ith level of A and the jth level of B, and e_{ij} is the random error with the usual assumptions. The other terms in the model are unknown constants; μ represents a grand mean, α_i is the main effect of the ith level of A, and β_j is the main effect of the jth level of B.

The design matrix $\mathbf{X}$ has $N = ab$ rows and $1 + a + b$ columns. The row corresponding to the observation y_{ij} has unit elements in the first, the $(i + 1)$th and the $(1 + a + j)$th columns, and zeros elsewhere. The first column is $\mathbf{1}_N$. The sum of the next a columns is also $\mathbf{1}_N$, and the sum of the last b columns is also $\mathbf{1}_N$; it follows that $r(\mathbf{X}) \leq a + b - 1$. This implies that we have two parameters too many. We could get rid of one parameter by putting $\mu = 0$, which is the same as leaving μ out of the model, but we would still be faced with a singular matrix $\mathbf{X}'\mathbf{X}$. The most common procedure is to leave μ in the model and to impose the two side conditions $\mathbf{1}'\boldsymbol{\alpha} = 0$ and $\mathbf{1}'\boldsymbol{\beta} = 0$, where $\boldsymbol{\alpha}$ and $\boldsymbol{\beta}$ are the vectors of the main effects α_i and β_j. The parameters μ, α_i, and β_j are not estimable. The expected values of observations, contrasts in the parameters α_i, and contrasts in the β_j are estimable.

As far as the derivation of the sums of squares is concerned, there is no change from the randomized complete block design. We write $T_{i.} = \sum_j y_{ij} = by_{i.}$ and $T_{.j} = \sum_i y_{ij} = ay_{.j}$. Then $\hat{\mu} = \bar{y}$, $\hat{\alpha}_i = y_{i.} - \bar{y}$ and $\hat{\beta}_j = y_{.j} - \bar{y}$. Writing $\alpha.$ for $\sum_i \alpha_i/a$ and $\beta.$ for $\sum_j \beta_j/b$, we have

$$E(\hat{\alpha}_i) = \mu + \alpha_i + \beta. - \mu - \alpha. - \beta. = \alpha_i - \alpha..$$

This brings out the concept of the main effect of the ith level of A as the difference between the average response at the ith level of A and the average response at all levels of A.

The comparison $\alpha_i - \alpha_{i'}$ is estimated by $y_{i.} - y_{i'.}$. We point out again that just as was the case in the randomized block experiment, the B effects are canceled out in the expectation of the contrast because of the balance, or orthogonality, that was built into the experiment by having each level of A appear exactly once with each level of B. Similarly, the comparison $\beta_j - \beta_{j'}$ is estimated by $y_{.j} - y_{.j'}$. Table 4.1 follows:

TABLE 4.1

ANALYSIS OF VARIANCE

Source	SS	d.f.
Total	$\sum_i \sum_j y_{ij}^2 - G^2/N$	$ab - 1$
Between levels of A	$\sum_i T_{i.}^2/b - G^2/N$	$a - 1$
Between levels of B	$\sum_j T_{.j}^2/a - G^2/N$	$b - 1$
Residual	$\sum_i \sum_j y_{ij}^2 - \sum_i T_{i.}^2/b - \sum_j T_{.j}^2/a + G^2/N$	$(a - 1)(b - 1)$

Tests of hypotheses about the α_i and about the β_j are carried out just as in a randomized complete block design.

EXAMPLE. In their studies on the habits of Zebu cattle, Rollison et al. observed a group of cattle and recorded every minute which of the cattle were grazing. The same group of five observers was used throughout the work carried out at Entebbe, Uganda. The authors (1956) reported upon an experiment carried out to compare these observers. They watched a group of ten cattle in the same paddock for eighty-eight minutes one afternoon. The numbers recorded are the numbers of minutes in which observer i reported that he saw animal j grazing.

Observers	Animals										
	1	2	3	4	5	6	7	8	9	10	Total
A	34	76	75	31	61	82	82	67	72	38	618
B	33	76	72	29	60	82	84	67	72	36	611
C	35	78	76	30	65	86	88	66	76	37	637
D	34	77	71	29	60	78	83	67	72	37	608
E	33	77	70	27	59	81	82	67	70	33	599
Total	169	384	364	146	305	409	419	334	362	181	3073

TABLE 4.2
ANALYSIS OF VARIANCE

Source	d.f.	MS
Animals	9	2149.42
Observers	4	20.33
Residual	36	2.06
Total	49	

4.2 Orthogonality with More Than One Observation per Cell

Suppose that we relax the requirement that there be exactly one observation at each level of A and at each level of B, and allow several observations. The model remains the same, except that we could write y_{ijm} to denote the mth observation at the ith level of A and the jth level of B, i.e., in the (ij)th cell. If there are an equal number of observations in each cell, the analysis will be a trivial extension of that of the previous section. If the cell numbers are not all the same, the solution of the normal equations is complicated by the fact that we cannot find two side conditions that will separate the unknowns in a simple fashion.

We are not going to discuss the solution of the equations at this time. We have already spoken in Chapter 2 about methods of finding pseudo-inverses, and the matter will come up again in more detail in our discussion of incomplete block designs. We address ourselves in this section to the concept of orthogonality. What do we mean when we say that the two factors A and B are orthogonal, and what are necessary and sufficient conditions for orthogonality? Two definitions of orthogonality of factors (or, in block designs, orthogonality of blocks and treatments) commonly occur in conversation among practitioners of analysis of variance. We shall show that they are equivalent.

Definition 1. The factors A and B are said to be orthogonal if, and only if, any comparison between averages of observations at two levels of A is orthogonal (when considered as a linear combination of the observations) to any comparison between the averages of observations at two levels of B.

Definition 2. A and B are orthogonal if, and only if, the estimate of any contrast in the parameters α_i is the same whether or not the parameters β_j are included in the model, and vice versa.

We shall consider in particular two levels, h and i, of A and two levels, j and k, of B. The corresponding totals of the observations at these levels are $T_{h\cdot}$, $T_{i\cdot}$, $T_{\cdot j}$, $T_{\cdot k}$; the sum of the observations at the hth level of A and the jth level of B is T_{hj}. Let there be n_{hj} observations in the (hj)th cell with marginal totals $n_{h\cdot}$ and $n_{\cdot j}$, and N observations altogether.

Suppose that A and B are orthogonal by Definition 1. Then $y_{h\cdot} - y_{i\cdot}$ is orthogonal to $y_{\cdot j} - y_{\cdot k}$. Consider the equivalent contrasts $n_{i\cdot}T_{h\cdot} - n_{h\cdot}T_{i\cdot}$ and $n_{\cdot k}T_{\cdot j} - n_{\cdot j}T_{\cdot k}$. The only observations that appear in both contrasts are those in the cell totals T_{hj}, T_{hk}, T_{ij}, T_{ik}, and we rewrite the contrasts

$$n_{i\cdot}T_{hj} + n_{i\cdot}T_{hk} - n_{h\cdot}T_{ij} - n_{h\cdot}T_{ik},$$

and so on, and

$$n_{\cdot k}T_{hj} - n_{\cdot j}T_{hk} + n_{\cdot k}T_{ij} - n_{\cdot j}T_{ik}$$

and so on.

These are orthogonal as contrasts in the observations if, and only if,

$$n_{i\cdot}n_{\cdot k}n_{hj} - n_{i\cdot}n_{\cdot j}n_{hk} - n_{h\cdot}n_{\cdot k}n_{ij} + n_{h\cdot}n_{\cdot j}n_{ik} = 0,$$

or,

$$n_{i\cdot}(n_{\cdot k}n_{hj} - n_{\cdot j}n_{hk}) = n_{h\cdot}(n_{\cdot k}n_{ij} - n_{\cdot j}n_{ik}).$$

Such equations hold for all pairs h, i and j, k. Summing over all values of k, including $k = j$, gives

$$n_{i\cdot}(Nn_{hj} - n_{\cdot j}n_{h\cdot}) = n_{h\cdot}(Nn_{ij} - n_{\cdot j}n_{i\cdot});$$

then, summing over all values of h,

$$n_{i\cdot}(Nn_{\cdot j} - n_{\cdot j}N) = 0 = (Nn_{ij} - n_{i\cdot}n_{\cdot j})N,$$

so that $Nn_{ij} - n_{\cdot j}n_{i\cdot} = 0$, or $n_{ij} = n_{i\cdot}n_{\cdot j}/N$ for each i and each j.

The converse, that this condition implies the orthogonality of the contrasts, is easily shown. We call this condition the condition of proportional frequencies.

Suppose now that A and B are orthogonal by Definition 2. We show that the condition of proportional frequencies is again a necessary and sufficient condition. There are now two models under consideration:

$$E(\mathbf{Y}) = \mu\mathbf{1} + \mathbf{X}_1\boldsymbol{\alpha} + \mathbf{X}_2\boldsymbol{\beta}, \tag{4.1}$$

$$E(\mathbf{Y}) = \mu\mathbf{1} + \mathbf{X}_1\boldsymbol{\alpha}, \tag{4.2}$$

where $\mathbf{X} = (\mathbf{1}, \mathbf{X}_1, \mathbf{X}_2)$, and $\boldsymbol{\alpha}$ and $\boldsymbol{\beta}$ are the vectors of main effects.

Under model (4.2) we have a one-way layout and the least squares estimate of $\alpha_h - \alpha_i$ is $y_{h\cdot} - y_{i\cdot}$. Consider the estimable function $\psi = n_{h\cdot}n_{i\cdot}(\alpha_h - \alpha_i)$. Its least squares estimate is $\hat{\psi} = n_{i\cdot}T_{h\cdot} - n_{h\cdot}T_{i\cdot}$. Since the least squares estimates are the unique minimum variance unbiased estimates, and since any linear combination of the observations which is an unbiased estimate of ψ under model (4.1) will also be unbiased under model (4.2), it follows that $\hat{\psi}$ is the least squares estimate of ψ under (4.1) if, and only if, it is unbiased. Under model (4.1)

$$E(\hat{\psi}) = n_{h\cdot}n_{i\cdot}(\alpha_h - \alpha_j) + \sum_j (n_{i\cdot}n_{hj} - n_{h\cdot}n_{ij})\beta_j,$$

and $\hat{\psi}$ is unbiased if, and only if, $n_{i\cdot}n_{hj} - n_{h\cdot}n_{ij} = 0$ for each j. Summing over all values of h leads to the condition of proportional frequencies.

Conversely, if the condition of proportional frequencies is satisfied and the side conditions $\sum_i n_{i\cdot}\alpha_i = 0$ and $\sum_j n_{\cdot j}\beta_j = 0$ are used in the standard least squares procedure, it is easily seen that the least squares estimates of μ and α_i are $\bar{y}$ and $y_{i\cdot} - \bar{y}$ under both models.

Under model (4.2) we may subdivide the sum of squares for regression into two components $S(\mu)$ and $S(\alpha|\mu)$, where $S(\mu) = G^2/N$. Under model (4.1) we can have three components $S(\mu)$, and $S(\alpha|\mu)$, as before, and the extra portion $S(\beta|\mu, \alpha)$. If A and B are orthogonal, $S(\beta|\mu) = S(\beta|\mu, \alpha)$.

4.3 The Two-Way Layout with Interaction

An important and restrictive implication of the simple additive model used in the previous sections for the two-factor experiment is that the expected value of the difference between the responses at two levels of A is the same at each level of B. In terms of the gasoline experiment, this amounts to saying that apart from the random error if the Plymouth rates gasoline 1 to have an octane number

0.5 ON higher than gasoline 2, so will every other car. Similarly, if the Plymouth gives a rating to gasoline 1 that is one ON higher than the rating given by the Ford, then the Plymouth will rate each gasoline (apart from the random error) at one ON higher than the Ford rates them. This simple additive model may not be adequate. The failure of the differences between the responses at the different levels of A to remain constant over the different levels of B is attributed to interaction between the two factors. To take this into account, we add an extra term to the model.

We now consider the situation in which there are n observations at each combination of levels of A and B, i.e., n observations in each cell. The new model is

$$y_{ijk} = \mu + \alpha_i + \beta_j + (\alpha\beta)_{ij} + e_{ijk}, \qquad 1 \le i \le a, \quad 1 \le j \le b, \quad 1 \le k \le n,$$

where y_{ijk} is the kth observation at the ith level of A and the jth level of B and e_{ijk} is the error term; μ, α_i, β_j are as we defined them before, and $(\alpha\beta)_{ij}$ is the interaction term. There are ab such interaction terms. The design matrix $\mathbf{X}$ has abn rows and $ab + a + b + 1$ columns. The rank of $\mathbf{X}$ is ab.

We can approach the subdivision of the total sum of squares into component sums of squares in two equivalent ways. One approach is to argue that basically we have a one-way layout with n observations on each of ab treatments, a treatment being in this context a combination of a level of A and a level of B. Analyzing the experiment in this way gives a sum of squares for treatments $S_t = \sum_i \sum_j T_{ij.}^2/n - C$ where $T_{ij.} = \sum_k y_{ijk}$ and $C = G^2/N$. (We shall now write $T_{i..}$, $T_{.j.}$ for the sums $\sum_j \sum_k y_{ijk}$, $\sum_i \sum_k y_{ijk}$.) Differences between levels of A account for an amount S_A; differences between the levels of B account for another S_B; the remainder $S_t - S_A - S_B$ must result from interaction.

The other approach is to consider three regression models:

$$y_{ijk} = \mu + \alpha_i + \beta_j + (\alpha\beta)_{ij} + e_{ijk}. \tag{4.3}$$

(If we impose the side conditions $\alpha_i = 0$ for each i and $\beta_j = 0$ for each j, we obtain the oneway layout, and $S_R = S_t + C$):

$$y_{ijk} = \mu + \alpha_i + \beta_j + e_{ijk}; \tag{4.4}$$

$$y_{ijk} = \mu + \alpha_i + e_{ijk}. \tag{4.5}$$

The reduction in S_e at each stage gives us the desired sum of squares, and, with the appropriate divisor (σ^2), the conditions of Cochran's Theorem are satisfied when normality is assumed so that the forms have independent χ^2 distributions.

In solving the normal equations under model (4.3), we impose the side conditions $\sum_i \alpha_i = 0$, $\sum_j \beta_j = 0$, $\sum_j (\alpha\beta)_{ij} = 0$ for each i, and $\sum_i (\alpha\beta)_{ij} = 0$ for each j. This actually amounts to only $a + b + 1$ conditions because the conditions imposed upon the interaction terms are not independent; each of the two sets implies $\sum_i \sum_j (\alpha\beta)_{ij} = 0$. With these side conditions, the estimates of the parameters are $\hat{\mu} = \bar{y}$, $\hat{\alpha}_i = y_{i..} - \bar{y}$ where $y_{i..} = T_{i..}/bn$, $\hat{\beta}_j = y_{.j.} - \bar{y}$, and $\widehat{(\alpha\beta)}_{ij} = y_{ij.} - y_{i..} - y_{.j.} + \bar{y}$ where $y_{ij.} = T_{ij.}/n$.

TABLE 4.3
ANALYSIS OF VARIANCE

Source	SS	d.f.	EMS
Total	$\sum_i \sum_j \sum_k y_{ijk}^2 - C$	$abn - 1$	
A	$\sum_i T_{i..}^2/bn - C = S_A$	$a - 1$	$\sigma^2 + bn\sigma_A^2$
B	$\sum_j T_{.j.}^2/an - C = S_B$	$b - 1$	$\sigma^2 + an\sigma_B^2$
Interaction	$\sum_i \sum_j T_{ij.}^2/n - C - S_A - S_B = S_{AB}$	$(a - 1)(b - 1)$	$\sigma^2 + n\sigma_{AB}^2$
Error	$\sum_i \sum_j \sum_k y_{ijk}^2 - \sum_i \sum_j T_{ij.}^2/n$	$ab(n - 1)$	σ^2

The sum of squares for interaction may be derived in two ways. It is the difference between the sum of squares for regression with the model of a one-way layout with ab treatments and the sums of squares attributable to A and B, i.e., $n^{-1}\sum_i \sum_j - C - S_A - S_B$. Alternatively, the equation $S_R = \hat{\beta}X'Y$ derived in Chapter 2 may be rewritten as $S_R = \sum_i \beta_i(\sum_j x_{ij}y_j)$ where $\sum x_{ij}y_j$ is the single term on the left side of the normal equation corresponding to β_i. With all the x_{ij} being zero or one, $\mathbf{X'Y}$ is the vector of totals, and so we have

$$S_A = \sum_i T_{i..}\hat{\alpha}_i, \qquad S_B = \sum_j T_{.j.}\hat{\beta}_j, \qquad S_{AB} = \sum_i \sum_j T_{ij.}\widehat{(\alpha\beta)}_{ij}.$$

The expression for S_{AB} can also be written

$$S_{AB} = n \sum_i \sum_j (y_{ij.} - y_{i..} - y_{.j.} + \bar{y})^2 = n \sum_i \sum_j \widehat{(\alpha\beta)}_{ij}^2,$$

or else

$$S_{AB} = n \sum_j \left(\sum_i (y_{ij.} - y_{.j.})^2\right) - an \sum_j (y_{.j.} - \bar{y})^2,$$

which expresses S_{AB} in terms of the variance of the estimates of α_i over the several levels of B.

The expectations of the mean squares require a word of explanation. Our side conditions simplify the computations. Thus we have

$$G = N\mu + bn \sum_i \alpha_i + an \sum_j \beta_j + n\sum_i \sum_j (\alpha\beta)_{ij} + \sum_i \sum_j \sum_k e_{ijk},$$

and this simplifies to $G = N\mu + \sum_i \sum_j \sum_k e_{ijk}$ so that $E(G^2/N) = N\mu^2 + \sigma^2$.

Similarly, $y_{i..} - \bar{y}$ reduces from $\mu + \alpha_i + \beta. + (\alpha\beta)_{i.} + e_{i..} - \mu - \alpha. - \beta. - (\alpha\beta).. - e...$, where $e_{i..} = aN^{-1} \sum_j \sum_k e_{ijk}$ and $e... = N^{-1} \sum_i \sum_j \sum_k e_{ijk}$ to

$$y_i - \bar{y} = \alpha_i + e_{i..} - e...;$$

thus, $E(S_A) = bn \sum_i E(y_i - \bar{y})^2 = bn \sum_i \alpha_i^2 + (a - 1)\sigma^2$, and we shall, in an obvious notation, write $\sum_i \alpha_i^2$ as $(a - 1)\sigma_A^2$.

Finally, we can proceed in the same way and obtain

$$\begin{aligned} E(S_{AB}) &= n \sum_i \sum_j E(y_{ij.} - y_{i..} - y_{.j.} + \bar{y})^2 \\ &= n \sum_i \sum_j (\alpha\beta)_{ij}^2 + (a - 1)(b - 1)\sigma^2. \end{aligned}$$

We then write $\sum_i \sum_j (\alpha\beta)_{ij}^2$ as $(a - 1)(b - 1)\sigma_{AB}^2$.

4.4 The Subdivision of the Sum of Squares for Interaction

We discussed in the previous chapter the subdivision of the sum of squares for treatments in the one-way layout. This process can be extended to the factorial experiment, and the sums of squares for A, B, and AB interaction can all be subdivided into single degrees of freedom. We can take $ab - 1$ orthogonal contrasts in the cell totals, of which $(a - 1)$ are contrasts in the A totals and $(b - 1)$ are contrasts in the B totals. The squares of the remaining contrasts with the appropriate divisors each correspond to one degree of freedom, and their total is S_{AB}. We mention two uses of this procedure, one of which, Tukey's one degree of freedom for interaction, will be discussed in the next section.

We have already seen that when A is a quantitative factor, preferably with its levels equally spaced, S_A can be divided into such components as linear, quadratic, and cubic. If both A and B are quantitative, this approach can be extended to fit the model.

$$E(y) = \beta_0 + \sum_{i=1}^{a} \beta_{i0} x_1^i + \sum_{j=1}^{b} b_{0j} x_2^j + \sum_{i=1}^{a} \sum_{j=1}^{b} \beta_{ij} x_1^i x_2^j,$$

where x_1 and x_2 are coordinates for the levels of A and B respectively. Replacing x_1^i by the ith degree orthogonal polynomial for a levels, u_i, and x_2^j by the corresponding polynomial v_j, we obtain

$$E(y) = \gamma_0 + \sum \gamma_{i0} u_i + \sum \gamma_{0j} v_j + \sum \sum \gamma_{ij} u_i v_j.$$

The estimates of the γ_{ij} are orthogonal contrasts. The contrasts corresponding to $\hat{\gamma}_{i0}$ and $\hat{\gamma}_{0j}$ are the linear, quadratic, cubic, and so on contrasts for A and B and give the subdivision of S_A and S_B into single degrees of freedom as before. The cross-product contrasts, $\gamma_{ij}(i > 0, j > 0)$, give a subdivision of the interaction sum of squares into single degrees of freedom. The contrasts corresponding to $\hat{\gamma}_{11}$, $\hat{\gamma}_{12}$, $\hat{\gamma}_{21}$, and $\hat{\gamma}_{22}$ are usually called the lin A × lin B, lin A × quad B, quad A × lin B, quad A × quad B contrasts respectively. All $(a - 1)(b - 1)$ degrees of freedom for interaction can be accounted for in this way by associating them with estimates of binomial coefficients. The physical interpretation of the quad A × quad B and higher order interactions is by no means straightforward. A large higher order interaction can sometimes be the result of bizarre behavior in a single cell.

Let $\mathbf{T}$ be the matrix of cell totals and let $\mathbf{L}_A$, $\mathbf{L}_B$, $\mathbf{Q}_A$, $\mathbf{Q}_B$ be the vectors of the coefficients for the linear and quadratic contrasts for A and B. The lin A × lin B, lin A × quad B, quad A × lin B, and quad A × quad B contrasts are $\mathbf{L}_A'\mathbf{T}\mathbf{L}_B$, $\mathbf{L}_A'\mathbf{T}\mathbf{Q}_B$, $\mathbf{Q}_A'\mathbf{T}\mathbf{L}_B$, and $\mathbf{Q}_A'\mathbf{T}\mathbf{Q}_B$. The divisors for the sums of squares are $n\mathbf{L}_A'\mathbf{L}_A\mathbf{L}_B'\mathbf{L}_B$, $n\mathbf{L}_A'\mathbf{L}_A\mathbf{Q}_B'\mathbf{Q}_B$, $n\mathbf{Q}_A'\mathbf{Q}_A\mathbf{L}_B'\mathbf{L}_B$, and $n\mathbf{Q}_A'\mathbf{Q}_A\mathbf{Q}_B'\mathbf{Q}_B$.

Yates (1937, p. 36) gives the data for an experiment carried out at Rothamsted with two factors, varieties of oats (A) and quantity of manure (B). There were three varieties of oats; four levels of manure were used, 0, 0.2, 0.4, 0.6 cwt per acre. The figures given are the cell totals for six plots per cell; the recorded responses are yields measured in quarter pounds per plot. Although the oats factor (A) was not quantitative, it is nevertheless instructive to compute the sums of squares for the several components of interaction.

	B_1	B_2	B_3	B_4	Total
A_1	429	538	665	711	2343
A_2	480	591	688	749	2508
A_3	520	651	703	761	2635
Total	1429	1780	2056	2221	7486

It is important to remember that these are cell totals, and the factor $n = 6$ must appear in the denominator of each sum of squares; C, for example, is $7486^2/72 = 778336.06$.

The eleven degrees of freedom between cells are partitioned in the ordinary analysis of variance table into

TABLE 4.4

ANALYSIS OF VARIANCE

Source	SS	d.f.
Total	22128.61	11
A	1786.36	2
B	20020.50	3
$A \times B$	321.75	6

To obtain the interaction contrasts, we first obtain the B contrasts at each level of A by using $\mathbf{L}'_B = (-3, -1, 1, 3)$, $\mathbf{Q}'_B = (1, -1, -1, 1)$, $\mathbf{C}'_B = (1, -3, 3, -1)$. Then we multiply by $\mathbf{L}'_A = (-1, 0, 1)$, $\mathbf{Q}'_A = (-1, 2, -1)$. Thus lin A × lin B = 775 − 973 = −198, and quad A × quad B = 63 − 2(50) + 73 = 36. The B contrasts are

	lin B	quad B	cub B	A divisor
A_1	973	−63	99	
A_2	904	−50	22	
A_3	775	−73	−85	
lin A	−198	−10	−184	2
quad A	60	36	30	6
B divisors	20	4	20	

The individual sums of squares are

lin A × lin B	$(-198)^2/(2 \times 20 \times 6) =$	163.35
lin A × quad B	$(-10)^2/(2 \times 4 \times 6) =$	2.08
lin A × cub B	$(-184)^2/(2 \times 20 \times 6) =$	141.07
quad A × lin B	$(60)^2/(6 \times 20 \times 6) =$	5.00
quad A × quad B	$(36)^2/(6 \times 4 \times 6) =$	9.00
quad A × cub B	$(30)^2/(6 \times 20 \times 6) =$	1.25
		321.75

The value of the lin A × lin B term is large, and we note that the lin B contrast decreases as we change levels of A: 973, 904, 775. (If we had interchanged varieties 2 and 3, this inconsistency of the lin B contrast would have been manifested as a large quad A × lin B interaction; however, A is not really a quantitative factor.) It would appear that the increase in the response caused by using extra manure is greater for A_1 than for A_3.

4.5 Interaction with One Observation per Cell

Our earlier discussion of the two-way layout with one observation per cell came before we introduced the idea of interaction. If there is interaction present, the sum of squares called S_e is actually S_{AB} with $n = 1$. We have, therefore, two problems:

(i) Our estimate of the error variance is biased inasmuch as $E(M_e) = \sigma^2 + \sigma^2_{AB}$;
(ii) We have lost our F test for the hypothesis $\sigma^2_{AB} = 0$.

There is not much that we can do about problem (i) except to assume that $\sigma_{AB}^2 = 0$ and to comfort ourselves with the thought that if our assumption were incorrect we should be erring on the conservative side because s^2 would tend to be larger than σ^2. Thus, it would take a larger value of $\sum \alpha_i^2$ or of $\sum \beta_j^2$ than otherwise to provide a significant F ratio.

In regard to problem (ii), Tukey (1949) has suggested a method of obtaining a sum of squares with one degree of freedom to test the hypothesis $\sigma_{AB}^2 = 0$. He takes the sum of squares for one of the interaction contrasts, $\mathbf{a'Tb}$ where $\mathbf{a}' = (\ldots, y_{i..} - \bar{y}, \ldots)$ and $\mathbf{b}' = (\ldots, y_{.j.} - \bar{y}, \ldots)$. He gives an example in his paper in which he replaces the response y_{ijk} in a layout where σ_{AB}^2 is small by $z_{ijk} = y_{ijk}^2$. He thereby obtains a situation in which there is a large interaction sum of squares, much of which is absorbed in this single degree of freedom. In practical applications we can more easily compute Tukey's statistic as

$$T^* = \left\{\sum_i \sum_j y_{ij} T_{i.} T_{.j} - G(S_A + S_B + C)\right\}^2 \Big/ abS_A S_B.$$

The distribution of T^* is discussed at some length by Graybill (1961).

4.6 The Random Effects Model

The model that we considered at the beginning of the chapter had parameters α_i and β_j which were assumed to be unknown constants corresponding to fixed effects. We discussed in the previous chapter the situation in which the levels of the single factor were random. We proceed now to consider the two-way layout with both factors random. The procedure that we shall adopt is to take the same breakdown into sums of squares that we used in the fixed effects case, and then to investigate the expectations of the mean squares under the changed model. The new model is

$$y_{ijk} = \mu + a_i + b_j + (ab)_{ij} + e_{ijk},$$

where y_{ijk}, μ, and e_{ijk} are defined as they were for the fixed effects case; a_i, b_j, and $(ab)_{ij}$ are uncorrelated random variables (as usual, we require normality, and, hence, independence, when we test hypotheses) with zero means and variances σ_A^2, σ_B^2, and σ_{AB}^2 respectively.

We write

$$a_. = a^{-1} \sum_i a_i, \qquad b_. = \sum_j b^{-1} b_j, \qquad (ab)_{i.} = b^{-1} \sum_j (ab)_{ij},$$

$$(ab)_{.j} = a^{-1} \sum_i (ab)_{ij}, \qquad (ab)_{..} = a^{-1} \sum_i (ab)_{i.} = b^{-1} \sum_j (ab)_{.j},$$

$$e_{ij.} = n^{-1} \sum_k e_{ijk}, \qquad e_{i..} = b^{-1} \sum_j e_{ij.}, \qquad e_{...} = N^{-1} \sum_i \sum_j \sum_k e_{ijk}.$$

Then we also have $y_{ij\cdot} = T_{ij\cdot}/n$, $y_{i\cdot\cdot} = (bn)^{-1}T_{i\cdot\cdot}$, $y_{\cdot j\cdot} = (an)^{-1}T_{\cdot j\cdot}$ and $\bar{y} = G/N$.
In computing $E(S_A)$, we again note that $S_A = nb \sum_i (y_{i\cdot\cdot} - \bar{y})^2$;

$$\begin{aligned} y_{i\cdot\cdot} - \bar{y} &= \mu + a_i + b. + (ab)_{i\cdot} + e_{i\cdot\cdot} - \mu - a. - b. - (ab).. - e... \\ &= (a_i - a.) + ((ab)_{i\cdot} - (ab)..) + (e_{i\cdot\cdot} - e...). \end{aligned}$$

When we square these expressions and take the expected values, the cross-product terms drop out because the random variables concerned are uncorrelated, and we have

$$\begin{aligned} E(S_A) &= (a - 1)bn\{V(a_i) + V(ab)_{i\cdot} + V(e_{i\cdot\cdot})\} \\ &= (a - 1)bn\{\sigma_A^2 + b^{-1}\sigma_{AB}^2 + b^{-1}n^{-1}\sigma^2\} \\ &= (a - 1)(\sigma^2 + n\sigma_{AB}^2 + bn\sigma_A^2). \end{aligned}$$

Similarly, we obtain

$$E(S_B) = (b - 1)(\sigma^2 + n\sigma_{AB}^2 + an\sigma_B^2),$$

$$E(S_{AB}) = (a - 1)(b - 1)(\sigma^2 + n\sigma_{AB}^2).$$

This points out the crucial difference in hypothesis testing with the two models. In the fixed effects case, the hypothesis $H_0: \alpha_1 = \alpha_2 = \cdots = 0$ was tested by the ratio $\mathscr{F} = M_A/M_e$ where $M_A = S_A/(a - 1)$. In the random effects case, the hypothesis $H_0: \sigma_{AB}^2 = 0$ is tested by the ratio $\mathscr{F} = M_{AB}/M_e$. If this hypothesis is rejected and the experimenter decides to act as if $\sigma_{AB}^2 > 0$, then the correct ratio for testing the hypothesis $\sigma_A^2 = 0$ is the ratio $\mathscr{F} = M_A/M_{AB}$.

If the hypothesis $\sigma_{AB}^2 = 0$ is not rejected, the experimenter has now three choices for a test statistic for the hypothesis $\sigma_A^2 = 0$.

(i) He can argue that the failure to reject $\sigma_{AB}^2 = 0$ is not the same as proving $\sigma_{AB}^2 = 0$ and that M_{AB} is still his best estimate of $\sigma^2 + n\sigma_{AB}^2$.
(ii) He can act as if $\sigma_{AB}^2 = 0$, in which case $E(M_A) = \sigma^2 + bn\sigma_A^2$, and use the test statistic $\mathscr{F} = M_A/M_e$.
(iii) He can pool S_{AB} and S_e to obtain a joint estimate of σ^2, and use the ratio $\mathscr{F} = (abn - a - b + 1)M_A/(S_e + S_{AB})$.

4.7 The Mixed Model

We proceed now to the situation in which one factor is fixed and the other is random. If there is no interaction between the factors, this situation poses no problems; the experiment is similar to the randomized block experiment of the previous chapter. If, however, interaction is present, difficulties arise. The question of a suitable model for this situation has been widely discussed, particularly by Wilk and Kempthorne (1955) and (1956), by Scheffé (1956*a* and *b*), and by Cornfield and Tukey (1956). The approaches of Cornfield and Tukey and of Wilk and Kempthorne involve deriving the expected values on the assumption that the levels of the fixed factor A are a random sample of size a from a finite population of size p, and then putting p equal to a. We shall follow Scheffé's derivation. Again, we take the breakdown into sums of squares derived for the fixed effects case, and concern ourselves with the expected values of those sums of squares under the new conditions.

The new model is

$$y_{ijk} = \mu + \alpha_i + b_j + c_{ij} + e_{ijk};$$

y_{ijk}, μ, b_j, and e_{ijk} are defined in the same way as in the random effects model, and the b_j and the e_{ijk} are all uncorrelated; the effects α_i are unknown constants subject to the side condition $\sum \alpha_i = 0$; c_{ij} is the interaction term.

The interactions c_{ij} are random variables: they involve a random factor B. They are not correlated with any of the b_j or the e_{ijk}. However, bearing in mind that for the fixed effects α_i we have $\sum \alpha_i = 0$, we impose a similar condition $\sum_i c_{ij} = 0$ for each j upon the interaction terms. Consequently, the c_{ij} are, for any given j, correlated with each other. It is convenient to write $V(c_{ij})$ as $(a - 1)\sigma_{AB}^2/a$.

Lemma. *Let $x_1, x_2, \ldots, x_n$ be a set of random variables having the same variance σ^2 and the same covariances $\rho\sigma^2$, and being subject to the condition $\sum_i x_i =$ const; then $\rho = -(n - 1)^{-1}$.*

PROOF. We have $V(\sum x_i) = n\sigma^2 + n(n - 1)\rho\sigma^2$. But $\sum x_i =$ const implies that $V(\sum x_i) = 0$. Hence, $n(1 + (n - 1)\rho)\sigma^2 = 0$, and so $\rho = -(n - 1)^{-1}$. ∎

Applying this lemma to the c_{ij}, with $n = a$ and $\sigma^2 = (a - 1)\sigma_{AB}^2/a$, we have for $i \neq i', j \neq j'$

$$\operatorname{cov}(c_{ij}, c_{i'j}) = -a^{-1}\sigma_{AB}^2, \qquad \operatorname{cov}(c_{ij}, c_{ij'}) = \operatorname{cov}(c_{ij}, c_{i'j'}) = 0,$$

$$V(c_{i\cdot}) = b^{-1}V(c_{ij}) = b^{-1}a^{-1}(a - 1)\sigma_{AB}^2.$$

We proceed now to derive the expectations of the sums of squares.

$$E(S_e) = ab(n - 1)\sigma^2, \tag{4.6}$$

as before.

$$S_A = bn \sum_i (y_{i\cdot\cdot} - \bar{y})^2.$$

$$(y_{i\cdot\cdot} - \bar{y}) = \alpha_i - \alpha_\cdot + b_\cdot - b_\cdot + c_{i\cdot} - c_{\cdot\cdot} + e_{i\cdot\cdot} - e_{\cdot\cdot\cdot}$$

$$= \alpha_i + c_{i\cdot} + e_{i\cdot\cdot} - e_{\cdot\cdot\cdot},$$

since $\alpha_. = 0$ and $c_{..} = 0$. Thus, $E(S_A) = bn \sum \alpha_i^2 + n(a - 1)\sigma_{AB}^2 + (a - 1)\sigma^2$, and writing $\sum \alpha_i^2 = (a - 1)\sigma_A^2$,

$$E(M_A) = \sigma^2 + n\sigma_{AB}^2 + bn\sigma_A^2. \tag{4.7}$$

$$S_B = an \sum_j (y_{.j.} - \bar{y})^2.$$

$$\begin{aligned}(y_{.j.} - \bar{y}) &= \alpha_. - \alpha_. + b_j - b_. + c_{.j} - c_{..} + e_{.j.} - e_{...}\\ &= b_j - b_. + e_{.j.} - e_{...},\end{aligned}$$

since $c_{.j} = c_{..} = 0$. Then

$$E(S_B) = an(b - 1)\{V(b_j) + V(e_{.j.})\} = (b - 1)(\sigma^2 + an\sigma_B^2),$$

and

$$E(M_B) = \sigma^2 + an\sigma_B^2. \tag{4.8}$$

$$S_{AB} = n \sum_i \sum_j (y_{ij.} - y_{i..} - y_{.j.} + \bar{y})^2,$$

where

$$(y_{ij.} - y_{i..} - y_{.j.} + \bar{y}) = c_{ij} - c_{i.} + (e_{ij.} - e_{i..} - e_{.j.} + e_{...}).$$

The error term is the same term that appears in the earlier models; its contribution to $E(S_{AB})$ is $(a - 1)(b - 1)\sigma^2$. $E\{\sum_j (c_{ij} - c_{i.})^2\} = (b - 1)V(c_{ij})$ for each i. Hence, $E(S_{AB}) = (a - 1)(b - 1)(\sigma^2 + n\sigma_{AB}^2)$, and

$$E(M_{AB}) = \sigma^2 + n\sigma_{AB}^2. \tag{4.9}$$

From this, we see that if $\sigma_{AB}^2 \neq 0$, the hypothesis $\sigma_A^2 = 0$ is tested by the ratio $\mathscr{F} = M_A/M_{AB}$, while the hypothesis $\sigma_B^2 = 0$ is tested by $\mathscr{F} = M_B/M_e$.

If there is only one observation per cell, S_e does not exist, and we have only the three sums of squares S_A, S_B, and S_{AB}. In the random effects case, this presents no problem in testing $\sigma_A^2 = 0$ or $\sigma_B^2 = 0$ because M_{AB} is the appropriate denominator for the test statistic anyway. However, in the fixed effects case (and in the mixed case considered in this section when we test $\sigma_B^2 = 0$), there is no test for the main effects unless it can be assumed a priori that $\sigma_{AB}^2 = 0$.

Some people find it surprising that although A is a fixed effect, its expected mean square contains the interaction term σ_{AB}^2, which would appear at first glance to go with A random rather than with A fixed. The approaches of Wilk and Kempthorne and of Cornfield and Tukey throw more light on this. They obtain $E(M_A)$ as

$$E(M_A) = \sigma^2 + (1 - b/q)\sigma_{AB}^2 + bn\sigma_A^2,$$

where q denotes the number of levels of B in the population of levels from which the experimenter chose the b levels actually used.

Then, if B is fixed, $b = q$ and $(1 - b/q) = 0$; but, if B is random, $q = \infty$ and $(1 - b/q) = 1$. We may also picture ourselves, when looking at the main effects of A, as averaging the interactions c_{ij} over all the levels of B that were used. If B is fixed, then we have averaged over all possible levels of B, and the interaction terms "cancel out"; on the other hand, if B is a random factor, we are averaging over only a fraction of the levels of B, and it would not be reasonable to assume that the interactions average out in any given experiment.

4.8 The Two-Stage Nested or Hierarchic Design

We illustrate this design by an example. Consider b batches of gasoline produced by some process. From each batch, s samples are taken, and then n observations are made on the octane number of each sample. If y_{ijk} denotes the kth observation on the jth sample from the ith batch, we have

$$y_{ijk} = \mu + b_i + s_{j(i)} + e_{k(ij)},$$

where b_i is the effect of the ith batch, $s_{j(i)}$ is the effect of the jth sample in the ith batch, and $e_{k(ij)}$ is the observational error.

If we assume to begin with that μ, b_i, and $s_{j(i)}$ are all fixed, we have bs samples and $1 + b + bs$ parameters. We impose side conditions $\sum_i b_i = 0$ and $\sum_j s_{j(i)} = 0$ for each i. Then, if $y_{ij.}$ and $y_{i..}$ denote the sample and batch means,

$$\hat{\mu} = \bar{y}, \qquad \hat{b}_i = y_{i..} - \bar{y}, \qquad \hat{s}_{j(i)} = y_{ij.} - y_{i..}.$$

The sum of squares attributable to batch differences is obtained by treating the setup as just a one-way layout with b batches and sn observations per batch, which gives

$$S_B = \sum_i T_{i..}^2/ns - G^2/N,$$

where $N = bsn$. Regarding the set-up as a one-way layout with bs samples and n observations per sample gives a sum of squares

$$S_R = \sum_i \sum_j T_{ij.}^2/n - G^2/N.$$

Then, subtracting, we have the sum of squares for samples within batches, as

$$S_{S \text{ in } B} = S_{S(B)} = \sum_i \sum_j T_{ij.}^2/n - \sum_i T_{i..}^2/ns;$$

and also

$$S_e = \sum_i \sum_j \sum_k y_{ijk}^2 - \sum_i \sum_j T_{ij.}^2/n.$$

Alternatively, we may argue that $S_B = \sum_i \hat{b}_i T_{i..}$ and $S_{S(B)} = \sum_i \sum_j \hat{s}_{j(i)} T_{ij.}$. The extension to more than two stages is obvious.

If either of the two factors is random, we take the same breakdown into sums of squares. The most common situation is that in which S is a random effect, the $s_{j(i)}$ being (normally) distributed with mean zero and variance σ_1^2. We write σ_0^2 for $V(e_{ijk})$. We shall assume that the batches are distributed with mean zero and variance σ_2^2. An example from genetics is the case of s sires being chosen and each of them being mated to d dams (there being ds dams altogether); then n siblings are chosen from each of the ds litters. The expectations of the mean squares are given in Table 4.5.

TABLE 4.5

ANALYSIS OF VARIANCE

Source	SS	d.f.	EMS
Total	$\sum_i \sum_j \sum_k y_{ijk}^2 - C$	$N - 1$	
Between sires	$\sum_i T_{i..}^2/dn - C$	$s - 1$	$\sigma_0^2 + n\sigma_1^2 + dn\sigma_2^2$
Dams in sires	$\sum_i \sum_j T_{ij.}^2/n - \sum_i T_{i..}^2/dn$	$s(d - 1)$	$\sigma_0^2 + n\sigma_1^2$
Between siblings	$\sum_i \sum_j \sum_k y_{ijk}^2 - \sum_i \sum_j T_{ij.}^2/n$	$sd(n - 1)$	σ_0^2

Under normality, the testing of the hypotheses $\sigma_1^2 = 0$ and $\sigma_2^2 = 0$ is carried out by the F test in the usual way. The principal interest in the experiment usually centers, however, on the problems of estimating the components of variance σ_0^2, σ_1^2, and σ_2^2.

4.9 The Diallel Cross Experiment

This kind of experiment is commonly performed by geneticists who are interested in selecting lines and strains of plants or animals for further breeding. The structure of the model used is similar to that of the two-way layout with interaction. Observations are made on the offspring of the crosses of pairs of inbred lines. The main effect of a given line is called its *general combining ability*, a term originally defined by Sprague and Tatum (1942). Corresponding to the interaction terms in the two-way layout are the *specific combining abilities*. The reader is referred for a more detailed discussion to the papers by Griffing (1956) and Kempthorne (1956*b*).

We consider here a situation in which p inbred lines are chosen and some crosses are made. Griffing discusses four different experiments. In methods 1 and 2 the p parental lines are also included; in methods 3 and 4 they are omitted. In methods 1 and 3 the reciprocal crosses are included, i.e., both the cross in which the ith line is the male line and the jth line is the female line and the reciprocal cross with j as the male line and i as the female line. In methods 2 and 4 the differences between reciprocal crosses are ignored. We shall consider method 4, which is the simplest one, in detail, and confine ourselves to only a few remarks about the others.

In Griffing's method 4 all $p(p - 1)/2$ crosses are made. From each cross n offspring are taken and some response, such as the yield or the number of bristles, is observed; y_{ijk} denotes the kth observation on the cross between the ith and jth lines. The model is

$$y_{ijk} = m + g_i + g_j + s_{ij} + e_{ijk}, \qquad 1 \leq i < j \leq p, \quad 1 \leq k \leq n,$$

where g_i and g_j are the general combining abilities of the ith and jth lines, s_{ij} is the specific combining ability of the pair of lines, m is a general mean, and e_{ijk} is a random error term with the usual assumptions.

We consider first the fixed effects model in which the parameters g_i, s_{ij}, and m are unknown constants. Let Y_i denote the sum of all the observations on all offspring from crosses made with the ith line; Y_{ij} denotes the sum of all the observations on the $i \times j$ cross, i.e., $Y_{ij} = \sum_k y_{ijk}$. Let G denote the sum of all the observations. Since each observation y_{ijk} appears in the two totals Y_i and Y_j, we have $2G = \sum_i Y_i$. When we write $\sum \sum s_{ij}$ the summation will be over all i and j with $i < j$; there is no term s_{ii}.

The normal equations are

$$G = N\hat{m} + n(p - 1) \sum_i \hat{g}_i + n \sum \sum \hat{s}_{ij},$$

where $2N = np(p - 1)$,

$$Y_i = n(p - 1)\hat{m} + n(p - 1)\hat{g}_i + n \sum_j{}' \hat{g}_j + n \sum_j \hat{s}_{ij},$$

where the prime denotes summation over all values $1 \leq j \leq p$, except $i = j$,

$$Y_i = n(p - 1)\hat{m} + n(p - 2)\hat{g}_i + n \sum \hat{g}_j + n \sum_j \hat{s}_{ij},$$

and

$$Y_{ij} = n\hat{m} + n\hat{g}_i + n\hat{g}_j + n\hat{s}_{ij}.$$

Imposing the side conditions $\sum_j g_j = 0$, $\sum_j s_{ij} = 0$ for each i, we obtain the estimates

$$\hat{m} = G/N, \qquad \hat{g}_i = (Y_i - n(p - 1)\hat{m})/n(p - 2),$$

$$\hat{s}_{ij} = Y_{ij}/n - \hat{g}_i - \hat{g}_j - \hat{m}.$$

The corresponding sums of squares in the analysis of variance table are

(i) The mean, G^2/N, with one degree of freedom,
(ii) Between lines, or the sum of squares for general combining ability,

$$S_g = \sum_i Y_i \hat{g}_i = \frac{\sum Y_i^2}{n(p-2)} - \frac{4G^2}{np(p-2)}$$

with $(p - 1)$ degrees of freedom,
(iii) Specific combining ability,

$$S_s = \sum\sum Y_{ij}\hat{s}_{ij} = \frac{\sum\sum Y_{ij}^2}{n} - \frac{\sum Y_i^2}{n(p-2)} + \frac{2G^2}{n(p-1)(p-2)}$$

with $p(p - 3)/2$ degrees of freedom,
(iv) S_e, obtained by subtraction with $p(p - 1)(n - 1)/2$ degrees of freedom.

In the random effects case we assume that g_i, s_{ij}, and e_{ijk} are uncorrelated random variables with zero means and variances σ_g^2, σ_s^2, and σ_e^2, respectively. The geneticist is interested in obtaining estimates of these components of variance. We take the same sums of squares as in the fixed effects case, and we now derive their expectations. There is no difficulty in showing that $E(S_e) = p(p - 1) \times (n - 1)\sigma_e^2/2$. We also have

$$\begin{aligned} E(G^2) &= N^2m^2 + n^2(p-1)^2p\sigma_g^2 + n^2p(p-1)\sigma_s^2/2 + N\sigma_e^2, \\ E(Y_i^2) &= n^2(p-1)^2m^2 + n^2(p-1)^2\sigma_g^2 \\ &\quad + n^2(p-1)\sigma_g^2 + n^2(p-1)\sigma_s^2 + n(p-1)\sigma_e^2 \\ &= n^2(p-1)^2m^2 + n^2p(p-1)\sigma_g^2 + n^2(p-1)\sigma_s^2 + n(p-1)\sigma_e^2, \\ E(Y_{ij}^2) &= n^2m^2 + 2n^2\sigma_g^2 + n^2\sigma_s^2 + n\sigma_e^2. \end{aligned}$$

It follows that

$$\begin{aligned} np(p-2)E(S_g) &= p\sum E(Y_i^2) - 4E(G^2) = p^2E(Y_i^2) - 4E(G^2) \\ &= n^2p(p-1)(p-2)^2\sigma_g^2 + n^2p(p-1)(p-2)\sigma_s^2 \\ &\quad + np(p-1)(p-2)\sigma_e^2, \end{aligned}$$

so that the expectation of the mean square is

$$E(M_g) = \sigma_e^2 + n\sigma_s^2 + n(p-2)\sigma_g^2.$$

We may similarly show that $E(M_s) = \sigma_e^2 + n\sigma_s^2$.

In the general case of method 1, in which the reciprocal crosses and the parental lines are included, the model assumed by Griffing is

$$y_{ijk} = m + g_i + g_j + s_{ij} + r_{ij} + e_{ijk},$$
$$1 \le i \le p, \quad 1 \le j \le p, \quad 1 \le k \le n.$$

Here y_{ijk} denotes the kth observation on the cross with i as the male line and j as the female line; the specific combining ability s_{ij} is symmetric, $s_{ij} = s_{ji}$. The reciprocal effect, r_{ij}, accounts for the difference between the $i \times j$ and $j \times i$ crosses, and is antisymmetric, $r_{ij} = -r_{ji}$.

We define the totals $Y_{i\cdot} = \sum_j \sum_k y_{ijk}$, $Y_{\cdot j} = \sum_i \sum_k y_{ijk}$, and $Y_{ij\cdot} = \sum_k y_{ijk}$. Then the estimates of the parameters in the fixed effects case are given by the equations

$$np^2\hat{m} = G, \qquad 2np^2\hat{g}_i = 2p(Y_{i\cdot} + Y_{\cdot i}) - 2G, \qquad 2n\hat{r}_{ij} = Y_{ij\cdot} - Y_{ji\cdot},$$
$$2np^2\hat{s}_{ij} = p^2(Y_{ij\cdot} + Y_{ji\cdot}) - p(Y_{i\cdot} + Y_{\cdot i} + Y_{j\cdot} + Y_{\cdot j}) + 2G, \qquad (i \neq j),$$
$$np^2\hat{s}_{ii} = p^2 Y_{ii\cdot} - p(Y_{i\cdot} + Y_{\cdot i}) + G.$$

The models for the other two methods may be derived from that of method 1 by dropping the unnecessary terms. The analyses follow the same lines.

Exercises

1. An experiment is carried out to investigate the deterioration in a product after storage for different lengths of time at different temperatures. The experimental design is a two-way crossed classification with factor A being time at a levels and factor B being storage temperature at b levels. However, one of the levels of A is zero, and zero storage time is the same at all temperatures. Thus there are $b(a - 1) + 1$ cells and n observations in each cell. How would you analyze the data?
2. In the two-stage nested design we assumed that there were n siblings chosen from each litter. In practice, however, some litters may have fewer than n siblings. Modify the analysis for the case in which the cell corresponding to the jth litter by the ith sire contains n_{ij} members where $n_{ij} > 0$ for all i and j.
3. (Continuation). Another possibility is that one or more litters might be completely lost. Modify the analysis for the situation in which there are only $d - 1$ litters by one of the sires; each nonempty cell has n observations.
4. (Continuation). In the industrial context with batches, samples in batches, and observations or analyses on samples, the hierarchical design can result in a large bill for chemical analysis. Suppose that we decide to make two observations on the first sample from each batch, and only one observation on each of the subsequent samples. Modify the analysis accordingly.
5. Find the expected values of the mean squares in Griffing's method 1 for the diallel cross experiment.
6. (Continuation). Derive the analysis for methods 2 and 3.

7. In a two-way layout the factors A and B are fixed and the cell frequencies satisfy the condition of proportional frequencies. Derive the analysis with AB interaction terms.

8. Batchelder et al. (1966) investigated the growth of tobacco. They grew tobacco in four different soils (blocks) with three different mulches, A, B, and C, and continued the experiment for three years. The data were analyzed as a two-factor experiment, years Y and mulches M in four randomized blocks. The data which follow are yields in pounds per acre. A was a control (no mulch at all); blocks I and II were a different type of soil from blocks III and IV. Show that the sums of squares for mulches, years, and blocks each have one large contrast that accounts for most of the sum. What conclusions would you draw from this?

		Blocks			
Year	Mulch	I	II	III	IV
	A	984	421	563	862
1962	B	1440	1276	1016	880
	C	1324	1232	1064	1016
	A	1539	1039	914	1032
1963	B	2141	1894	1819	1675
	C	1939	1833	2394	2055
	A	1607	1461	1593	1507
1964	B	1860	1912	1703	1378
	C	1846	1973	1043	1319

9. The following data are yields, in grams per plot, in diallel crosses between inbred lines of winter beans [Bond (1966)]. He planted all crosses between six inbred lines, and also the inbred lines themselves, a total of 36 varieties in all. The yields reported are the mean values over two years. Analyze the data under Griffing's model 1, both with fixed effects and with random effects.

Female Parent	Male Parent					
	24	31	36	55	64	67
24	172.8	279.5	277.2	278.2	279.8	315.0
31	247.7	177.8	258.5	263.3	303.2	313.2
36	267.5	274.0	236.0	250.7	258.5	256.7
55	301.7	278.5	269.5	224.2	260.5	285.5
64	267.5	253.0	248.0	267.2	208.2	287.5
67	262.0	274.3	254.5	259.8	281.8	221.3

CHAPTER 5

Experiments with Several Factors

The results obtained in the previous chapter can be extended to larger experiments involving more than two factors. The simplest situation is the complete factorial experiment in which all the factors are fixed and crossed. This is a routine extension of the two-way crossed classification. The term *crossed classification* is used to denote that each factor is crossed with the others: factor A cuts across factor B in the sense that each level of factor A appears with each level of factor B.

We shall see that in the crossed classification with some of the factors random complications can arise about the F tests. In some cases, no one mean square will provide a suitable denominator for the F statistic to test a hypothesis, and a denominator has to be put together from several mean squares. We shall also consider designs that contain both crossed and nested factors. A set of rules will be given for deriving the degrees of freedom, sums of squares, and the expectations of the mean squares. The chapter ends with a short discussion, including some examples, of split-plot and split-split-plot designs. Before proceeding further, it is appropriate to make a few remarks about error terms, pooling, and replicates.

5.1 Error Terms and Pooling

We have seen in the previous chapter that in a two-way layout with a single observation per cell we do not have an estimate of the error variance unless we are prepared to assume, a priori, that there is no interaction between the two factors. This situation persists when there are more than two factors, and it is a general operating procedure that if there is only one observation per cell, the highest order interaction (the interaction between all the factors) is taken to be zero, a priori. If the experimenter does not want to assume this in any case except that in which all the effects are random, he will have to find some way of duplicating his observations to obtain an unbiased estimate of error. If several of the factors are at more than two levels, this procedure of ignoring the highest order interaction will provide an error term with a reasonable number of degrees of freedom.

If an experiment involves four factors, each at four levels, the $ABCD$ interaction will have $3^4 = 81$ d.f. Assuming the effects to be fixed, the critical value of F for testing the hypothesis that the main effects for A are zero, using $\alpha = 0.05$, is $F^*(3, 81)$, which is approximately 2.7. If the factors are at only three levels each, we should have 16 d.f. for $ABCD$ and $F^*(3, 16) = 3.24$, which means that we should need a 20 per cent larger F value in order to detect differences between the levels of A, a considerable decrease in power. If all the factors are at two levels each, there will only be a single degree of freedom for each of the mean squares. In the latter circumstances, the experimenter is going to be rather desperate for degrees of freedom in the denominators of his F statistics, and it is customary in such experiments to assume, in the absence of information to the contrary, that all interactions involving more than two factors are negligible, a priori, and to pool their sums of squares to obtain an error term. We shall tacitly make that assumption in the later chapters when we discuss experiments with factors at two or three levels.

The pooling of interaction sums of squares to provide an error term is a subjective matter. On the one hand, purists can deplore the use of a possibly biased denominator and a possibly invalid F statistic. On the other hand, the practical man can hardly be condemned for making use of reasonable assumptions to obtain enough degrees of freedom in the denominator for his F test to have enough power to be useful. What is important is that the experimenter should use his common sense. There used to be a popular procedure, which has hopefully passed into disuse, by which the experimenter would start his F tests at the foot of the table, testing the mean square in the next to last line against the mean square for the error. If the ratio was larger than the critical value of F, he recorded the fact in the right-hand column and moved on to the next line. If the ratio was not significant, he pooled the sum of squares with the error term and went on to the next line. So he continued up the table automatically pooling every sum of squares that was not significant, oblivious to the fact that just because the F ratio is not quite big enough for significance it does not necessarily follow that the hypothesis is true and that if you pool a sum of squares that is almost large enough to be significant you will inflate your error estimate and bias it upward.

5.2 Replicates

The word *replicate* (or replication) commonly occurs in the jargon of experimental design, but its meaning is not always the same. When in later chapters we come to discuss incomplete block designs, we shall consider a design in which there are five treatments, each tested four times for a total of twenty plots; we shall then say that each of the treatments is replicated four times, or has four replications. In the context of the factorial experiment, however, a replicate or a replication means a complete replica, repetition, or copy of the basic design with one observation per cell.

Suppose, for example, that we wish to test five varieties of beans using four different fertilizers, and that there are to be three plots with each combination of bean and fertilizer. In the completely randomized experiment there would be sixty plots assigned at random, and it would be improper to speak of replications. To say that the experiment has three replications implies that there are three separate groups of twenty plots each. Each group constitutes a replicate, or replication, inasmuch as it consists of a single plot with each bean-fertilizer combination. If the three replicates are in the same field, they may just as well be called blocks; it seems sometimes to be a matter of individual taste. The replicates need not, however, be in the same field. It would be possible to have them on different farms or, in an industrial experiment, to run them on different days.

The following operating rule is commonly accepted, often without being stated: replicates and blocks do not interact with treatments or factors. As a result, the reader will often see in the literature analysis of variance tables that contain a sum of squares for replicates but in which there is no mention of interactions involving replicates because they have been automatically pooled into the error term.

Henceforth in the expected values of sums of squares we shall use, for example, the notation of σ_A^2 whether A is a fixed effect or a random effect. If A is random, σ_A^2 is the usual component of variance. If A is fixed, σ_A^2 is what might be called a pseudocomponent of variance and represents an expression such as $\sum_i (\alpha_i - \alpha.)^2/(a - 1)$.

5.3 The Complete Three-Factor Experiment

We consider now a complete factorial experiment with three factors A at a levels, B at b levels, and C at c levels, There are n observations in each cell, for a total of $N = abcn$ data points. The complete model has terms for two-factor interactions and for a three-factor interaction:

$$y_{ijkm} = \mu + a_i + b_j + c_k + (ab)_{ij} + (ac)_{ik} + (bc)_{jk} + (abc)_{ijk} + e_{ijkm}.$$

The three-factor interaction essentially takes in all the scatter in the cell means that is not explained by the main effects and the two-factor interactions. Its sum of squares can also be derived by considering the variation in the sums of squares for AB interaction at the several levels of C (or equivalently the AC interaction at the levels of B, or the BC interaction at the levels of A).

In the fixed effects case, the analysis follows the same lines as that of the two-factor experiment. The side conditions imposed are

$$\sum_i a_i = 0, \qquad \sum_j b_j = 0, \qquad \sum_k c_k = 0;$$
$$\sum_j (ab)_{ij} = 0, \qquad \sum_i (ab)_{ij} = 0, \qquad \sum_k (ac)_{ik} = 0;$$
$$\sum_i (ac)_{ik} = 0, \qquad \sum_k (bc)_{jk} = 0, \qquad \sum_j (bc)_{jk} = 0;$$
$$\sum_i (abc)_{ijk} = 0, \qquad \sum_j (abc)_{ijk} = 0, \qquad \sum_k (abc)_{ijk} = 0.$$

Typical estimates of parameters are

$$\mu = \bar{y}, \qquad a_i = y_{i\cdots} - \bar{y}, \qquad (ab)_{ij} = y_{ij\cdot\cdot} - y_{i\cdots} - y_{\cdot j\cdot\cdot} + \bar{y},$$
$$(abc)_{ijk} = y_{ijk\cdot} - y_{ij\cdot\cdot} - y_{i\cdot k\cdot} - y_{\cdot jk\cdot} + y_{i\cdots} + y_{\cdot j\cdot\cdot} + y_{\cdot\cdot k\cdot} - \bar{y},$$

with typical sums of squares

$$S_A = \sum_i T_{i\cdots}^2/bcn - C$$

with $a - 1$ d.f.,

$$S_{AB} = \sum_i \sum_j T_{ij\cdot\cdot}^2/cn - C - S_A - S_B,$$

with $(a - 1)(b - 1)$ d.f.,

$$S_{ABC} = \sum_i \sum_j \sum_k T_{ijk\cdot}^2/n - C - S_A - S_B - S_C - S_{AB} - S_{AC} - S_{BC}$$

with $(a - 1)(b - 1)(c - 1)$ d.f..

The expectations of the mean squares are $E(M_A) = \sigma^2 + bcn\sigma_A^2$, $E(M_{AB}) = \sigma^2 + cn\sigma_{AB}^2$, $E(M_{ABC}) = \sigma^2 + n\sigma_{ABC}^2$ where $\sigma_A^2 = \sum a_i^2/(a - 1)$, and so on. The mean square for error $M_e = s^2$ serves as the denominator for each of the F tests. The extension to four or more factors is straightforward.

When one or more of the factors is random, matters become somewhat more complex. In each case we take the breakdown into sums of squares for the fixed effects case, but there are differences in the expectations of the mean squares. We may follow the development in the two-factor case to derive these expectations. If all three factors are random, we have

$$E(M_A) = \sigma^2 + n\sigma_{ABC}^2 + cn\sigma_{AB}^2 + bn\sigma_{AC}^2 + bcn\sigma_A^2,$$
$$E(M_{AB}) = \sigma^2 + n\sigma_{ABC}^2 + cn\sigma_{AB}^2,$$
$$E(M_{ABC}) = \sigma^2 + n\sigma_{ABC}^2, \qquad E(M_e) = \sigma^2.$$

For the mixed model, we recall that in the two-factor case with A fixed and B random, the term σ_{AB}^2 did not appear in $E(M_B)$ because for each level of B the interactions were averaged over the entire set of levels of the fixed factor A. On the other hand, σ_{AB}^2 did appear in $E(M_A)$ because for each level of A the interactions were averaged over only those b levels of B that were (randomly) chosen from the infinite population.

With this in mind, we can formulate a set of rules for writing down the expectations of the mean squares in the complete factorial experiment which are applicable to any number of factors, fixed or random.

1. Each expected mean square contains a term σ^2.
2. Each component of variance has some letters as subscripts. The coefficient of a component in any expectation is the product of all the letters that are not among its subscripts. Thus, in the three-factor case we have the four letters a, b, c, and n. The coefficient of σ_{AB}^2 whenever it appears is the product cn; similarly, we have $n\sigma_{ABC}^2$ and $bcn\sigma_A^2$. The coefficient is the number of observations contained in the corresponding total, such as $T_{ij\cdots}$.
3. If all the factors in an experiment are random, the expectation of a mean square consists of σ^2 together with all the terms for effects and interactions containing all the letters in the name of the mean square in question. Thus, $E(M_{AB})$ contains terms for all components with both A and B among the subscripts, e.g., $\sigma^2 + n\sigma_{ABC}^2 + cn\sigma_{AB}^2$.
4. If some of the factors are fixed, we strike out in the expectation of any mean square all those components which have among their subscripts any letters, other than those in the name of the mean square, which correspond to fixed factors. For example, in the three-factor case we have, when all three factors are random,

$$E(M_A) = \sigma^2 + n\sigma_{ABC}^2 + cn\sigma_{AB}^2 + bn\sigma_{AC}^2 + bcn\sigma_A^2 .$$

Whether A is fixed or random has no bearing on $E(M_A)$. However, if C were fixed, we should strike out all terms with C as a subscript and obtain

$$E(M_A) = \sigma^2 + cn\sigma_{AB}^2 + bcn\sigma_A^2.$$

If B were also fixed, we should have $E(M_A) = \sigma^2 + bcn\sigma_A^2$.

We give in Table 5.1 the expected mean squares for a three-factor experiment in three cases:

(i) A, B, and C are all random;
(ii) A and B are random, C is fixed;
(iii) A is random, B and C are fixed.

In the right-hand column we list the mean square that is the denominator in the F test for the hypothesis that the corresponding effects or interactions or components of variance are zero. An asterisk denotes that we have to construct a suitable denominator. That topic is discussed in the next section. If $n = 1$, we have to assume that $\sigma^2_{ABC} = 0$ and use M_{ABC} to estimate σ^2.

TABLE 5.1
EXPECTED MEAN SQUARES

I. All Factors Random		
A	$\sigma^2 + n\sigma^2_{ABC} + cn\sigma^2_{AB} + bn\sigma^2_{AC} + bcn\sigma^2_A$	*
B	$\sigma^2 + n\sigma^2_{ABC} + cn\sigma^2_{AB} + an\sigma^2_{BC} + acn\sigma^2_B$	*
C	$\sigma^2 + n\sigma^2_{ABC} + bn\sigma^2_{AC} + cn\sigma^2_{BC} + abn\sigma^2_C$	*
AB	$\sigma^2 + n\sigma^2_{ABC} + cn\sigma^2_{AB}$	M_{ABC}
AC	$\sigma^2 + n\sigma^2_{ABC} + bn\sigma^2_{AC}$	M_{ABC}
BC	$\sigma^2 + n\sigma^2_{ABC} + an\sigma^2_{BC}$	M_{ABC}
ABC	$\sigma^2 + n\sigma^2_{ABC}$	M_e
Error	σ^2	

II. A and B Random, C Fixed		
A	$\sigma^2 + cn\sigma^2_{AB} + bcn\sigma^2_A$	M_{AB}
B	$\sigma^2 + cn\sigma^2_{AB} + acn\sigma^2_B$	M_{AB}
C	$\sigma^2 + n\sigma^2_{ABC} + bn\sigma^2_{AC} + an\sigma^2_{BC} + abn\sigma^2_C$	*
AB	$\sigma^2 + cn\sigma^2_{AB}$	M_e
AC	$\sigma^2 + n\sigma^2_{ABC} + bn\sigma^2_{AC}$	M_{ABC}
BC	$\sigma^2 + n\sigma^2_{ABC} + an\sigma^2_{BC}$	M_{ABC}
ABC	$\sigma^2 + n\sigma^2_{ABC}$	M_e
Error	σ^2	

III. A Random, B and C Fixed		
A	$\sigma^2 + bcn\sigma^2_A$	M_e
B	$\sigma^2 + cn\sigma^2_{AB} + acn\sigma^2_B$	M_{AB}
C	$\sigma^2 + bn\sigma^2_{AC} + abn\sigma^2_C$	M_{AC}
AB	$\sigma^2 + cn\sigma^2_{AB}$	M_e
AC	$\sigma^2 + bn\sigma^2_{AC}$	M_e
BC	$\sigma^2 + n\sigma^2_{ABC} + an\sigma^2_{BC}$	M_{ABC}
ABC	$\sigma^2 + n\sigma^2_{ABC}$	M_e
Error	σ^2	

EXAMPLE. The following data are taken from an experiment reported by Woodman and Johnson (1946). They were investigating the effects of various fertilizers on the growth of carrots. The three factors were nitrogen (ammonium sulphate), phosphorus (monocalcium phosphate), and potassium (potassium sulphate). The data are means of the weights (in grams) of the roots of five plants in each group grown under the various conditions. We have multiplied the data by 100.

	N_0			N_1			N_2		
	K_0	K_1	K_2	K_0	K_1	K_2	K_0	K_1	K_2
P_0	8876	9141	9785	9483	10049	9975	9990	10023	10451
P_1	8745	9827	9585	8457	9720	11230	9298	10777	11094
P_2	8601	10420	9009	8106	12080	10877	9472	11839	10287

Table 5.2 follows:

TABLE 5.2
ANALYSIS OF VARIANCE

Source	SS	d.f.
Nitrogen	488.3675	2
Potassium	1090.6564	2
Phosphorus	49.1485	2
$N \times K$	142.5844	4
$N \times P$	32.3475	4
$K \times P$	592.6238	4
$N \times K \times P$	185.7762	8

5.4 Approximate F Tests

We consider now the problem of testing the hypothesis $\sigma_A^2 = 0$ when all three factors are random. The difficulty lies in the fact that there is no mean square that has for its expectation $E(M_A) - bcn\sigma_A^2$ or $\sigma^2 + n\sigma_{ABC}^2 + cn\sigma_{AB}^2 + bn\sigma_{AC}^2$. If either σ_{AB}^2 or σ_{AC}^2 can be taken as zero, the problem is resolved. If we decide that $\sigma_{AC}^2 = 0$, M_{AB} becomes a suitable denominator for the F test; if $\sigma_{AB}^2 = 0$, we can use M_{AC}. Otherwise, a large value for the ratio M_A/M_{AB} merely indicates that the sum $bn\sigma_{AC}^2 + bcn\sigma_A^2$ is not zero, and, if the hypothesis $\sigma_{AC}^2 = 0$ has already been rejected, this throws little light on the size of σ_A^2. We have, therefore, to construct our own denominator from the several denominators available. (We are assuming now that the appropriate tests have already rejected the hypotheses $\sigma_{ABC}^2 = 0$, $\sigma_{AB}^2 = 0$, and $\sigma_{AC}^2 = 0$).

We shall take as denominator the linear combination

$$u = M_{AB} + M_{AC} - M_{ABC},$$

and take as our test statistic $\mathscr{F}' = M_A/u$. $M_A(a-1)/E(M_A)$ has a $\chi^2(a-1)$ distribution, and M_A and u are independent. We now need to find the approximate distribution of u.

It will be recalled that in the simpler case of the one-way layout with n observations per cell and random effects our test statistic was

$$\mathscr{F} = M_A/M_e.$$

Under normality, $S_A/(\sigma^2 + n\sigma_A^2)$ and S_e/σ^2 had independent χ^2 distributions with $(a-1) = \phi_A$ and $a(n-1) = \phi_e$ d.f. respectively. We then argue that

$$w = \frac{\phi_A^{-1}S_A/(\sigma^2 + n\sigma_A^2)}{\phi_e^{-1}S_e/\sigma^2} = \frac{\sigma^2}{\sigma^2 + n\sigma_A^2} \cdot \frac{M_A}{M_e}$$

has the $F[\phi_A, \phi_e]$ distribution and that

$$\mathscr{F} \sim \frac{\sigma^2 + n\sigma_A^2}{\sigma^2} F(\phi_A, \phi_e).$$

This argument depends upon M_A and M_e being multiples of independent χ^2 variates. In the present case M_A and u are certainly independent, but u does not have a χ^2 distribution. Indeed, it has not yet been possible to derive the distribution of u. We seek therefore to approximate the distribution of u by regarding u as a multiple of a χ^2 variate, i.e., by obtaining numbers p and ϕ such that $z = u/p$ has the same mean and variance as a χ^2 variate with ϕ d.f.

For the more general case of the distribution of linear combinations of χ^2 variates, the reader is referred to Satterthwaite (1946) and Welch (1956). We shall confine our discussion to the present example.

If $z \sim \chi^2_\phi$, then $E(z) = \phi$ and $V(z) = 2\phi$. The random variables $\phi_{AB}M_{AB}/[E(M_{AB})]$, $\phi_{AC}M_{AC}/[E(M_{AC})]$, and $\phi_{ABC}M_{ABC}/[E(M_{ABC})]$ where $\phi_{AB} = (a-1) \times (b-1)$, and so on, have independent χ^2 distributions. Then, for example

$$V(M_{AB}) = \frac{[E(M_{AB})]^2}{\phi_{AB}^2} \cdot 2\phi_{AB} = 2\frac{[E(M_{AB})]^2}{\phi_{AB}}.$$

If we fit p and ϕ by letting $E(u) = E(pz) = p\phi$ and $V(u) = p^2V(z) = 2p^2\phi$, then $E(M_{AB} + M_{AC} - M_{ABC}) = p\phi = E(u)$ and $V(M_{AB}) + V(M_{AC}) + V(M_{ABC}) = 2p^2\phi = V(u)$, so that $\phi = 2[E(u)]^2/V(u)$ and $p = \phi^{-1}E(u)$. Unfortunately, we are in the difficult position of not knowing any of the expectations. We therefore replace $E(M_{AB})$ by the observed value M_{AB}, and so on. We then obtain $p\phi = u$, and

$$p^2\phi = \frac{M_{AB}^2}{\phi_{AB}} + \frac{M_{AC}^2}{\phi_{AC}} + \frac{M_{ABC}^2}{\phi_{ABC}},$$

so that,

$$\phi = \frac{(a-1)(b-1)(c-1)u^2}{(c-1)M_{AB}^2 + (b-1)M_{AC}^2 + M^2{}_{ABC}}.$$

With this approximation we have

$$\frac{M_A}{u} = \frac{S_A}{(a-1)zp} = \frac{E(M_A)}{p\phi} \cdot \frac{S_A}{(a-1)E(M_A)} \cdot \frac{\phi}{z} = \frac{E(M_A)}{E(u)} \cdot F(a-1, \phi).$$

The other cases are handled in a similar fashion.

5.5 An Experiment with Crossed and Nested Factors

Suppose that we have a chemical process in which a feed stock is pretreated in one of several ways to form an intermediate, and then the intermediate is processed in a second stage at one of several temperatures. We suppose that there are a methods of pretreatment. We prepare b batches by each method. From each batch we take cn samples; n of these are then treated at each of c temperatures. If y_{ijkm} denotes the response observed on the mth of those samples which were treated at the kth temperature and taken from the jth of the batches that were prepared by the ith method, we assume a model

$$y_{ijkm} = \mu + a_i + b_{j(i)} + c_k + (ac)_{ik} + bc_{jk(i)} + e_{ijkm}.$$

The batches are nested within the methods (of pretreatment). The temperatures are crossed with the batches and, therefore, with the methods.

In the formulation of the model (we assume to begin with that all the effects are fixed), μ is a grand mean and a_i is the main effect of the ith level of the first factor; a_i represents the difference between the expected average response on all bcn runs made on stock pretreated by the ith method and the grand mean μ, i.e., $a_i = E(y_{i\cdots}) - \mu$. We have the side condition $\sum_i a_i = 0$. The b batches made by the ith method may differ; $b_{j(i)}$ denotes the main effect of the jth batch in the ith method. It is the difference between the expected average response of all cn runs made on stock taken from the (ij)th batch and the expected average for all the batches made by the ith method, i.e., $b_{j(i)} = E(y_{ij\cdot\cdot} - y_{i\cdots})$. It follows that $\sum_j b_{j(i)} = 0$ for each i. The main effects of the levels of temperature are denoted by c_k; we have $c_k = E(y_{\cdot\cdot k\cdot}) - \mu$, the averages being taken over every batch from every method, and $\sum_k c_k = 0$.

There are two interaction terms. The method $\times$ temperature interaction $(ac)_{ik}$ represents nonadditivity between methods and temperatures. Perhaps changing from temperature 1 to temperature 2 produces a greater change in response on feed stock pretreated by method 1 than on feed stock pretreated by method 2; we have $(ac)_{ik} = E(y_{i \cdot k \cdot}) - a_i - c_k - \mu = E(y_{i \cdot k \cdot} - y_{i \cdots} - y_{\cdot \cdot k \cdot} + y_{\cdots \cdot})$. The side conditions are $\sum_i (ac)_{ik} = 0$, and $\sum_k (ac)_{ik} = 0$. The temperature × batch within methods interaction, $(bc)_{jk(i)}$, represents nonadditivity between temperatures and different batches made by the same method, and, in essence, accommodates all the scatter between cell means that has not been taken into account so far. The side conditions are $\sum_k (bc)_{jk(i)} = 0$ for each batch, and $\sum_j (bc)_{jk(i)} = 0$ for each combination of temperature and method.

There is no AB interaction term. We cannot talk about the main effect of batches changing from method to method because batches are nested in methods. A run made at temperature 1 on a stock pretreated by method 1 and a run at temperature 1 on a stock pretreated by method 2 have in common the (possibly important) fact that they were both made at the same temperature. The first batch made by method 1 and the first batch made by method 2 do not have a batch factor in common; as we have formulated the experiment there are just b batches made by each method, and being the first or second batch is just a label. Similarly, there is no ABC interaction. That would have corresponded to an AB interaction varying from temperature to temperature.

We could now proceed in two ways. We could go back to the linear model, obtain the normal equations, solve them, and derive sums of squares by routine algebraic procedures. We shall, however, take instead a more practical approach, which can be formalized by considering a set of nested hypotheses.

Our experiment involves ab batches and c temperatures. We can analyze it as a two-way layout with factors B (batches), C (temperatures), and the usual sums of squares.

TABLE 5.3

ANALYSIS OF VARIANCE

Source	SS	d.f.
Total	$\mathbf{Y}'\mathbf{Y} - C$	$abcn - 1$
B	$\sum_i \sum_j T_{ij\cdot\cdot}^2/cn - C = S_B$	$ab - 1$
C	$\sum_k T_{\cdot\cdot k \cdot}^2/abn - C = S_T$	$c - 1$
BC	$\sum_i \sum_j \sum_k T_{ijk\cdot}^2/n - C - S_B - S_T$	$(c - 1)(ab - 1)$
S_e	By subtraction	$abc(n - 1)$

The sum of squares for batches falls into two components, just as in the ordinary nested design.

$S_A = \sum_i T_{i\cdots}^2/bcn - C$ is the sum of squares between (or for) methods. $S_{B(A)} = \sum_i \sum_j T_{ij\cdot\cdot}^2/cn - \sum_i T_{i\cdots}^2/bcn$ is the remainder, and is attributable to differences between batches made by the same method.

A similar process of subdivision can now be applied to the interaction sum of squares. We divide S_{BC} into two components; S_{AC} is the sum of squares for interaction that we would obtain were we to ignore batches and to consider a two-way layout with a methods, c temperatures, and bn observations in each cell. Thus $S_{AC} = \sum_i \sum_k T^2_{i \cdot k}/bn - \sum_i T_i^2{}_{\cdot\cdot}/bcn - \sum T^2{}_{\cdot\cdot k}/abn + C$ with $(a-1)(b-1)$ d.f.; the rest of S_{ABC} with $a(b-1)(c-1)$ d.f. is the sum of squares for interaction between temperature and batches made by the same method (or more concisely temperature $\times$ batches in methods).

In practice, it is more reasonable to regard the $b_{j(i)}$ as random variables, uncorrelated (or independent under the normality assumption) with mean zero and variance σ_B^2. The interactions $(bc)_{jk(i)}$ will also be random variables. Suppose that we continue to take methods (A) and temperatures (C) as fixed. The expectations of the sums of squares may be derived by following the same procedures that we used for the mixed model in the chapter on the two-way layout. This can, however, be a lengthy process and we shall give in the next section some rules for writing down the expected mean squares more easily. In our example the table of expected mean squares with A and C fixed and B random is

TABLE 5.4
EXPECTED MEAN SQUARES

SS	d.f.	EMS
S_A	$a - 1$	$\sigma^2 + cn\sigma^2_{B(A)} + bcn\sigma^2_A$
$S_{B(A)}$	$a(b - 1)$	$\sigma^2 + cn\sigma^2_{B(A)}$
S_C	$c - 1$	$\sigma^2 + n\sigma^2_{BC(A)} + abn\sigma^2_C$
S_{AC}	$(a - 1)(c - 1)$	$\sigma^2 + n\sigma^2_{BC(A)} + bn\sigma^2_{AC}$
$S_{BC(A)}$	$a(b - 1)(c - 1)$	$\sigma^2 + n\sigma^2_{BC(A)}$
S_e	$abc(n - 1)$	σ^2

5.6 General Rules for Sums of Squares and EMS

We present now a set of rules for writing down the form of the model, formulae for the sums of squares, their degrees of freedom, and the expectations of the mean squares. These rules are also given by Scheffé (1959, pp. 282–289) and by Bennett and Franklin (1954, Sec. 7.6). We shall use the example of the last section as an illustration.

1. The model contains, in addition to th^ mean and the error term, all main effects and interactions corresponding to sets of 2, 3, . . ., factors, except that there are no interaction terms containing two factors, one of which is nested in the other. Thus, in the example there was no AB interaction and no ABC interaction. The subscripts attached to a term are those corresponding to the factors appearing in the term. If one of the factors in a term is a nested factor, the letters corresponding to the factors in which it is nested are added in brackets as subscripts. Thus, we had for the BC in A interaction, $(bc)_{jk(i)}$. The error term is written in the form $e_{m(ijk)}$ for our present purposes.

2. For each term or effect in the model (we let the word *effect* now embrace both main effects and interactions), we divide the subscripts into three classes: (*a*) live—those subscripts that are present in the term and are not bracketed; (*b*) dead—those that are present and bracketed; (*c*) absent—those subscripts that appear in the model but not in the particular term. Thus, in $bc_{jk(i)}$, j and k are live, i is dead, and m is absent. We then form the symbolic product for the term, in which each dead subscript is represented by its letter and each live subscript by its letter minus one. For $(bc)_{jk(i)}$ the symbolic product is $i(j - 1) \times (k - 1)$.

3. The corresponding product in the numbers of levels, e.g., $a(b - 1)(c - 1)$, gives the number of degrees of freedom for the sum of squares.

4. Expanding the symbolic product, we obtain $(ijk - ij - ik + i)$. The corresponding sum of squares is $\sum (y_{ijk\cdot} - y_{ij\cdot\cdot} - y_{i\cdot k\cdot} + y_{i\cdot\cdot\cdot})^2$ or

$$\sum_i \sum_j \sum_k \frac{T_{ijk\cdot}^2}{n} - \sum_i \sum_j \frac{T_{ij\cdot\cdot}^2}{cn} - \sum_i \sum_k \frac{T_{i\cdot k\cdot}^2}{bn} + \sum_i \frac{T_{i\cdot\cdot\cdot}^2}{bcn}.$$

Similarly, we have for the main effect A the product $(i - 1)$, which gives

$$S_A = (bcn)^{-1} \sum_i T_{i\cdot\cdot\cdot}^2 - C.$$

5. To obtain the EMS we prepare the following auxiliary table. There is a row for each mean square in the analysis of variance and a column for each letter. At the head of each column we write the subscript letter, the number of levels of the factor, and whether it is random (R) or fixed (F). At the beginning of each row we write the subscripts for that term in the model. The rules for filling in the table follow the table itself.

TABLE 5.5
AUXILIARY TABLE

	i_a^F	j_b^R	k_c^F	m_n^R
i	0	b	c	n
$j(i)$	1	1	c	n
k	a	b	0	n
ik	0	b	0	n
$jk(i)$	1	1	0	n
$m(ijk)$	1	1	1	1

(i) In any row, write the number of levels in any column headed by a letter which is an absent subscript.

(ii) Write 1 if the column is headed by a dead subscript.

(iii) In any column corresponding to a random effect, write 1 in the remaining cells.

(iv) In any column corresponding to a fixed effect, write 0 in the remaining cells.

To obtain, for example, $E(M_A)$, we cover the ith column. In each row for which i is a live or dead subscript, we take the product of the visible numbers and multiply by the corresponding component of variance. Thus, $E(M_A) = \sigma^2 + cn\sigma^2_{B(A)} + bcn\sigma^2_A$.

For $E(M_{AC})$, we cover the i column and the k column and obtain

$$E(M_{AC}) = \sigma^2 + n\sigma^2_{BC(A)} + bn\sigma^2_{AC}.$$

5.7 Split-Plot Designs

The experiment that we have just considered is an example of a split-plot design. In the agricultural context from which the name is derived, we picture a piece of land divided into several plots, perhaps strips of land or fields. Each of these plots is then given a single treatment. It is convenient to call these plots *whole plots* and to designate the treatments that are applied, one to each whole plot, as *main treatments* or levels of a main factor. The whole plots are then split into subplots, and one or more subtreatments are applied to each subplot. We might, for example, take several varieties of a crop and sow several fields (whole plots) with each variety, one variety to a field. Then each field could be divided into six subplots, and one subplot from each field could be harvested each week over a six-week period. The varieties would be the main treatments, and the harvesting times the subtreatments. In comparing the varieties, we must take into account the field-to-field variation. In comparing harvest times, we can compare yields within each field. It follows that the main treatments are tested against a larger error term (the whole-plot error) than the subtreatments.

In our example the methods of preparing the feedstock are the main treatments, the batches are the whole plots, the temperatures are the subtreatments, and the samples drawn from the batches to be run at the various temperatures are the subplots. It will be recalled that the main effect C is tested against $M_{BC(A)}$, as is the interaction AC; however, the main effect A is tested against $M_{B(A)}$. The most common situation is to have one subplot in each whole plot for each subtreatment, i.e., $n = 1$. The main treatments are sometimes said to be confounded with the whole plots.

The experiment reported by Yates (1937) involving oats and manure (nitrogen), to which we referred in the previous chapter, was a split-plot experiment. Six blocks of three plots each were taken, and one plot in each block was sown with each of the three varieties of oats chosen for the experiment. Each plot was divided into four subplots, one of which was assigned at random to each level of the nitrogen factor. The four levels were no manure and 0.01, 0.02, and 0.03 tons per acre. The main treatments were the varieties of oats and the subtreatments were the levels of nitrogen. The data and the analysis of variance table are given below. The sum of squares for whole-plot error is the same as the sum of squares for block × varieties interaction. In practice, the necessary sums of squares can be obtained on a computer by using a program for a three-factor experiment with factors A (oats), N (nitrogen), C (blocks), and one observation per cell. Then $S_e = S_{NC} + S_{ANC}$.

Block		n_1	n_2	n_3	n_4	Block	n_1	n_2	n_3	n_4
	A_1	111	130	157	174		74	89	81	122
I	A_2	117	114	161	141	II	64	103	132	133
	A_3	105	140	118	156		70	89	104	117
	A_1	61	91	97	100		62	90	100	116
III	A_2	70	108	126	149	IV	80	82	94	126
	A_3	96	124	121	144		63	70	109	99
	A_1	68	64	112	86		53	74	118	113
V	A_2	60	102	89	96	VI	89	82	86	104
	A_3	89	129	132	124		97	99	119	121

TABLE 5.6
ANALYSIS OF VARIANCE

Source	SS	d.f.	MS
Total	51985.95	71	
Blocks	15875.28	5	3175.06
Varieties	1786.36	2	893.18
Whole plot error	6013.30	10	601.33
Between whole plots	23674.94	17	
Nitrogen	20020.50	3	6673.50
$N \times V$	321.75	6	53.63
Subplot error	7968.76	45	177.08

The following situation occurs occasionally in industrial experimentation and is similar to the split-plot experiment. We shall illustrate it by an example. Consider a chemical process in which propylene is to be polymerized in a reactor using an acid catalyst, and pilot plant experiments are to be conducted using acid strength and temperature as the two factors with a and t levels respectively. The response measured is the amount of propylene converted into polymer. At each experimental run, a quantity of the product is taken from the plant, and the amount of conversion is determined by a laboratory analysis.

Suppose now that it is decided to have four observations at each combination of levels of temperature and acid strength. There are two possibilities that come to mind. The first is to make 4 *at* runs in random order, and then to make an analysis of the product of each run. The second is to make only one run at each combination of levels, but to make four analyses on the product of each run. In either case, we end up with four numbers for each cell in the acid-temperature table.

In the first case, the scatter among the numbers in each cell reflects both the precision of the laboratory analysis and also the 'noise' in the whole system. It measures the reproducibility of the results and throws some light upon the question: if we make another run at the ith acid strength and the jth temperature tomorrow, how well shall we be able to reproduce the results that were obtained last week? If y_{ijk} denotes the response on the kth run at the ith level of acid and the jth level of temperature, we could write

$$y_{ijk} = m + a_i + b_j + (ab)_{ij} + e_{ijk} + f_{ijk},$$

where e_{ijk} is a term for the general noise, such as the inability to reproduce exactly the same temperature and perhaps the same acid strength, on different runs; f_{ijk} is the analytic error. They are independent random variables with zero means and variances σ_e^2 and σ_f^2 respectively, and $V(y_{ijk}) = (\sigma_e^2 + \sigma_f^2)$, $V(y_{ij\cdot}) = (\sigma_e^2 + \sigma_f^2)/4$. The ordinary two-way analysis of variance procedure produces a mean square for error s^2 with 3 *at* d.f. and expectation $\sigma_e^2 + \sigma_f^2$, and we have the situation of the previous chapter.

On the other hand, in the second experimental plan, y_{ijk} denotes the result of the kth analysis on the (ij)th run. We write

$$y_{ijk} = m + a_i + b_j + (ab)_{ij} + e_{ij} + f_{ijk},$$

because for any given i and j, e_{ijk} is the same for all four values of k. We have $V(y_{ijk}) = (\sigma_e^2 + \sigma_f^2)$ as before, but now $V(y_{ij\cdot}) = \sigma_e^2 + (\sigma_f^2/4)$.

The sum of squares between duplicate analyses provides an estimate of σ_f^2 with 3 *at* degrees of freedom. It does not, however, provide any information about σ_e^2. In consequence, the experimenter who uses the mean square between duplicates as his error term in comparing temperatures is using too small an error estimate. He is using only the analytic variance and is tacitly assuming that his experimental conditions are perfectly reproducible. This will result in his making too many wrong decisions to reject null hypotheses.

How, then, do we test for temperature differences in this case? The answer is that we use for the denominator the mean square for $A \times T$ interaction. We essentially have a two-factor experiment with one run at each combination of levels; nested within each run are the four analyses. The setup is the same as if there were *at* whole plots, each split into four subplots to which no subtreatment is applied. If either A or T is fixed, as such factors usually would be, we have the same problem that we faced in the last chapter when there was one observation per cell and interaction was present. We may sum up with a useful reminder. If, in a factorial experiment, all the main effects and all the interactions turn out to be highly significant in the analysis of variance, go back and look at the experiment again. It may actually be a split-plot situation, and you may be testing everything against the repeatability of a chemical balance in the hands of an expert technician!

The following example illustrates a slightly different aspect of the topic that has just been discussed. Chanda, et al. (1952) fed carotene to goats and cows over consecutive two-day periods and used two methods to measure the digestibility of the carotene. The data shown are for four goats over five periods. In each cell, the first number given is the apparent digestibility measured by method 1. The second number is the measurement by method 2.

The goat-period combinations can be regarded as $4 \times 5 = 20$ whole plots. Each whole plot is split into two subplots. Method 1 is applied to one subplot and method 2 to the other. The goat and period main effects are tested against the mean square for goat $\times$ period interaction. The $M \times G \times P$ mean square is used to test M, $M \times G$ and $M \times P$. It will be noted that if the $M \times G \times P$ mean square had been used to test G and P, each of those effects would have been deemed significant.

	Periods				
Goats	I	II	III	IV	V
Anna	75.1, 75. 2	69.0, 63.5	74.3, 80.7	51.2, 46.7	72.6, 71.0
Betty	65.4, 62.7	63.3, 61.2	59.4, 65.3	62.9, 62.1	63.8, 63.1
Bluebell	61.9, 60.1	57.7, 59.6	74.4, 75.8	62.9, 55.3	54.3, 58.3
Diana	69.4, 70.1	60.7, 57.0	57.0, 54.5	56.6, 49.5	64.3, 63.3

TABLE 5.7
ANALYSIS OF VARIANCE

Source	SS	d.f.	MS	F values
Methods (M)	11.03	1	11.03	2.28
Goats (G)	328.44	3	109.48	1.20
Periods (P)	760.13	4	190.03	2.09
$M \times G$	10.23	3	3.41	0.71
$M \times P$	67.27	4	16.82	3.48
$G \times P$	1090.77	12	90.90	18.82
$M \times G \times P$	57.99	12	4.83	
Total	2325.86	39		

5.8 Split-Split-Plot Designs

In some experimental situations, further subdivision of the subplots is used. An example is the experiment conducted by Tandon (1949) to investigate the response of flax to various rates of application and formulations of the herbicide 2,4-D.

He chose seven varieties of flax with three formulations of 2,4-D and four rates of application. There were four replications in the experiment. Each replicate involved twelve blocks, each with seven plots, so that there were 336 plots altogether. One plot in each block was sown with each variety of flax. The three formulations of 2,4-D were made up, and four blocks in each replicate were assigned to each formulation. The 2,4-D was then applied to one of the blocks at the lowest rate, to another at the second rate, and so on. In any block, each of the seven plots received the same formulation at the same rate. In the split-split-plot context there are twelve whole plots with formulations as the whole-plot factor. Each of the whole plots is split into four subplots (the blocks), with rates of application as the subplot factor. Each subplot is split into seven sub-subplots with flax as the sub-subplot factor. The sources of variation and the degrees of freedom for their sums of squares are

Replications	3
F = formulations	2
Reps $\times$ F = Error (a)	6
R = Rates	3
$R \times F$	6
Error (b)	27
V	6
$V \times F$	12
$V \times R$	18
$V \times R \times F$	36
Error (c)	216

He obtained his error (b) term for testing R and $R \times F$ by pooling the sums of squares for reps $\times$ R and reps $\times$ $R \times F$. The sum of squares for error (c) is obtained by subtraction and consists of everything that has not been used already. As usual, it is assumed that replications do not interact with any of the other factors.

5.9 Computer Programs

The analysis of variance calculations for all these designs can readily be carried out on a computer. It is relatively easy to write a program for a complete factorial design with one observation per cell, and most computer centers have one available. It is not necessary to write a special program for each design when some factors are nested or when there are split plots. A few examples will suffice to show how the simple program can be used for the more complex designs with the help of a few additions with pencil and paper. We have already mentioned this in regard to the Yates experiment with oats and manure.

1. *A two-factor design with r observations per cell.* We have merely to pretend to ourselves that there are three factors. If y_{ijk} is the kth observation at the ith

level of A and the jth level of B, we suppose it to be at the kth level of a third factor, C. The computer program then gives the correct sums of squares S_A, S_B, and S_{AB}. The error term is obtained by adding S_C, S_{AC}, S_{BC}, and S_{ABC}. If the experiment is blocked, S_C is the sum of squares for blocks.

2. *The two-stage nested design.* Suppose that we have s sires, d dams per sire, and n offspring per dam. Take three factors A, B, and C so that y_{ijk}, the observation on the kth offspring from the jth dam by the ith sire, becomes the observation at the ith level of A, the jth level of B, and the kth level of C. The sums of squares appearing in the analysis table are then:

Between sires	S_A
Between dams in sires	$S_{AB} + S_B$
Between offspring in the same litter	$S_C + S_{AC} + S_{BC} + S_{ABC}$.

The idea involved here is that assuming a sires and b dams per sire, there are $a(b - 1)$ d.f. for dams in sires. We write this in terms of $(a - 1)$ and $(b - 1)$ as $a(b - 1) = ab - a = (a - 1)(b - 1) + (b - 1)$. Hence, $S_{B(A)} = S_{AB} + S_B$.

3. *The example with methods, batches, and temperatures.* We add a fourth factor, D, for pretended replications and then present the data to the computer as a four-factor experiment in A, B, C, and D. We have $S_{B(A)} = S_B + S_{AB}$, $S_{BC(A)} = S_{ABC} + S_{BC}$ and $S_e = S_D + S_{AD} + S_{BD} + S_{CD} + S_{ABD} + S_{ACD} + S_{BCD} + S_{ABCD}$.

Exercises

1. The following data are yields of dry herbage (cwt/acre) from an experiment on grazed grass [Widdowson et al., (1966)]. There were three fertilizers, nitrogen at three levels, phosphorus at two levels, and potassium at two levels. Yields are given for four years, 1961–1964.

Years	1961	1962	1963	1964
$N_0P_0K_0$	56.6	46.7	62.4	69.1
$N_0P_1K_0$	77.5	50.8	65.6	70.8
$N_0P_0K_1$	63.3	46.8	65.4	87.8
$N_0P_1K_1$	65.5	61.8	78.5	88.7
$N_1P_0K_0$	82.9	79.0	89.8	100.6
$N_1P_1K_0$	116.8	85.4	104.4	107.8
$N_1P_0K_1$	99.4	80.0	88.0	105.8
$N_1P_1K_1$	111.7	84.3	106.2	118.7
$N_2P_0K_0$	78.7	73.3	78.2	98.6
$N_2P_1K_0$	100.0	84.6	98.8	109.4
$N_2P_0K_1$	102.6	74.8	100.4	106.0
$N_2P_1K_1$	123.8	95.2	100.7	119.5

We denote the levels of the three factors by N_0, N_1, N_2, P_0, P_1, and K_0, K_1. Analyze the data, assuming all factors to be fixed.

2. Peart (1968) reports lactation studies with ewes. In one of his experiments nine ewes were divided into groups of three. In group 1 the ewes and their lambs were fed freely during the lactation period; in the other groups two different methods of restricting the food intake were used. He records the yields of residual milk in grams for each ewe during the fourth through tenth weeks of lactation.

Group	Ewe	Lactation Week						
		4	5	6	7	8	9	10
1	184	190	240	270	280	280	280	260
	284	220	370	390	400	390	360	410
	294	200	200	220	230	260	290	300
2	243	580	580	570	560	570	490	370
	272	230	200	200	170	240	290	270
	236	190	110	100	100	80	110	120
3	266	160	110	90	110	100	100	80
	245	140	130	130	130	130	100	120
	280	390	210	200	130	140	120	80

Analyze the data.

3. Gross et al. (1953) tested twelve fertilizer treatments (T) and four varieties of alfalfa (V). They took four blocks; in each block there were forty-eight plots, one for each alfalfa-fertilizer combination. Assuming that V and T are fixed but that the blocks are random, set up the analysis of variance table and show which ratio would be used to test the three hypotheses

$$\sigma_V^2 = 0, \qquad \sigma_T^2 = 0, \qquad \sigma_{VT}^2 = 0.$$

4. Aircraft are now used to sow seed and to spread fertilizer. Suppose that in a large field we set up a grid pattern in the following way. Flying east and west, a plane sows av strips, that is, a strips with each of v varieties. Flying north and south, the pilot spreads fertilizer, b strips for each of t fertilizer treatments. The intersection of a sowing strip and a fertilizer strip is a plot. There are $avbt$ of these, and n observations are made from each of them. In both directions, the varieties or the fertilizers are assigned to the strips at random. Set up the analysis of variance table.

5. Lessman and Nyquist (1966) investigated the growth of an oil seed crop called *crambe*. They tested eleven varieties in two locations, and in both locations they made five replicates, each of which was divided into two blocks. In each block four plants of each variety were sown. In one block the plants were closely spaced; in the other they were widely spaced. There were thus 880 plants in all. Set up the analysis of variance table with the expectations of mean squares regarding locations and plants as random, and regarding replicates, spacings, and varieties as fixed.

CHAPTER 6

Latin Square Designs

In Chapter 3 we discussed the experiment in which five gasolines were tested in cars and their road octane numbers observed. Four observations were to be made on each gasoline. We began with the completely randomized experiment in which twenty cars were chosen and four of them were assigned by some random procedure to each gasoline. Later in the chapter we introduced the randomized complete block design. There we took only four cars. Each gasoline was tested once in each car. The car differences canceled out when gasolines were compared, and the precision of the experiment for comparing gasolines was thereby improved. In an agricultural experiment, where we have reason to think that the land becomes more fertile as we move from north to south, suitable blocks might be strips of land running from east to west with one plot per block for each of the treatments or varieties being tested. In either case, the treatments are randomized within the blocks. In the case of the gasolines, the order in which they are to be tested is chosen randomly for each car; the agronomist assigns the plots within each block at random to the different varieties.

Let us change the experiment slightly so that the response we observe is the rate of consumption in miles per gallon, with the added restriction that each car can only make one run, i.e., test one gasoline, per day. It may be that conditions vary during the working week. Perhaps early in the week the weather is cold and damp, but as the week wears on the sun appears and the weather becomes warmer and drier. This could conceivably have some influence upon the performance of the cars and upon the ratings that they give the gasolines. Unless something is done about it, this day-to-day variation may ruin the whole experiment. In an extreme situation suppose that all the cars use only gasoline A on Monday, which is bright and sunny, and that all the cars make their test of gasoline B on Friday when the temperature is twenty degrees lower and it is raining steadily. We might gain some insight into the relative merits of the cars used under a variety of conditions, but we should not be able to tell whether a higher rating in miles per gallon of gasoline A over gasoline B results from the differences between the gasolines or from the differences in the weather. We were able to increase the precision of the earlier experiment when we removed the car effects by blocking. We should be able to make a further improvement in the precision if we could balance out the day effects. This calls for an experimental scheme in which each gasoline is tested once in each car and once on each day (or in each position in the block). This cannot be done with five days and only four observations on each gasoline. Suppose that we allow ourselves a fifth car. We can then use a Latin square design for the experiment.

A Latin square of side p, or a $p \times p$ Latin square, is an arrangement of p letters, each repeated p times, in a square array of side p in such a manner that each letter appears exactly once in each row and in each column. Four Latin squares of side five are

$$
\begin{array}{ccccc}
A\,B\,C\,D\,E & \quad & A\,B\,C\,D\,E & \quad & A\,B\,C\,D\,E & \quad & A\,B\,C\,D\,E \\
B\,C\,D\,E\,A & & C\,D\,E\,A\,B & & D\,E\,A\,B\,C & & E\,A\,B\,C\,D \\
C\,D\,E\,A\,B & & E\,A\,B\,C\,D & & B\,C\,D\,E\,A & & D\,E\,A\,B\,C \\
D\,E\,A\,B\,C & & B\,C\,D\,E\,A & & E\,A\,B\,C\,D & & C\,D\,E\,A\,B \\
E\,A\,B\,C\,D & & D\,E\,A\,B\,C & & C\,D\,E\,A\,B & & B\,C\,D\,E\,A.
\end{array}
$$

Suppose that we were to use the first of these squares and to let the rows denote the cars, the columns the days, and the letters the gasolines. Then, on the second day (column 2) the first car (row 1) uses gasoline B, the second car (row 2) uses gasoline C, and so on. As we have written it, this square has a more systematic pattern than is needed or is desirable since for every car the gasolines follow one another in alphabetical order. In practice, the experimenter should assign the cars at random to the rows and the days at random to the columns (i.e., order the columns by a random procedure).

6.1 Analysis

We assume a completely additive model; more will be said about this assumption later. Let y_{ijk} denote the observation corresponding to the kth letter in the ith row and the jth column, if there is one, i.e., to the observation on the kth gasoline in the ith car on the jth day where $1 \leq i, j, k \leq p$. The row, column, and letter effects are denoted by ρ_i, γ_j, and τ_k, respectively, and for the present we assume them to be fixed effects; μ is the grand mean and e_{ijk} is the random error with the usual assumptions about homoscedasticity and normality. We then have

$$y_{ijk} = \mu + \rho_i + \gamma_j + \tau_k + e_{ijk}.$$

We shall denote the row, column, and letter totals by $T_{i\cdot\cdot}$, $T_{\cdot j\cdot}$, and $T_{\cdot\cdot k}$, and the corresponding means by $y_{i\cdot\cdot}$, $y_{\cdot j\cdot}$, $y_{\cdot\cdot k}$; the grand mean will be denoted by $y_{\cdots}$. In addition to the side conditions $\sum_i \rho_i = 0$ and $\sum_j \gamma_j = 0$, we impose the condition $\sum_k \tau_k = 0$. The normal equations are then easily solved, and the least squares estimates of the parameters are

$$\hat{\mu} = y_{\cdots}, \qquad \hat{\rho}_i = y_{i\cdot\cdot} - y_{\cdots}, \qquad \hat{\gamma}_j = y_{\cdot j\cdot} - y_{\cdots}, \qquad \hat{\tau}_k = y_{\cdot\cdot k} - y_{\cdots}.$$

The analysis of variance table follows. Under normality, the sums of squares are distributed independently. The appropriate test statistics for the hypotheses of no row effects, no column effects, or no letter effects are the ratios of the corresponding mean squares to the mean square for error. Under the null hypotheses, these ratios will each have the central F distribution with $(p - 1)$ and $(p - 1)(p - 2)$ degrees of freedom. If, for example, the row effects are random variables with mean zero and variance σ_R^2, we use the same breakdown into sums of squares. The expected value of the mean square for rows is then $\sigma^2 + p\sigma_R^2$.

TABLE 6.1
ANALYSIS OF VARIANCE

Source	SS	d.f.	EMS
Total	$\sum_i \sum_j \sum_k y_{ijk}^2 - C$	$p^2 - 1$	
Rows	$p^{-1} \sum_i T_{i\cdot\cdot}^2 - C$	$p - 1$	$\sigma^2 + p \sum_i \rho_i^2/(p - 1)$
Columns	$p^{-1} \sum_j T_{\cdot j\cdot}^2 - C$	$p - 1$	$\sigma^2 + p \sum_j \gamma_j^2/(p - 1)$
Letters	$p^{-1} \sum_k T_{\cdot\cdot k}^2 - C$	$p - 1$	$\sigma^2 + p \sum_k \tau_k^2/(p - 1)$
Error	By subtraction	$(p - 1)(p - 2)$	σ^2

In the example of the gasoline experiment, the purpose of using the Latin square design was to improve the precision of the gasoline comparisons by eliminating the car and day effects. In agricultural experiments we could use such designs to eliminate fertility trends in two perpendicular directions, east-west as well as north-south; this was the context in which they were first introduced, and they are sometimes called designs for the elimination of two-way heterogeneity. It is customary to refer to the three classifications as rows, columns, and treatments. The requirement that there should be the same number of rows and columns as there are treatments can present a difficulty, and alternative designs which involve incomplete blocks can be used. That, however, is a topic for later discussion.

The Latin square can also be used in factorial experiments. A complete factorial experiment with three factors each at five levels, which we may call a 5^3 experiment, calls for 125 points. If we take one of the Latin squares we can let the rows denote the levels of the first factor, the columns the levels of the second factor, and the letters the levels of the third factor. Then we have a set of twenty-five out of the 125 points with the property that for any two of the factors each level of one appears exactly once at each level of the other. In the general case we have a $(1/p)$th fraction of the complete factorial design. This reduction in the number of points can be very attractive. When $p = 5$, it represents an 80 per cent reduction in the number of observations that have to be made. The additive model that we have assumed may be rewritten

$$y_{ijk} = \mu + \alpha_i + \beta_j + \gamma_k + e_{ijk},$$

where y_{ijk} is the observation, if any, made at the ith level of the first factor, the jth level of the second factor, and the kth level of the third factor, and α_i, β_j, and γ_k are the corresponding main effects. The analysis follows the same lines, and we have tests for the main effects of the factors. Contrasts may be tested by any of the usual multiple comparison methods. The reduction in the number of points obliges us to assume in our model that there are no interactions. This restriction may be serious; it is discussed in the next section.

This type of fractional factorial design may become necessary because of budget restrictions. It may also be necessary if time is one of the factors, as it is for the dairy cattle and crop rotation experiments that we shall discuss later in this chapter. Consider, for example, an experiment to investigate the consumption of lubricating oil in buses. Suppose that there are three brands of oil, X, Y, and Z. We could take three groups of new buses (A, B, and C) and conduct a test over three months (I, II, and III), in which case an observation would show the average consumption of a test oil by a group of buses in a month. We might have group A use oil X for the first month, followed by oil Y for the second month, and oil Z for the third month. We cannot use two oils simultaneously in the same bus and so, even though we might think of a factorial experiment in the three factors (oils, groups, and months, each at three levels), we can observe only one of the three points AIX, AIY, and AIZ. We are obliged to use one-third fractions of the three-factor design, and the use of one or more Latin squares would be appropriate.

6.2 Limitations of the Model

In the analysis of the Latin square design we assumed a completely additive model with no interaction terms. As long as this model is justified, there is no problem about the validity of the analysis. However, if there is, for example, interaction between rows and columns, this interaction biases the estimates of the treatment effects. This difficulty can be illustrated in the gasoline experiment that has just been discussed. Using the first square, suppose that there is indeed a row-column interaction $(\rho\gamma)_{ij}$ to be added to the model. Over the set of rows and the set of columns we have $\sum_i (\rho\gamma)_{ij} = 0$ for each j and $\sum_j (\rho\gamma)_{ij} = 0$ for each i. These side conditions ensure that the interactions cancel out of $E(\mu)$.

Let us consider, in particular, the gasoline that had been labeled treatment A. We shall use subscripts 1, 2, 3, 4, and 5 for gasolines A, B, C, D, and E. We then have

$$T_{..1} = 5\mu + \sum_{i=1}^{5} \rho_i + \sum_{j=1}^{5} \gamma_j + 5\tau_1 + \sum_i \sum_j (\rho\gamma)_{ij} + 5e_{..1}.$$

The terms $\sum_j \gamma_j$ and $\sum_i \rho_i$ vanish by virtue of the side conditions but the interaction terms that are actually involved are only a subset of the p^2 possible $(\rho\gamma)$ interactions, and there is no reason to expect them to vanish. They remain as an unwanted and unknown nuisance. In this case the bias is

$$(\rho\gamma)_{11} + (\rho\gamma)_{25} + (\rho\gamma)_{34} + (\rho\gamma)_{43} + (\rho\gamma)_{52}.$$

We might have chosen to label the gasoline in question treatment B rather than treatment A, in which case the bias would have been

$$(\rho\gamma)_{12} + (\rho\gamma)_{21} + (\rho\gamma)_{35} + (\rho\gamma)_{44} + (\rho\gamma)_{53},$$

or we might have retained the label A and used the second square, in which case the bias would have been

$$(\rho\gamma)_{11} + (\rho\gamma)_{24} + (\rho\gamma)_{32} + (\rho\gamma)_{45} + (\rho\gamma)_{53},$$

and so on. Thus, summing, or averaging, over all the labels and over all four squares, it can be seen that the bias cancels out. In any particular square and for any particular label there is a bias, but over the whole set it averages out. This leads to the argument that the experimenter should choose his Latin square at random from the set of all squares of side p and, having done so, assign the letters at random to the treatments; by this procedure he would eliminate the unwanted bias.

It is, alas, cold comfort to the experimenter, who has for his particular experiment one particular square, to know that if he could repeat his experiment many times using randomly selected squares, the bias would on the average disappear and to that extent, even in the presence of row-column interactions, treatment differences would be estimable. He is saddled with the single square that has been chosen, and his results are conditional upon the choice of that square.

Nor is the error term free from bias. Scheffé (1959) shows that in the fixed effects case the expectation of the mean square for error contains a term involving the interactions, which is nonnegative and may or may not be appreciable. The problem of bias in Latin squares was first noted by Neyman et al. (1935). Wilk and Kempthorne (1957) consider a random effects model in a similar way. In their model, rows, columns, and treatments are all considered as random effects; they show that in this case, too, the mean squares for rows, columns, treatments, and error are all biased if the simple additive model does not hold. Kempthorne (1952) and Scheffé (1959) discuss randomization models as alternatives to the models presented here.

6.3 Graeco-Latin Squares

Suppose that we take a Latin square and superimpose upon it a second square with the treatments denoted by Greek letters. If the two squares have the property that each Latin letter coincides exactly once with each Greek letter when the squares are superimposed, then they are said to be orthogonal. The combined squares are said to form a Graeco-Latin square. Adding a third square orthogonal to each of the others gives a hyper-Graeco-Latin square. The four squares of side five that were given at the beginning of this chapter form a set of mutually orthogonal squares.

Graeco-Latin squares can be used to provide designs for four factors (rows, columns, Latin letters, Greek letters), each at p levels in p^2 runs. Thus, in our example of the gasoline experiment, a fourth factor, drivers, might be introduced by combining the first two squares and using the Greek letters to represent the drivers:

$$\begin{array}{ccccc} A\alpha & B\beta & C\gamma & D\delta & E\epsilon \\ B\gamma & C\delta & D\epsilon & E\alpha & A\beta \\ C\epsilon & D\alpha & E\beta & A\gamma & B\delta \\ D\beta & E\gamma & A\delta & B\epsilon & C\alpha \\ E\delta & A\epsilon & B\alpha & C\beta & D\gamma \end{array}$$

On the fourth day of the test, driver α uses gasoline E in car 2, driver β uses gasoline C in car 5, driver γ uses gasoline A in car 3, driver δ uses gasoline D in car 1, and driver ϵ uses gasoline B in car 4. The analysis with the additive model follows the same procedure as the analysis of the Latin square design. There is an additional sum of squares for Greek letters with $(p - 1)$ degrees of freedom and there are only $(p - 1)(p - 3)$ degrees of freedom for error.

Indeed, we could go further by having five different test courses and by having each of the cars driven over a different course each day. Courses could be added as a fifth factor by incorporating the third Latin square and using lower case letters a, b, c, d, and e to denote the five courses. That would leave only $(p - 1) \times (p - 4) = 4$ degrees of freedom for error. A sixth factor, perhaps predetermined average driving speeds, could be added by incorporating the fourth square and using Arabic numerals to denote the levels. There would then be no degrees of freedom left for error, and, without some prior estimate of the variance, testing hypotheses would present some difficulties. However, there is a chance that perhaps one or two of the main effects might stand out much larger than the others, which would point out the way to further experimentation. It may be observed that if the experiment is going to need five cars, each on all-day trips, we are certainly going to need at least five drivers. It does not, however, follow that we are obliged to add drivers as a fourth factor and to use a Graeco-Latin square. It depends upon the nature of the interest in the cars and drivers.

If we wish to investigate whether there are car differences and whether there are driver differences, then we must add drivers as a fourth factor and, unless we are prepared to increase the size of the experiment, the Graeco-Latin square would be appropriate, with only eight degrees of freedom for error. However, if we are interested only in comparing gasolines, and thus regard car and driver differences, if any, as nuisances to be eliminated, we should be better advised to use the same driver with the same car throughout the experiment. Then the driver effects and the car effects would both be absorbed in the row effects, which would now denote car-driver combinations, and we should have twelve degrees of freedom for error.

The Graeco-Latin squares and the hyper-Graeco-Latin squares are, like the Latin squares, fractional factorial designs. The Graeco-Latin square is a $(1/p^2)$ fraction of the p^4 design, i.e., the factorial with four factors, each at p levels. The square with no degrees of freedom left for error is a $(1/p^{p-1})$ fraction of the p^{p+1} factorial. Such a fraction, in which every degree of freedom is used for estimation, is said to be a saturated fraction.

This ability to add extra factors without increasing the size of the experiment appears attractive on the surface, but the experimenter should note that not only does the number of degrees of freedom for error diminish as more factors are added, but also the problems with nonadditivity, which are so bothersome in the Latin square layout, become even more troublesome. Graeco-Latin squares exist whenever p is a prime, or a power of a prime, and for some other values. Complete sets of $p - 1$ mutually orthogonal squares for $p = 3, 4, 5, 7, 8$, and 9 are given in the Fisher and Yates tables.

6.4 Sets of Orthogonal Latin Squares

Now we present a method, due to Bose (1938), of obtaining sets of $p - 1$ mutually orthogonal Latin squares of side p when p is a prime or a power of a prime. We associate each of the treatments with an element of the Galois field of $p = s^n$ elements, $GF(s^n)$, in a one-to-one correspondence. The elements of the field are ordered as $g_0 = 0$, $g_1 = 1$, $g_2 = x$, $g_3 = x^2, \ldots, g_{p-1} = x^{p-2}$ where x is a primitive element of the field, i.e., x is an element such that $x^{p-1} = 1$ and there is no other power $x^q = 1$, $0 < q < p$. Then the addition table forms a Latin square, and the other squares are obtained by rotating cyclically all the rows except the first.

EXAMPLE 1. $p = 4$. $GF(2^2)$ has elements $0, 1, x$, and $1 + x$. Arithmetic is carried out mod 2, and x is a primitive element. The irreducible polynomial of the second degree in the field is $x^2 + x + 1$. Equating this polynomial to zero, and recalling that $-1 \equiv 1 \pmod 2$, we have $x^2 = 1 + x$, and $x^3 = x + x^2 = 1$. In the addition table, the first row consists of the elements $0, 1, x$, and $1 + x$;

the second row is $0 + 1 = 1$, $1 + 1 = 0$, $x + 1 = 1 + x$, $1 + x + 1 = x$; and so on. The complete table is

0	1	x	$1 + x$
1	0	$1 + x$	x
x	$1 + x$	0	1
$1 + x$	x	1	0

Writing A, B, C, and D for the elements 0, 1, x, and $1 + x$, respectively, and rotating the rows other than the first cyclically, we obtain the following three squares. They form a mutually orthogonal set of squares of side 4.

$$\begin{matrix} A & B & C & D \\ B & A & D & C \\ C & D & A & B \\ D & C & B & A \end{matrix} \qquad \begin{matrix} A & B & C & D \\ C & D & A & B \\ D & C & B & A \\ B & A & D & C \end{matrix} \qquad \begin{matrix} A & B & C & D \\ D & C & B & A \\ B & A & D & C \\ C & D & A & B \end{matrix}$$

The first row of each square is the same. The second row of the second square is the same as the third row of the first square. The third row of the second square is the fourth row of the first square. The last row of the second square is the second row of the first square. The third square is obtained from the second square in the same way.

We now show that Bose's method is valid in the general case. We define the ith square of the set to be the square that has $g_i = x^{i-1}$ as the first element of the second row. The first element of the $(m + 1)$th row is $g_i.x^{m-1} = g_i g_m$. We write the ith square, therefore, as

$$\begin{matrix} 0 & 1 & \cdots & g_{p-1} \\ g_i & g_i + 1 & \cdots & g_i + g_{p-1} \\ g_i g_2 & g_i g_2 + 1 & \cdots & g_i g_2 + g_{p-1} \\ \vdots & \vdots & \cdots & \vdots \\ g_i g_{p-1} & g_i g_{p-1} + 1 & \cdots & g_i g_{p-1} + g_{p-1} \end{matrix}$$

We need to establish two properties:

Property 1. Such a square is a Latin square.

PROOF. Suppose that it is not a Latin square and that two of the elements that appear in the $(q + 1)$th row are identical, i.e., for some pair t, u, $(t \neq u)$, $g_i g_q + g_t = g_i g_q + g_u$. This implies $g_t = g_u$, which is false because the elements of the field are distinct. ∎

Property 2. The squares in the set are mutually orthogonal.

PROOF. Suppose that the ith and jth squares are not orthogonal. This means that when the jth square is superimposed on the ith square, one pair of elements occurs together in two of the cells. Suppose that these cells are in the $(q + 1)$th row and the $(t + 1)$th column and in the $(r + 1)$th row and the $(u + 1)$th column where $q \neq r$, $t \neq u$. In these two cells the elements of the ith square coincide to give

$$g_i g_q + g_t = g_i g_r + g_u,$$

and the elements of the jth square coincide to give

$$g_j g_q + g_t = g_j g_r + g_u.$$

Subtracting, we have $(g_i - g_j)g_q = (g_i - g_j)g_r$. This would imply that either $g_i = g_j$ or $g_q = g_r$ in contradiction. ∎

This method was also presented by Stevens (1939). In his paper, Stevens points out that the orthogonal set for $p = 9$ given by Yates in the 1948 edition of the Fisher and Yates tables is not isomorphic to the set that Stevens himself obtained using Galois fields, i.e., the one set cannot be obtained from the other by a reassignment of the letters to the treatments.

It is easily verified that if p is a prime, rather than s^n where $n > 1$, it is not necessary to find a primitive element of $GF(p)$ in order to obtain an initial square. Bose points out that we can, instead, work with the residue classes (mod p) and write the rows of the initial square in any order we wish.

EXAMPLE 2. $p = 5$. We shall now give two possible initial squares of side 5. The first of these is derived by the previous procedure. The elements of $GF(5)$ are 0, 1, 2, 3, and 4. Arithmetic is mod 5 and $x = 2$ is a primitive element. Its powers are $x^2 = 4$, $x^3 = 3$, and $x^4 = 1$. The first row (and column) of the addition table as we write it to form the initial square is, therefore, 0, 1, 2, 4, and 3.

The second initial square is obtained by writing the residue classes (mod 5) in the usual order 0, 1, 2, 3, and 4 and forming the addition table. The ith square of this set is obtained by re-ordering the rows of the initial square so that the first column is $0, i, 2i, 3i, \ldots$. Writing A, B, C, D, and E for 0, 1, 2, 3, and 4, respectively, produces the set of mutually orthogonal squares of side 5 given at the beginning of the chapter.

```
0 1 2 4 3     0 1 2 3 4
1 2 3 0 4     1 2 3 4 0
2 3 4 1 0     2 3 4 0 1
4 0 1 3 2     3 4 0 1 2
3 4 0 2 1     4 0 1 2 3
```

There are no similar methods of obtaining sets of $p - 1$ mutually orthogonal squares of side p when p is not a prime or a power of a prime. Indeed, Bruck and Ryser (1949) have shown that if $p \equiv 1$ or $2 \pmod 4$, and if the square free part of p contains at least one prime factor $q \equiv 3 \pmod 4$, then no orthogonal set of $p - 1$ squares exists. It had been conjectured by Euler that if $p \equiv 2 \pmod 4$, there did not exist a pair of orthogonal squares, and this conjecture was verified for the case $p = 6$ by Tarry (1901). However, Bose, Shrikhande, and Parker, in a series of papers, have shown that if $p \equiv 2 \pmod 4$ and $p > 6$, a pair of orthogonal squares always exists. A pair of orthogonal squares found by them (1960) for $p = 10$ follows.

A pair of orthogonal Latin squares of side 10.

0 0	4 7	1 8	7 6	2 9	9 3	8 5	3 4	6 1	5 2
8 6	1 1	5 7	2 8	7 0	3 9	9 4	4 5	0 2	6 3
9 5	8 0	2 2	6 7	3 8	7 1	4 9	5 6	1 3	0 4
5 9	9 6	8 1	3 3	0 7	4 8	7 2	6 0	2 4	1 5
7 3	6 9	9 0	8 2	4 4	1 7	5 8	0 1	3 5	2 6
6 8	7 4	0 9	9 1	8 3	5 5	2 7	1 2	4 6	3 0
3 7	0 8	7 5	1 9	9 2	8 4	6 6	2 3	5 0	4 1
1 4	2 5	3 6	4 0	5 1	6 2	0 3	7 7	8 8	9 9
2 1	3 2	4 3	5 4	6 5	0 6	1 0	8 9	9 7	7 8
4 2	5 3	6 4	0 5	1 6	2 0	3 1	9 8	7 9	8 7

6.5 Experiments Involving Several Squares

The single Latin square contains only p observations with each treatment. If the experimenter wishes to have more replications of each treatment, he may find it desirable to use several squares. The following analysis is valid whether the squares chosen are the same or different. We suppose that there are s squares.

Let y_{hijk} denote the observation, if there is one, made upon the kth treatment in the ith row and the jth column of the hth square. The model taken is

$$y_{hijk} = \mu + \pi_h + \rho_{i(h)} + \gamma_{j(h)} + \tau_k + (\pi\tau)_{hk} + e_{hijk},$$

where π_h is the effect of the hth square and $(\pi\tau)_{hk}$ represents the interaction between squares and treatments; $\rho_{i(h)}$ and $\gamma_{j(h)}$ are the row and column effects in the hth square. The side conditions on the parameters are $\sum_h \pi_h = 0$, $\sum_i \rho_{i(h)} = 0$ for each h, $\sum_j \gamma_{j(h)} = 0$ for each h, $\sum_k \tau_k = 0$, $\sum_k (\pi\tau)_{hk} = 0$ for each h, and $\sum_h (\pi\tau)_{hk} = 0$ for each k.

With these side conditions, the estimates of the parameters are

$$\hat{\mu} = y_{....}, \qquad \hat{\pi}_h = y_{h...} - y_{....}, \qquad \hat{\tau}_k = y_{...k} - y_{....},$$

$$\hat{\rho}_{i(h)} = y_{hi..} - y_{h...}, \qquad \hat{\gamma}_{j(h)} = y_{h.j.} - y_{h...},$$

$$(\widehat{\pi\tau})_{hk} = y_{h..k} - y_{h...} - y_{...k} + y_{....}.$$

The analysis of variance table is

TABLE 6.2
ANALYSIS OF VARIANCE

Source	SS	d.f.
Total	$\sum_h \sum_i \sum_j \sum_k y_{hijk}^2 - C$	$sp^2 - 1$
Squares	$\sum_h T_{h\cdots}^2/p^2 - C = S_s$	$s - 1$
Treatments	$\sum_k T_{\cdot\cdot\cdot k}^2/(sp) - C = S_t$	$p - 1$
$T \times S$	$\sum_h \sum_k T_{h\cdot\cdot k}^2/p - C - S_s - S_t$	$(s - 1)(p - 1)$
Rows in squares	$\sum_h \sum_i T_{hi\cdot\cdot}^2/p - \sum_h T_{h\cdots}^2/p^2$	$s(p - 1)$
Columns in squares	$\sum_h \sum_j T_{h\cdot j\cdot}^2/p - \sum_h T_{h\cdots}^2/p^2$	$s(p - 1)$
Residual	By subtraction	$s(p - 1)(p - 2)$

6.6 Change-Over Designs

In the previous section we considered the s squares as separate squares. We may also consider them as being arranged in a rectangle with p rows and sp columns. This situation typically occurs when time, in the form of periods, is a factor in the experiment. In experiments with dairy cattle, for example, we may test p treatments (such as kinds of feed) over p periods using sp cows. Each cow has each feed for one of the p periods, and the schedules are arranged so that during each period each feed is given to s cows. In the rectangular arrangement, the periods correspond to rows and the cows to columns. Alternatively, we may have p subjects receiving p treatments over sp periods; each subject receives each treatment for s of the periods; during each period each of the treatments is used on one of the subjects. In either case, we may let

$$y_{ijk} = \mu + \rho_i + \gamma_j + \tau_k + e_{ijk},$$

where $1 \leq i \leq p$, $1 \leq j \leq sp$, and $1 \leq k \leq p$.

TABLE 6.3
ANALYSIS OF VARIANCE

Source	SS	d.f.	EMS
Total	$\sum_i \sum_j \sum_k y_{ijk}^2 - C$	$sp^2 - 1$	
Rows	$\sum_i T_{i\cdot\cdot}^2/(sp) - C$	$p - 1$	$\sigma^2 + sp \sum_i \rho_i^2/(p - 1)$
Columns	$\sum_j T_{\cdot j\cdot}^2/p - C$	$sp - 1$	$\sigma^2 + p \sum_j \gamma_j^2/(sp - 1)$
Treatments	$\sum_k T_{\cdot\cdot k}^2/(sp) - C$	$p - 1$	$\sigma^2 + sp \sum_k \tau_k^2/(p - 1)$
Residual	By subtraction	$(sp - 2)(p - 1)$	σ^2

EXAMPLE. The following example is taken from Patterson (1950). In the experiment eighteen cows were fed, for consecutive five week periods, three rations: A, good hay; B, poor hay; and C, straw. The cows were divided into six blocks of three cows each, and a Latin square design was used for each block. The response recorded was the yield of milk in pounds for each five-week period. The analysis will be developed further in the next section.

We take for our model an adaptation of the previous model to include the blocking, and we shall now use Latin letters rather than Greek letters to denote the parameters, including the mean. (This is done merely for convenience and there is no deeper significance to the change.) We have

$$y_{hijk} = m + p_h + c_{i(j)} + b_j + t_k + e_{hijk},$$

where y_{hijk} is the yield during the hth period of the ith cow in the jth block; during that time the cow was receiving the kth treatment; p_h, $c_{i(j)}$, b_j, and t_k are the period, cow, block, and treatment effects.

Period	Block 1			Total of Period	Block 2			Total of Period
1	*A* 768	*B* 662	*C* 731	2161	*A* 669	*B* 459	*C* 624	1752
2	*B* 600	*C* 515	*A* 680	1795	*C* 550	*A* 409	*B* 462	1421
3	*C* 411	*A* 506	*B* 525	1442	*B* 416	*C* 222	*A* 426	1064
Total of Cow	1779	1683	1936	5398	1635	1090	1512	4237
	Block 3				Block 4			
1	*A* 1091	*B* 1234	*C* 1300	3625	*A* 1105	*B* 891	*C* 859	2855
2	*B* 798	*C* 902	*A* 1297	2997	*C* 712	*A* 830	*B* 617	2159
3	*C* 534	*A* 869	*B* 962	2365	*B* 453	*C* 629	*A* 597	1679
Total of Cow	2423	3005	3559	8987	2270	2350	2073	6693
	Block 5				Block 6			
1	*A* 941	*B* 794	*C* 779	2514	*A* 933	*B* 724	*C* 749	2406
2	*B* 718	*C* 603	*A* 718	2039	*C* 658	*A* 649	*B* 594	1901
3	*C* 548	*A* 613	*B* 515	1676	*B* 576	*C* 496	*A* 612	1684
Total of Cow	2207	2010	2012	6229	2167	1869	1955	5991

TABLE 6.4

ANALYSIS OF VARIANCE

Source	SS	d.f.
Total	2,756,705	53
Blocks	1,392,534	5
Cows in blocks	318,242	12
Periods	814,223	2
$P \times B$	57,791	10
Treatments	121,147	2
Errors	52,768	22

6.7 Change-Over Designs Balanced for Residual Effects

The change-over design that has just been discussed in the example is balanced for treatments. It is also a balanced design for the comparison or elimination of residual effects, and it was as an illustration of this that Patterson (1950) presented the data. When a cow that received treatment A for the first period is switched to treatment B for the second period, the yield in the second period may reflect not just the main effect of treatment B, but also the residual effect, if there is one, of treatment A.

In the design, blocks I and II together form a pair of orthogonal Latin squares,

$$\begin{matrix} A & B & C \\ B & C & A \\ C & A & B \end{matrix} \qquad \begin{matrix} A & B & C \\ C & A & B \\ B & C & A \end{matrix}$$

and the same is true for blocks III and IV and for blocks V and VI. In the pair of squares, each treatment is used once on each of the six cows. There are no residual effects during the first period. In the subsequent periods, the residual effect of each treatment appears four times. In each of the last two periods, the main (or direct) effect of each treatment is observed in two cows, and the residual effect of each treatment is also observed in two cows. Since no treatment can follow itself, no yield reflects both the main effect and the residual effect of the same treatment. This problem was first considered by Cochran et al. (1941). Further developments are due to Patterson (1950) and (1952), and Lucas (1951) and (1957). The designs are also applicable to experiments, such as those involving the rotation of crops, in which several treatments are applied in succession to the same experimental units.

In the general case with the dairy cattle, we make use of sp cows over p periods, and the feeding schedule is arranged in such a way that

(i) Each cow has each treatment for one period;
(ii) In each period each treatment is used on s cows;
(iii) Each treatment follows every other treatment s times.

We shall not consider blocks in our analysis; they present only a minor complication. The discussion given here is based on that of Patterson (1950).

A set of s of the cows receive the qth treatment in the last period, and therefore never experience the residual effect of that treatment. We call the kth of that subset of s cows the (kq)th cow. Let y_{hijkq} denote the yield of the (kq)th cow during the hth period in which she received the ith treatment, preceded, if $h > 1$, by the jth treatment. Then

$$y_{hijkq} = m + d_h + t_i + r_j + c_{kq} + e_{hijkq},$$

where d_h, t_i, r_j, and c_{kq} are the period, (direct) treatment, residual treatment, and cow effects, respectively. The side conditions on the parameters are

$$\sum_h d_h = 0, \qquad \sum_i t_i = 0, \qquad \sum_j r_j = 0, \qquad \sum_k \sum_q c_{kq} = 0.$$

In the normal equations we denote the various totals by G, D_h, T_i, R_j, and C_{kq}, and we have

$$G = sp^2\hat{m}, \qquad D_h = sp\hat{m} + sp\hat{d}_h,$$

$$T_i = sp\hat{m} + sp\hat{t}_i - s\hat{r}_i,$$

$$R_j = s(p-1)\hat{m} + s(p-1)\hat{r}_j - s\hat{d}_1 - \sum_k \hat{c}_{kj} - s\hat{t}_j,$$

$$C_{kq} = p\hat{m} + p\hat{c}_{kq} - \hat{r}_q.$$

We now write $\{C\}_i$ for the quantity $G - \sum_k C_{ki}$; $\{C\}_i$ is the sum of all the yields on all the cows that experience the ith residual effect. We have

$$\{C\}_i = sp(p-1)\hat{m} - p\sum_k \hat{c}_{ki} - s\hat{r}_i.$$

Let T_i' and R_i' denote the adjusted totals

$$T_i' = T_i - p^{-1}G = sp\hat{t}_i - s\hat{r}_i,$$

$$R_i' = R_i - p^{-1}\{C\}_i + p^{-1}D_1 - p^{-2}G = p^{-1}s(p^2 - p - 1)\hat{r}_i - s\hat{t}_i.$$

Solving these equations we have

$$\hat{t}_i = \frac{(p^2 - p - 1)T_i' + pR_i'}{sp(p+1)(p-2)}, \qquad \hat{r}_i = \frac{T_i' + pR_i'}{s(p+1)(p-2)}.$$

We turn now to the analysis of variance table.

Whether or not there are direct or residual treatment effects included in the model, we have $\hat{m} = G/(sp^2)$ and $\hat{d}_h = (D_h - sp\hat{m})/(sp)$. If, however, there are no residual effects in the model, the least squares estimate of c_{kq} becomes $(C_{kq} - p\hat{m})/p$, which is not the same as the estimate with our present model. We shall consider four models:

$$\Omega_0: \quad E(y_{hijkq}) = m + d_h + c_{kq},$$

$$\Omega_1: \quad E(y_{hijkq}) = m + d_h + c_{kq} + t_i + r_j,$$

$$\Omega_2: \quad E(y_{hijkq}) = m + d_h + c_{kq} + t_i,$$

$$\Omega_3: \quad E(y_{hijkq}) = m + d_h + c_{kq} + r_j.$$

Let S_n denote the sum of squares for regression under the nth model including the correction for the mean, $G^2/(sp^2)$. Then

$$S_0 = \sum_h D_h^2/(sp) + \sum_k \sum_q C_{kq}^2/p - G^2/(sp^2),$$

$$S_1 = \sum_h D_h^2/(sp) + \sum_k \sum_q C_{kq}\hat{c}_{kq} + \sum_i T_i\hat{t}_i + \sum_i R_i\hat{r}_i,$$

and

$$\begin{aligned} S_1 - S_0 &= \sum_i T_i\hat{t}_i + \sum_i R_i\hat{r}_i + \sum_k \sum_q (C_{kq}\hat{c}_{kq} - C_{kq}^2/p) + G^2/N \\ &= \sum_i T_i\hat{t}_i + \sum_i R_i\hat{r}_i - \sum_i \hat{r}_i\{C\}_i/p = \sum_i T_i'\hat{t}_i + \sum_i R_i'\hat{r}_i \\ &= \sum_i \{(p^2 - p - 1)T_i'^2 + 2pT_i'R_i' + p^2R_i'^2\}/\{sp(p + 1)(p - 2)\}. \end{aligned}$$

Under Ω_2, we have for the least squares estimate of t_i

$$\tilde{t}_i = T_i'/(sp),$$

so that

$$S_2 = S_0 + \sum_i T_i\tilde{t}_i = S_0 + \sum_i T_iT_i'/(sp) = S_0 + \sum_i T_i'^2/(sp),$$

and the sum of squares for direct effects ignoring residual effects is

$$S_2 - S_0 = \sum_i T_i'^2/(sp)$$

with $(p - 1)$ d.f., and the sum of squares for residual effects eliminating direct effects (i.e., for residual effects over and above the sum for direct effects alone) is

$$S_1 - S_2 = \sum_i (T_i' + pR_i')^2/sp(p + 1)(p - 2) = s(p + 1)(p - 2) \sum_i \hat{r}_i^2/p.$$

Under Ω_3, we have $s(p^2 - p - 1)\tilde{r}_i = pR_i'$, and

$$pS_3 = \sum_h D_h^2/s + \sum_k \sum_q C_{kq}^2 - G^2/(sp) - \sum_i \tilde{r}_i\{C\}_i + p \sum_i \tilde{r}_iR_i,$$

so that the sum of squares for residual effects ignoring direct effects is

$$S_3 - S_0 = \sum_i \tilde{r}_i(R_i - p^{-1}\{C\}_i) = p \sum_i R_i'^2/s(p^2 - p - 1),$$

and the sum of squares for direct effects, eliminating residual effects is

$$S_1 - S_3 = \frac{\sum_i ((p^2 - p - 1)T_i' + pR_i')^2}{sp(p + 1)(p - 2)(p^2 - p - 1)} = \frac{sp(p + 1)(p - 2)\sum \hat{t}_i^2}{p^2 - p - 1}.$$

Thus, to test the hypothesis that there are no residual effects we take the test statistic $\mathscr{F} = (S_1 - S_2)/M_e$, which has under the null hypothesis the $F[(p - 1), \phi]$ distribution. In the particular example that we have been considering, a term for block × period interaction has been removed. In the general case there are $(p - 1)$ d.f. each for periods, direct effects, and residual effects, and $(sp - 1)$ d.f. between cows; this leaves $\phi = sp^2 - (s + 3)p + 3$. If the sum of squares for block × period interaction is also taken out there is a further reduction of $(s - 1)(p - 1)$ d.f.

In the example the sum of squares listed for treatments is the sum of squares for direct effects neglecting residual effects. To obtain the sum of squares for residual effects eliminating direct effects we have, where $D_1 = 15313$,

	T_i	T_i'	R_i	$\{C\}_i$	$3R_i'$	$T_i' + 3R_i'$
A	13713	1201.33	7385	25297	−340.67	860.67
B	12000	−511.67	7036	23956	−46.67	−558.33
C	11822	−689.67	7801	25817	387.33	−302.33
	37535		22222			

Then $(S_1 - S_2) = [(860.67)^2 + (558.33)^2 + (302.33)^2]/72 = 15{,}888$. The sum of squares for error is $52{,}768 - 15{,}888 = 36{,}880$, and $M_e = 1{,}844$.

6.8 Designs for Experiments with Residual Effects

The design in the example just considered was a pair of orthogonal Latin squares repeated three times. Williams (1949) has investigated designs for this type of experiment. He has shown that designs balanced for the residual effect of the preceding treatment and satisfying the requirements listed in the previous section can be obtained with only p cows, or replications, if p is even, and with $2p$ cows if p is odd, as was the case in the example. His designs are based upon Latin squares.

EXAMPLE 1. p even. We let the treatments be represented by the residue classes, mod p, and we illustrate Williams' method by the following example with $p = 6$. The rows correspond to periods and the columns to cows.

$$\begin{array}{cccccc}
0 & 1 & 2 & 3 & 4 & 5 \\
1 & 2 & 3 & 4 & 5 & 0 \\
5 & 0 & 1 & 2 & 3 & 4 \\
2 & 3 & 4 & 5 & 0 & 1 \\
4 & 5 & 0 & 1 & 2 & 3 \\
3 & 4 & 5 & 0 & 1 & 2
\end{array}$$

The first step in Williams' procedure is to find a suitable initial (first) column. Then, each subsequent column is obtained from the previous column by adding one to each entry and reducing mod p. The elements of the initial column in the example are, in order, 0, 1, 5, 2, 4, and 3. The differences between adjacent pairs, reduced mod 6, are 1, 4, 3, 2, and 5. Every number other than zero appears exactly once, and it follows that as we take the several columns each treatment will be preceded exactly once by each of the other treatments. Thus, treatment 3 will be preceded by treatment 2 in the column where 3 appears in the second row because the difference between the second and first entries in any column is $1 = 3 - 2$.

We seek, therefore, for the initial column a permutation of the elements $0, 1, \ldots, p - 1$ such that in the $p - 1$ differences between adjacent elements each nonzero member, mod p, appears exactly once. When p is even, such an arrangement is given by $0, 1, p - 1, 2, p - 2, 3, p - 3, \ldots, p/2$.

This is not, however, a unique solution. If we multiply each element of the initial column by an integer that is prime to p, and reduce mod p, we obtain another initial column. In the example, multiplying by 5 gives the new column 0, 5, 1, 4, 2, and 3. Another initial column for $p = 6$ is 0, 2, 1, 4, 5, and 3. Multiplying this by 5 gives 0, 4, 5, 2, 1, and 3.

EXAMPLE 2. p odd. When p is odd, the sum $p(p - 1)/2$ of the elements is divisible by p, and if each difference is to occur once in a column beginning with zero the last entry would also be zero, which is not satisfactory. We have, however, seen an example of a scheme for $p = 3$ with two Latin squares in which half the differences appeared in one square and half in the other.

Williams makes use of two Latin squares of side p. One of the squares is derived from an initial column in which each of the odd elements appears twice as a difference. The other initial column gives each even difference twice. To obtain the odd differences, we take the initial column $0, 1, p - 1, 2, p - 2, \ldots, (p - 1)/2, (p + 1)/2$. For the even differences, we take the same initial column, but we write it in reverse order.

Using this procedure we obtain for $p = 5$ the following design using ten cows.

```
0 1 2 3 4    3 4 0 1 2
1 2 3 4 0    2 3 4 0 1
4 0 1 2 3    4 0 1 2 3
2 3 4 0 1    1 2 3 4 0
3 4 0 1 2    0 1 2 3 4
```

In a subsequent paper Williams (1950) goes on to consider designs balanced with respect to the residual effects of the previous two treatments.

Exercises

1. Derive the following missing plot formula for the Latin square design; G', R', C', and L' denote the sum of the $(p^2 - 1)$ actual observations taken and $(p - 1)$

actual observations occurring in the same row, column, and with the same letter as the missing plot.

$$x = \frac{p(R' + C' + L') - 2G'}{(p - 1)(p - 2)}.$$

Show that if the missing plot received the ith treatment and is replaced by a value calculated by the previous formula then

$$V(\hat{\tau}_h - \hat{\tau}_i) = \sigma^2\left(\frac{2}{p} + \frac{1}{(p - 1)(p - 2)}\right).$$

2. Youden and Hunter (1955) suggested modifying the Latin square design in the following way to obtain an unbiased estimate of the error. They chose a square in which each variety appeared exactly once on the diagonal and duplicated those plots, thus obtaining an estimate of error between duplicates with p degrees of freedom. Because the design is no longer orthogonal the analysis calls for testing subhypotheses: rows, columns (after rows), and treatments (after columns and rows). Derive the analysis.
3. Suppose that one entire row is omitted from a Latin square design and modify the analysis accordingly.
4. For the data for the dairy cattle experiment, compute

   (i) the sum of squares for residual effects ignoring direct effects, and
   (ii) the sum of squares for direct effects eliminating the residual effects.

CHAPTER 7

Factors with Two or Three Levels; Confounding

In the previous chapters we have developed the analysis of variance for the general multifactorial experiment. We finished up with algorithms for obtaining degrees of freedom, sums of squares, and expected mean squares for fixed, mixed, or random models. The only blocking situation considered was that of the randomized complete block, in which case the blocks appeared in the algorithms as the levels of another (random) factor.

In this chapter we shall take another look at the factorial experiment. The 2^n and 3^n factorials will be introduced. We shall also consider blocking in the case in which the blocks are not large enough to contain all the treatments, a situation which we saw for the first time in the split-plot design where each whole plot contained only one level of each of the main treatments. We shall assume throughout that all the factors have fixed effects, that the block effects are random effects with mean zero and variance σ_B^2, uncorrelated with each other or with the error terms, and that there is no interaction between blocks and treatments. We mean by the last statement that if y_{ij} denotes an observation on the ith treatment in the jth block, and y_{hj} denotes an observation on the hth treatment in the same block, then, apart from random error, the difference $y_{hj} - y_{ij}$ remains the same for all values of j.

The reader will not fail to notice the enormous contribution made by Yates to the development of the topics covered in this chapter. In particular, numerous references will be found to his pioneer monograph, *The Design and Analysis of Factorial Experiments*, which was published in 1937 as Technical Communication 35 by the Imperial Bureau of Soil Science, Harpenden, England.

7.1 Factors at Two Levels

Consider an experiment with two factors A and B, each at two levels. There are four sets of experimental conditions: low A, low B; high A, low B; low A, high B; high A, high B. Yates calls these the four treatment combinations, and denotes them by (1), a, b, and ab; the letter a appears whenever the factor A is at its high level. The use of the words *high* and *low* is arbitrary and has no particular significance in the analysis. With a quantitative factor, one level is clearly higher than the other; with a qualitative factor, we arbitrarily call one level high and the other low. We shall also use the notation (1), a, b, and ab to denote the mean responses at those treatment combinations. Thus, the symbol a is used in two senses: to denote the set of conditions having A at its high level and B at its low level, and also the average of the observations taken under those conditions. The notation can be extended to any number of factors. With four factors, the treatment combination cd has C and D at their high levels and A and B at their low levels. In the earlier chapters, the response cd would have been denoted by something like y_{1122}.

The usual model has

$$y_{ijk} = \mu + \alpha_i + \beta_j + (\alpha\beta)_{ij} + e_{ijk}.$$

(We shall use β later to denote something else.)

The side condition $\alpha_1 + \alpha_2 = 0$ gives $\alpha_1 = -\alpha_2$; we write $\alpha_2 = \alpha$. Then $\hat{\mu} = y_{...}$ and $\hat{\alpha} = y_{2..} - y_{...} = (y_{2..} - y_{1..})/2$, so that

$$4\hat{\mu} = ab + a + b + (1),$$

$$4\hat{\alpha} = 2ab + 2a - ab - a - b - (1) = ab + a - b - (1).$$

The corresponding sums of squares are $C = G^2/N$, as usual, and $S_A = N\hat{\alpha}^2$ where N is the total number of observations and we assume that each of the treatment combinations is observed the same number of times. Similarly, writing $\beta = \beta_2 = -\beta_1$ and $(\alpha\beta)_{11} = (\alpha\beta)_{22} = -(\alpha\beta)_{12} = -(\alpha\beta)_{21} = (\alpha\beta)$, we also have $S_B = N\hat{\beta}^2$ and $S_{AB} = N(\widehat{\alpha\beta})^2$.

We call $ab + a - b - (1)$ the A contrast. The B and AB contrasts are

$$4\hat{\beta} = ab - a + b - (1), \qquad 4(\widehat{\alpha\beta}) = ab - a - b + (1).$$

Symbolically they can be written $(a - 1)(b + 1)$, $(a + 1)(b - 1)$, and $(a - 1) \times (b - 1)$. The three contrasts are mutually orthogonal. Henceforth we shall assume, in the absence of a statement to the contrary, that we have only one observation on each treatment combination.

An alternative approach, which we shall use later, is to associate with the factors A and B two coordinates, x_1 and x_2; x_1 takes the value $+1$ when A is at its high level and the value -1 when A is at its low level; similarly, $x_2 = +1$ at high B and $x_2 = -1$ at low B. The four treatment combinations are thus $(-1, -1)$, $(+1, -1)$, $(-1, +1)$, and $(+1, +1)$. Since the coordinates are ± 1, it is convenient to write $(- -)$, $(+ -)$, $(- +)$, $(+ +)$. We can now fit the model

$$y_{ij} = \beta_0 + \beta_1 x_1 + \beta_2 x_2 + \beta_{12} x_1 x_2 + e_{ij},$$

and we shall have $\beta_0 = \mu$, $\beta_1 = \alpha$, $\beta_2 = \beta$, $\beta_{12} = (\alpha\beta)$.

The interaction contrast may be written in two ways, both of which provide insights into its nature. Writing it as $\{ab - b - a + (1)\}$ we see it as the difference between the increase in response obtained by changing the levels of A at high B and the corresponding increase at low B. Writing it as $\{ab - a\} - \{b - (1)\}$ shows it as the difference between the B effects at the two levels of A. In either case, we have the same story: the interaction measures the failure of the main effect of A (or B) to remain constant over the levels of B (or A). An interesting question then arises, one that is also relevant when the factors have more than two levels. What does it mean to have a significant A effect when there is also a significant AB interaction?

EXAMPLE. Consider the following three situations:

(i)				(ii)				(iii)			
(1)	40	*a*	60	(1)	40	*a*	80	(1)	60	*a*	40
b	60	*ab*	80	*b*	60	*ab*	60	*b*	60	*ab*	80.

Situation (i) is simple. At either level of B, increasing the level of A from low to high causes an increase of 20 units in the response. We have $4\hat{\alpha} = 40$ and $4\widehat{(\alpha\beta)} = 0$. In situation (ii) we still have $4\hat{\alpha} = 40$, but now $4\widehat{(\alpha\beta)} = -40$. On the average, assuming that a larger response is more desirable than a smaller one, it is a good idea to work at the high level of A. Actually, however, there is no difference between the responses at the two levels of A at high B, and all the forty units difference that went into $4\hat{\alpha}$ was found at low B. In situation (iii) $\hat{\alpha} = 0$ and $4\widehat{(\alpha\beta)} = 40$. It is not correct to say that changing the level of A has no effect on the response. Changing the level of A from low to high produces a decrease in the response at low B and an equal increase in the response at high B. In general we must take the point of view that a significant value of $\hat{\alpha}$, or of the ratio M_A/M_e, indicates that the responses are, *on the average*, higher at one level of A than they are at the other. A significant interaction indicates that looking at the average change in response is not enough, and that the effect of changing levels of A is not the same at all levels of B; equivalently, the effect of changing levels of B is not the same at all levels of A.

If we add a third factor, C, also at two levels, we introduce a three-factor interaction. The model becomes

$$y_{ijk} = \mu + \alpha_i + \beta_j + \gamma_k + (\alpha\beta)_{ij} + (\alpha\gamma)_{ik} + (\beta\gamma)_{jk} + (\alpha\beta\gamma)_{ijk} + e_{ijk},$$

and, proceeding as before, we have

$$8\hat{\mu} = abc + ab + ac + bc + a + b + c + (1),$$

$$8\hat{\alpha} = abc + ab + ac - bc + a - b - c - (1),$$

$$8\widehat{(\alpha\beta)} = abc + ab - ac - bc - a - b + c + (1),$$

which may be written symbolically as $(a + 1)(b + 1)(c + 1)$, $(a - 1)(b + 1) \times (c + 1)$, and $(a - 1)(b - 1)(c + 1)$.

The eight interaction terms $(\alpha\beta\gamma)_{ijk}$ have the same absolute value. If they are not all zero, four of them are positive and the other four are negative. The side conditions

$$\sum_i \sum_j (\alpha\beta\gamma)_{ijk} = 0, \qquad \sum_i \sum_k (\alpha\beta\gamma)_{ijk} = 0, \qquad \sum_j \sum_k (\alpha\beta\gamma)_{ijk} = 0$$

lead to

$$(\alpha\beta\gamma)_{222} = -(\alpha\beta\gamma)_{221} = -(\alpha\beta\gamma)_{212} = -(\alpha\beta\gamma)_{122} = (\alpha\beta\gamma)_{211} = (\alpha\beta\gamma)_{121}$$
$$= (\alpha\beta\gamma)_{112} = -(\alpha\beta\gamma)_{111} = (\alpha\beta\gamma),$$

whence

$$8\widehat{(\alpha\beta\gamma)} = abc - bc - ac - ab + a + b + c - (1).$$

Alternatively, we may write the model as

$$E(y) = \beta_0 + \beta_1 x_1 + \beta_2 x_2 + \beta_3 x_3 + \beta_{12} x_1 x_2$$
$$+ \beta_{13} x_1 x_3 + \beta_{23} x_2 x_3 + \beta_{123} x_1 x_2 x_3.$$

Then $8\hat{\beta}_{123} = \sum$ (responses at those points with $x_1x_2x_3 = 1$) $- \sum$ (responses at those points with $x_1x_2x_3 = -1$) $= 8\widehat{(\alpha\beta\gamma)}$. Symbolically we have $8\widehat{(\alpha\beta\gamma)} = 8\hat{\beta}_{123} = (a - 1)(b - 1)(c - 1)$.

There are several notations that can be used to denote the ABC contrasts and, similarly, other contrasts. We shall use either $\{ABC\}$ or $\{x_1x_2x_3\}$. We shall also use $\{x_1x_2x_3 = +1\}$ and $\{x_1x_2x_3 = -1\}$ to denote the sums of the responses at the points at which $x_1x_2x_3 = +1$ or -1, respectively. Then $\{ABC\} = \{x_1x_2x_3\} = \{x_1x_2x_3 = +1\} - \{x_1x_2x_3 = -1\}$. Yates has denoted ABC by $A.B.C.$.

The ABC contrast can be written $(abc - ac - bc + c) - (ab - a - b + (1))$, which is the difference between the AB contrast at high C and the AB contrast at low C. It may also be considered as the difference between the AC contrasts at the two levels of B, or the difference between the BC contrasts at the two levels of A.

The variance of the estimate of any effect, either main effect or interaction, is $N^{-1}\sigma^2$. Each point contributes with the same weight to the estimate of each effect. In particular, so far as the estimation of α is concerned, we have lost nothing in precision by adding B and C to the experiment because the variance of $\hat{\beta}_1$ in this experiment is the same as it would have been if we had not had the factors B and C and had made instead only four runs at low A and four runs at high A. Alternatively, we may take the point of view that the addition of the extra factor C to the experiment also adds two more points at high A and at low A, and two more at high B and low B for estimating A, B, and AB. This is an example of what is sometimes referred to as the hidden replication property of the factorial experiment.

We have in the 2^3 experiment four direct comparisons to estimate the main effect of A, namely, $a - (1)$, $ab - b$, $ac - c$, and $abc - bc$. We obtain a vivid pictorial representation when the experimental points are portrayed as the vertices of a cube with coordinates $(\pm 1, \pm 1, \pm 1)$.

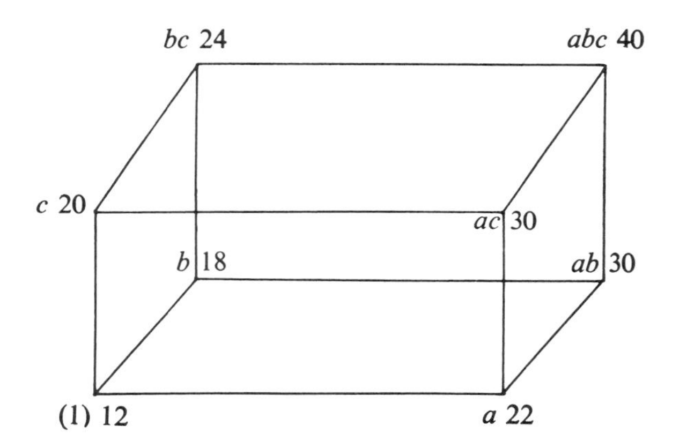

The four constituent A contrasts are 10, 12, 10, and 16. The overall A contrast is the sum, 48, with an average of 12. At low B we have A contrasts 10 and 10; at high B we have 12 and 16, giving an interaction contrast $28 - 20 = 8$. At low C the AB contrast is $12 - 10 = 2$; at high C the AB contrast is $16 - 10 = 6$; the ABC contrast is $6 - 2 = 4$.

7.2 Yates' Algorithm and the General 2^n Design

The 2^n design is the complete factorial for n factors, each at two levels. We have now seen enough to be able to tell how the derivation of the contrasts will proceed. If ψ is some effect (and we emphasize that we use the term *effect* to embrace both main effects and interactions), the ψ contrast will be the symbolic product given by

$$\{\psi\} = 2^n\hat{\psi} = (a \pm 1)(b \pm 1)(c \pm 1)\ldots,$$

where the sign of ± 1 is negative for all those letters appearing in ψ and positive elsewhere. Thus, in a 2^6 experiment we should have

$$64\hat{\beta}_{135} = 64(\widehat{ACE}) = (a - 1)(b + 1)(c - 1)(d + 1)(e - 1)(f + 1).$$

We shall, henceforth, use upper case letters such as ACE to denote either the effects or their estimates in a particular experiment. We shall no longer write the hat over ACE unless its omission would cause confusion. Some statisticians prefer to take as the effects twice the corresponding regression coefficients, e.g., $A = 2\beta_1$ rather than $A = \beta_1$ as we do. Their procedure corresponds to a change from $x_1 = -1$ to $x_1 = +1$ rather than a change from $x_1 = 0$ to $x_1 = +1$.

Based on the symbolic product rule, Yates developed an algorithm for computing the contrasts in a 2^n design. We now illustrate the algorithm with the data for the 2^3 design. There are several steps.

(i) The data are listed in the standard order (1), a; then b, ab; then c, ac, bc, abc, and so on; they are entered in this order in column 1.

(ii) The first 2^{n-1} entries in any of the subsequent columns are obtained by addition. The entry in the ith row of the $(k + 1)$th column is the sum of the $(2i - 1)$th and the $(2i)$th entries in the kth column.

(iii) The next 2^{n-1} entries are differences. The $(2^{n-1} + i)$th entry in the $(k + 1)$th column is obtained from the kth column by subtracting the $(2i - 1)$th entry from the $(2i)$th entry. It is easy to make a mistake in the sign at this stage.

(iv) The entries in the $(n + 1)$th column are the grand total M, followed by the contrasts in standard order.

(v) W, X, Y, and Z are part of the checking procedure. W is the sum of the entries in rows $1, 3, \ldots, 2^{n-1} - 1$, and X is the sum of the entries in rows $2, 4, \ldots, 2^{n-1}$. Y and Z are the corresponding sums for the second half of the table. The check in the calculations is that in any column $X + W$ is equal to $X + W + Z + Y$ in the previous column, and $Z + Y$ is equal to $X - W + Z - Y$ in the previous column.

(vi) We may add an extra column, the $(n + 2)$th, for the sum of the squares of the contrasts. The entries in this column are obtained by squaring the entries in the $(n + 1)$th column and dividing by 2^n. The sum of the entries in this column is equal to the sum of the squares of the original observations.

The calculations in this algorithm are notoriously error prone, especially in respect to wrong signs used in recording the differences. Three other methods of checking the calculations are given in the literature. Good (1960), in an addendum to his earlier paper (1958) on the topic, suggests reversing the order of the elements of the $(n + 1)$th column so as to begin with ABC and end with M. The Yates procedure is now carried out on these figures, and the new $(n + 1)$th column will have for its entries the original data, multiplied by 2^n, in the reverse order. For the 2^2 design with (1) = 12, a = 22, b = 18, ab = 30, we should have

(1)	12	34	82	M	2	16	120 = 4ab
a	22	48	22	A	14	104	72 = 4b
b	18	10	14	B	22	12	88 = 4a
ab	30	12	2	AB	82	60	48 = 4(1)

It is easily shown that the sum of the entries in the $(n + 1)$th column should be 2^n times the last entry in the data column, e.g., $82 + 22 + 14 + 2 = 4 \times 30 = 120$. Quenouille (1955) points out that this "check" does not assure that the calculations are correct. Consider again a 2^2 design and replace the first and third entries in the second column by any numbers x and y. The entries in the third column will be $ab + b + x$, $ab - b + y$, $ab + b - x$, and $ab - b - y$. No matter what values are given to x and y, the entries in the third column sum to $4ab$.

Quenouille (1953) suggests using the fact that the sum of the squares of the entries in the $(n + 1)$th column is 2^n times the sum of the squares of the data. It is obvious that this check is vulnerable to errors of sign in the $(n + 1)$th column. However, Rayner (1967) has shown that this "check" holds "true" in spite of mistakes in sign at any stage of the algorithm. To illustrate this we return to the 2^2 example and consider what happens if an error is made in the sign of the last entry in the second column. We should then have

(1)	12	34	82
a	22	48	−2
b	18	10	14
ab	30	−12	−22

and $82^2 + (-2)^2 + (14)^2 + (-22)^2 = 7408 = 4 \times 1852$. Yates' Algorithm for a 2^3 design follows.

	(1)	(2)	(3)	(4)		(5)
(1)	12	34	82	196	M	4802
a	22	48	114	48	A	288
b	18	50	22	28	B	98
ab	30	64	26	8	AB	8
c	20	10	14	32	C	128
ac	30	12	14	4	AC	2
bc	24	10	2	0	BC	0
abc	40	16	6	4	ABC	2

	(1)	(2)	(3)	(4)
W	30	84	104	
X	52	112	140	
Y	44	20	16	
Z	70	28	20	
$X + W$	82	196	244	280
$X - W$	22	28	36	
$Z + Y$	114	48	36	40
$Z - Y$	26	8	4	
$X + W + Z + Y$	196	244	280	
$X - W + Z - Y$	48	36	40	

7.3 The 3^n Series of Factorials

The 3^n factorial experiment has n factors, each at three levels. An example of a 3^3 experiment appeared in Chapter 5. We denote the levels of the factors by 0, 1, and 2. Then the treatment combination with A and C at level 1 and B at level 2 may be denoted by 121 or $a_1b_2c_1$ or ab^2c. We shall use either of the first two notations as convenient, and we shall sometimes not mention factors at their low levels, writing b_1c_2 for the point with A at level 0, B at level 1, and C at level 2.

To illustrate the 3^2 experiment we take some of the data from the 3^3 experiment of Chapter 5. Subtracting 80 and rounding off we obtain

(1)	9,	a_1	15,	a_2	20,
b_1	7,	a_1b_1	5,	a_2b_1	13,
b_2	6,	a_1b_2	1,	a_2b_2	15.

TABLE 7.1
Analysis of Variance

Total	8 d.f.	$1211 - 920.1 =$	290.9
S_A	2	$1076.3 - 920.1 =$	156.2
lin A	1		112.7
quad A	1		43.5
S_B	2	$1015 - 920.1 =$	94.9
lin B	1		80.7
quad B	1		14.2
S_{AB}	4		39.8

The interaction sum of squares is usually subdivided in one of two ways. The first method involves the use of orthogonal Latin squares. As was pointed out in

Chapter 6, a ($p \times p$) Latin square is a square array of side p in which p letters are arranged in such a way that each letter appears exactly once in each row and each column. Two particular Latin squares of side 3 are

$$\begin{matrix} P & Q & R \\ Q & R & P \\ R & P & Q \end{matrix} \qquad \text{and} \qquad \begin{matrix} P & Q & R \\ R & P & Q \\ Q & R & P. \end{matrix}$$

These two squares have an additional property. If the second square is superimposed on the first square, each letter in the first square will appear exactly once with each letter in the second square. Such a pair of squares is said to be orthogonal. If we denote the coordinates (levels) for A and B by x_1 and x_2, respectively, P occupies in the first square those places for which $x_1 + x_2 \equiv 0 \pmod 3$, while for Q we have $x_1 + x_2 \equiv 1 \pmod 3$, and for R, $x_1 + x_2 \equiv 2 \pmod 3$. The totals for the letters are P 23, Q 37, R 31. Any contrast between the letter totals is orthogonal to any contrast in the A totals or in the B totals. We can, therefore, calculate a sum of squares between the letters in square 1 with 2 d.f. as $(23^2 + 37^2 + 31^2)/3 - 920.1 = 32.9$. Alternatively, we could take two orthogonal contrasts in the totals such as $P - Q$ and $P + Q - 2R$ and add their individual sums of squares to obtain $(-14)^2/6 + (-2)^2/18 = 32.9$. For the second square, the letter totals are P 29, Q 34, R 28. Any contrast in these totals is orthogonal to any of the previous contrasts, and we have another sum of squares, also with 2 d.f.: $(29^2 + 34^2 + 28^2)/3 - 920.1 = 6.9$. The sum of the two components is $32.9 + 6.9 = 39.8 = S_{AB}$. Because their derivations involve the sets of points defined by $x_1 + x_2 \equiv 0, 1, 2 \pmod 3$ and by $x_1 + 2x_2 \equiv 0, 1, 2 \pmod 3$, respectively, they are sometimes called the sums of squares for AB and AB^2 interactions respectively.

For the 3^3 design we may subdivide S_{AB}, S_{AC}, and S_{BC} in the same way. S_{ABC} with 8 d.f. can be divided into four constituent sums of squares, each with 2 d.f., by breaking up the twenty-seven points into three sets of nine points each in four orthogonal ways. We then have

$$\begin{aligned} ABC&: \quad x_1 + x_2 + x_3 \equiv 0, 1, 2 \pmod 3, \\ AB^2C&: \quad x_1 + 2x_2 + x_3 \equiv 0, 1, 2 \pmod 3, \\ ABC^2&: \quad x_1 + x_2 + 2x_3 \equiv 0, 1, 2 \pmod 3, \\ AB^2C^2&: \quad x_1 + 2x_2 + 2x_3 \equiv 0, 1, 2 \pmod 3. \end{aligned}$$

Yates calls the sums of squares for AB and AB^2 in the 3^2 design the sums of squares for the J diagonals and the I diagonals, respectively. He calls the classifications that we have used for the 3^3 design Z, X, Y, and W, respectively.

The second approach is to split up the sum S_{AB} into single degrees of freedom for lin A lin B, and so on. We should then have the following four component sums of squares: lin A lin B 1.0; lin A quad B 5.3; quad A lin B 33.3; quad A

quad B 0.1. There is no connection between these sums of squares and the sums AB and AB^2. The four components that we have just derived can also be obtained by fitting the model

$$y_{ij} = \gamma_0 + \gamma_{10}u_1 + \gamma_{20}u_2 + \gamma_{01}v_1 + \gamma_{02}v_2 + \gamma_{11}u_1v_1 + \gamma_{12}u_1v_2 + \gamma_{21}u_2v_1 + \gamma_{22}u_2v_2 + e_{ij},$$

where $u_1 = x_1 - 1$, $u_2 = 3u_1^2 - 2$, $v_1 = x_2 - 1$, and $v_2 = 3v_1^2 - 2$.

Yates' algorithm for the 2^n designs may be adapted to the analysis of the 3^n designs. We illustrate the procedure by applying it to the 3^2 design. The procedure leads to the lin A, lin A quad B contrasts, and so on. The steps are similar to those of the original algorithm. We begin by arranging the data in standard order: (1), a_1, a_2, b_1, a_1b_1, a_2b_1, b_2, a_1b_2, a_2b_2, c_1, a_1c_1, $a_2c_2, \ldots$, and so on. The entries in the $(k + 1)$th column are obtained from those in the preceding column. Denote the entries in the kth column by k_i. The entries in the $(k + 1)$th column fall into three groups. In the first group the ith entry is $k_{3i} + k_{3i-1} + k_{3i-2}$; in the second group the ith entry is $k_{3i} - k_{3i-2}$; in the last group the ith entry is $k_{3i} - 2k_{3i-1} + k_{3i-2}$. The sums used in the check are $S_1 = k_1 + k_4 + k_7 + \cdots$, $S_2 = k_2 + k_5 + k_8 + \cdots$, $S_3 = k_3 + k_6 + k_9 + \cdots$, $X = S_1 + S_2 + S_3$, and $Y = S_1 - S_2 + 3S_3$. The check is that X in any column should equal Y in the preceding column.

(1)	9	44	91	M
a_1	15	25	26	lin A
a_2	20	22	28	quad A
b_1	7	11	−22	lin B
a_1b_1	5	6	−2	lin A lin B
a_2b_1	13	9	20	quad A lin B
b_2	6	−1	16	quad B
a_1b_2	1	10	8	lin A quad B
a_2b_2	15	19	−2	quad A quad B
S_1	22	54	85	
S_2	21	41	32	
S_3	48	50	46	
X	91	145	163	
Y	145	163		

7.4 Confounding

When we first discussed blocking we considered only the situation in which the blocks were complete, which meant that each block contained as many points or plots as there were treatments or treatment combinations. We turn our attention now to the situation which arises in a factorial experiment when the block size is too small to accomodate one plot with each of the treatment combinations, but is large enough for some fraction of them. We begin by discussing a 2^2 experiment in two blocks of two points each. We shall refer to the experimental points as points or runs; the latter is a common usage among chemical engineers who speak of making a run on their pilot plant at some set of conditions.

We have three choices for the division into blocks:

(i) (1), a; b, ab;
(ii) (1), b; a, ab;
(iii) (1), ab; a, b.

The revised model, including the block effects, may be written

$$y_{ijg} = \beta_0 + \beta_1 x_1 + \beta_2 x_2 + \beta_{12} x_1 x_2 + \delta_g + e_{ij},$$

where y_{ijg} is the response at the ith level of A and the jth level of B in the gth block, and δ_g is the gth block effect, which we now treat as fixed. The usual side condition $\sum_g \delta_g = 0$ becomes $\delta_1 + \delta_2 = 0$, and so we write $\delta_1 = -\delta$ and $\delta_2 = +\delta$.

Consider blocking scheme (i). The experimental points in block I are both at low B. The experimental points in block II are both at high B. Hence, the expectation of the B contrast is (using the suffix 1 to denote the low level of a factor and 2 for the high level)

$$E\{B\} = E(y_{22} + y_{12} - y_{21} - y_{11}) = 4\beta_2 + 4\delta.$$

On the other hand, $E\{A\} = 4\beta_1$ and $E\{AB\} = 4\beta_{12}$.

Another way of looking at the situation is to say that, since both points in one block are at high B and both points in the other block are at low B, we cannot, when we look at the B contrast, tell whether the observed difference results from the B effect, from block differences, or from a combination of both. On the other hand, each block contains one treatment combination at high A and one at low A, and so the block effects δ cancel out; they also cancel out for the AB contrast. We say, therefore, that the B effect is completely confounded, i.e., confused, with blocks. In scheme (ii), A is confounded but B and AB are not. In practice most experimenters would choose scheme (iii) because here it is the two-factor interaction AB that is confounded, while the main effects A and B are clear of the block effects.

When we consider dividing the 2^3 design into two blocks of four runs each, there are more options; indeed, there are thirty-five possible ways in which to choose three treatment combinations out of seven to be put in the same block as (1). One possibility is to take as the first block (1), a, b, c, which is the traditional one-at-a-time experiment (inasmuch as we start at the base point and vary

A, B, and C singly in turn). With this blocking the expected value of the contrast between block totals is

$$E\{abc + ab + ac + bc - a - b - c - (1)\}$$
$$= 8\delta + 4\beta_1 + 4\beta_2 + 4\beta_3 - 4\beta_{123}.$$

The expectation of the A contrast is $8\beta_1 + 4\delta$. All three main effects and the three-factor interaction are confounded with blocks.

We can, however, arrange the blocking so that one and only one of the effects is confounded. This we achieve by taking any one of the seven contrasts for estimating effects, and placing in one block all the treatment combinations that appear with a positive sign in the contrast and in the other block those treatment combinations that have a negative sign. The chosen contrast will then be identical with the difference between block totals, and the chosen effect will be completely confounded with block effects. However, the estimating contrasts are mutually orthogonal, and so the other contrasts are orthogonal to the contrast in the block totals; the other effects are clear of the block biases.

If we decided to confound ABC with blocks, the division would be

block I abc, a, b, c; block II $ab, ac, bc, (1)$.

We note that in each block there are exactly two points at the high level of A and two at the low level; the same is true for B and C. Block I consists of all the treatment combinations with positive signs in the ABC contrast; we shall call them the set of points defined by $\text{I} = +ABC$. They are also the set of points for which $x_1x_2x_3 = +1$. The other block contains the set of points defined by $\text{I} = -ABC$ or by $x_1x_2x_3 = -1$. In general, let P be any effect. We shall call the set of points that have positive (or negative) signs in the P contrast the set (or fraction of the 2^n points) defined by $\text{I} = +P$ (or by $\text{I} = -P$).

The chosen contrast is said to be the defining contrast of the confounding scheme. The analysis proceeds as before, using Yates' algorithm, except that the sum of squares for the defining contrast is discarded, since it is biased by the block effects. The extension to blocking the general 2^n design into two blocks of 2^{n-1} points each is obvious. When there are more than two blocks the situation is more interesting; this will be discussed later.

7.5 Partial Confounding

If the experiment contains several replications, i.e., if the set of treatment combinations is to be repeated several times instead of only once, we have two choices. Either we can repeat the same basic design each time, which would amount in our example to dividing each replicate into two blocks always confounding ABC, or else we can confound one or more of the other effects in the subsequent replications. Thus, if the experiment were to call for four replicates we might elect to confound ABC in the first replicate, AB in the second replicate, AC in the third, and BC in the fourth. In that case the four interactions are said to be partially confounded with blocks. Each of them is completely confounded with blocks in one of the replicates, but it is unconfounded in the other three replicates. We thus lose 25 per cent of the information about each of the interactions by dividing the thirty-two observations into eight blocks of four points each.

The experimenter has, therefore, two choices. He can either elect to learn nothing about the three-factor interaction while at the same time retaining 100 per cent information about each of the two-factor interactions, or he can obtain estimates of all four interactions at the cost of an increase in variance for the estimates of AB, AC, and BC. This is a choice which he himself must make in designing his experiment.

AN EXAMPLE OF PARTIAL CONFOUNDING. The following example, given by Yates (1937), deals with an experiment on potatoes. The factors are A, sulphate of ammonia; B, sulphate of potash; and C, nitrogen (manure). The responses are the yields in pounds per plot.

(1) 101	*c* 312	(1) 106	*b* 272
ac 373	*a* 106	*ab* 306	*a* 89
bc 398	*b* 265	*c* 324	*bc* 407
ab 291	*abc* 450	*abc* 449	*ac* 338
1163	1133	1185	1106
I*a*	I*b*	II*a*	II*b*
(1) 87	*c* 323	(1) 131	*c* 324
ac 324	*a* 128	*a* 103	*ac* 361
b 279	*bc* 423	*bc* 445	*b* 302
abc 471	*ab* 334	*abc* 437	*ab* 272
1161	1208	1116	1259
III*a*	III*b*	IV*a*	IV*b*

Using Yates' algorithm for each of the four replicates separately, we obtain the following totals and contrasts. The entries in the last column are the row totals. The underlined contrasts are those that are confounded with blocks, and they are not included in the corresponding row totals.

The estimates of the unconfounded effects are obtained by dividing the corresponding totals by 32; for the partially confounded interactions the divisor is 24. In Table 7.2 the sums of squares for the effects are obtained by squaring the row totals and dividing by 32 or 24, as appropriate, the divisor being the number of observations that went into the total.

The results of applying Yates' algorithm to the replicates follow.

	I	II	III	IV	Total
M	2296	2291	2369	2375	9331
A	144	73	145	−29	333
B	512	577	645	537	2271
AB	12	79	61	−47	26
C	770	745	713	759	2987
AC	82	39	−47	87	208
BC	−186	−189	−151	−143	−526
ABC	−30	−23	33	−43	−33.

TABLE 7.2

ANALYSIS OF VARIANCE

Source	SS		d.f.	MS
Total	$\sum y^2 - C$	$= 466779.7$	31	
Blocks	$(1163^2 + 1133^2 + \cdots) - C$	$= 4499.0$	7	642.7
Main effects	$(333^2 + 2271^2 + 2987^2)/32$	$= 443453.1$	3	147817.7
Interactions	$(26^2 + 208^2 + 526^2 + 33^2)/24$	$= 13404.4$	4	3351.1
Error		5423.2	17	319.0

7.6 Confounding 2^n Designs in 2^k Blocks of 2^{n-k} Points Each

Consider the set of all effects with the addition of a unit element I. They are $I, A, B, \ldots, AB, \ldots, ABC$, and, with the condition $A^2 = B^2 = \cdots = I$, they form a multiplicative Abelian group of order 2^n. The product of the effects ABC, ABD is A^2B^2CD, or CD. The product CD is said to be the generalized interaction of ABC and ABD. We note that we could also have said $(x_1x_2x_3) \times (x_1x_2x_4) = x_1^2x_2^2x_3x_4 = x_3x_4$ (since $x_1^2 = x_2^2 = 1$ at each experimental point).

A similar group is formed by the treatment combinations with (1) as the identity and $a^2 = b^2 = c^2 = \cdots = (1)$. Here, multiplication by a corresponds to multiplying by x_1 or to changing the level of factor A. The reduction of the 2^n experiment to an experiment on the vertices of a hypercube with coordinates ± 1, and the corresponding equivalence of $A^2 = I$ and $x_1^2 = 1$, appears to be due to Box. The group property for the sets of effects and of treatment combinations is due to Fisher (1942).

Suppose now that we elect to divide a 2^n experiment into four blocks, confounding ABC and ABD. We can place in the first block all points with $x_1x_2x_3 = +1$ and $x_1x_2x_4 = +1$, and in the second block all the points with $x_1x_2x_3 = +1$ and $x_1x_2x_4 = -1$, and so on. But whenever we have $x_1x_2x_3 = x_1x_2x_4 = +1$, we also have $x_3x_4 = +1$. The division into blocks is thus

	Block I	Block II	Block III	Block IV
$x_1x_2x_3$	$+1$	$+1$	-1	-1
$x_1x_2x_4$	$+1$	-1	$+1$	-1
x_3x_4	$+1$	-1	-1	$+1$

Each of the blocks contains 2^{n-2} points. The contrast in the block totals I + II − III − IV is the ABC contrast; I − II + III − IV is the ABD contrast. The third degree of freedom between blocks corresponds to I − II − III + IV, which is the CD contrast. No other effects are confounded.

In the general case, let P and Q be two effects. We may divide the 2^n design into four blocks, each of size 2^{n-2}, by assigning to block I all points which have positive signs in both the P and Q contrasts; block II contains the points which have positive signs in the P contrast and negative signs in the Q contrast, and so on. Then the generalized interaction PQ will also be confounded with blocks, and these will be the only three effects confounded.

If we wish to make eight blocks of size 2^{n-3}, we may take a third effect R (other than PQ). One block will contain all the 2^{n-3} points that have positive signs in the P, Q, and R contrasts. The seven effects confounded will be P, Q, PQ, R, PR, QR, and PQR. In the general case of 2^k blocks, we take for the defining contrasts a set of k independent generators $P, Q, R, S, \ldots$, by which we mean k effects such that no one of them is the generalized interaction of any set of the others. The blocks are then the 2^k sets obtained by taking the intersections of the sets defined by $I = \pm P, I = \pm Q, I = \pm R, I = \pm S, \ldots$. The complete set of confounded effects consists of the generators and all possible generalized interactions that can be derived from them.

If we add I we obtain a set that includes the identity and is closed under the group operation. The set of defining contrasts together with I is thus a subgroup of the group of effects. This subgroup is called the defining contrast subgroup. It is usually desirable to choose the defining contrasts, if possible, so that only higher order interactions are confounded. For example, in dividing a 2^5 experiment into four blocks of eight runs each, we might consider confounding the five-factor interaction and one of the four-factor interactions. However, confounding $ABCDE$ and $ABCD$ would be unwise because we should then automatically confound their generalized interaction, which is the main effect E. Similarly, choosing $ABCDE$ and ABC as defining contrasts also confounds the two-factor interaction DE. A preferable choice of defining contrasts would be two three-factor interactions and one four-factor interaction such as ABD, ACE, and $BCDE$. When the block effects are random, the design in which there are one or more main effects confounded is similar to a split-plot design.

7.7 The Allocation of Treatment Combinations to Blocks

We have already mentioned that the blocks are the intersections of the sets defined by $I = \pm P$ and $I = \pm Q, \ldots$. The actual assignment of points to blocks is simplified in practice by taking advantage of the fact that the treatment combinations form a group. The procedure used is to obtain the principal block, and then to derive the other blocks from it. The principal block is defined to be the block that contains the base point, or identity element, of the group, (1).

Theorem. *The treatment combinations in the principal block form a subgroup of the group of all treatment combinations.*

PROOF. Let P be a defining contrast and let S_P be the set of all treatment combinations having an even number of letters (or none) in common with P. Thus, if P were ABC, the set S_P would contain (1), ab, ac, bc, abd, and so on. The set S_P is closed and contains (1), for, if ab and ac are both in S_P, so is their product bc. Hence, S_P is a subgroup. If P contains an even number of letters, S_P is the set defined by $I = +P$. If P contains an odd number of letters, S_P is defined by $I = -P$. The principal block is the intersection of the sets S_P, S_Q, . . ., and is thus a subgroup. The principal block is sometimes called the intrablock subgroup. ∎

If a treatment combination has an even number of letters (or none) in common with both P and Q, it also has an even number (or none) in common with PQ. Hence, it is enough to find $n - k$ independent members of the principal block, each having an even number (or none) of letters in common with a set of k generators of the defining contrast subgroup, and then to complete the block by taking products.

The other blocks are obtained by changing the levels of one or more factors. Changing the level of A is equivalent to multiplying each member of the principal block by a. To change the levels of A and B simultaneously, we multiply each member by ab. Thus the other blocks are the cosets of the principal block.

EXAMPLE. Confounding a 2^5 design in four blocks of eight runs. The defining contrasts are ABD, ACE, and $BCDE$.

	Principal Block	Block II	Block III	Block IV
	(1)	*d*	*e*	*de*
	ade	*ae*	*ad*	*a*
	bd	*b*	*bde*	*be*
	abe	*abde*	*ab*	*abd*
	ce	*cde*	*c*	*cd*
	acd	*ac*	*acde*	*ace*
	bcde	*bce*	*bcd*	*bc*
	abc	*abcd*	*abce*	*abcde*
$x_4 =$	$-x_1x_2$	$+x_1x_2$	$-x_1x_2$	$+x_1x_2$,
$x_5 =$	$-x_1x_3$	$-x_1x_3$	$+x_1x_3$	$+x_1x_3$

The second block is derived by multiplying each member of the principal block by d, the third by multiplying by e, and the fourth by multiplying by de.

We could also approach the problem of allocation of points in the following way. The principal block consists of the points for which $x_1x_2x_4 = -1$ and $x_1x_3x_5 = -1$. Multiplying these equations by x_4 and x_5, respectively, and noting that $x_4^2 = x_5^2 = 1$, we have $x_4 = -x_1x_2$, $x_5 = -x_1x_3$. We can thus obtain the principal block by writing down a basic 2^3 design in the factors A, B, and C, and then adding the other two factors D and E by putting $x_4 = -x_1x_2$ and $x_5 = -x_1x_3$. Similarly, the second block could be obtained by letting $x_4 = +x_1x_2$ and $x_5 = -x_1x_3$.

7.8 The Composition of the Defining-Contrast Subgroup

In this section we shall temporarily refer to the effects in the defining contrast subgroup as words. The number of letters in an effect is the length of the word. A word with an odd (even) number of letters is called an odd (even) word. The product of two words is the word for the corresponding generalized interaction. The patterns of word lengths in confounding a 2^n design in 2^k blocks have been investigated by Brownlee et al. (1948), Burton and Connor (1957), and John (1966*a*). We shall assume in the remainder of this section that each of the n letters appears in at least one of the words in the subgroup. This eliminates from our consideration, for example, the design for the 2^3 factorial in two blocks of four runs confounding the main effect A.

The product of two even words, or of two odd words, is an even word. The product of an odd word and an even word is an odd word. The following necessary (but not sufficient) conditions can be deduced for the set of $2^k - 1$ words in the defining contrast subgroup:

(i) Each letter appears in exactly 2^{k-1} words. Hence, if w_i is the length of the ith word, $\sum_i w_i = 2^{k-1}n$;

(ii) Either all the words are even, or else exactly 2^{k-1} of them are odd. In either case, the even words together with the identity form a subgroup;

(iii) If one of the words contains all n letters, the remaining words occur in pairs such that the sum of the lengths of the words in any pair is n;

(iv) If one of the words consists of a single letter, the remaining words fall into pairs, such that the lengths of the words in any pair differ by one.

Suppose that we take any set of k generators. We define a set of parameters: t_i denotes the number of letters appearing only in the ith generator; t_{ij} is the number of letters that appears only in the ith and jth generators; and so on. The sum of these parameters is $\sum t = n$. We shall let $w(i)$ denote the length of the ith generator and $w(i, j)$ the length of the product of the ith and jth generators. Then

$$w(1) = t_1 + t_{12} + \cdots + t_{123} + \cdots,$$

$$w(1, 2) = w(1) + w(2) - 2(t_{12} + t_{123} + \cdots).$$

In general let $\sum_{O(S)} t$ denote the sum of all the t which contain in their subscripts an odd number of members from a set S of integers $i_1, i_2, \ldots, 1 \le i \le k$. Then, if $w(S)$ denotes the length of the generalized interaction of the i_1th, i_2th,... generators we have

$$w(S) = \sum_{O(S)} t.$$

Burton and Connor (1957) derived the formula

$$\sum w^2 = 2^{k-2}\left(\sum t^2 + n^2\right),$$

and showed that properties (i) and (ii) were not sufficient for the existence of a confounding scheme.

Their example involved trying to block a 2^9 design into 2^4 blocks, using seven words of length 4, six words of length 5, and two words of length 7. We have $\sum w = 72$ and so property (i) is satisfied. There is no difficulty in finding a set of seven four-letter words satisfying property (ii); *ABCD*, *ABEF*, and *ADFG* generate such a set.

We note next that $\sum w^2 = 360$ so that $\sum t^2 = \sum t = n = 9$. It follows that nine of the t must be unity and the others zero. If such a scheme existed we could take as generators the two seven-letter words and two of the five-letter words. We should then have

$$w(1) + w(2) + w(3) + w(4) = 24.$$

But

$$\sum w(i) = 4t_{1234} + 3\sum t_{hij} + 2\sum t_{ij} + \sum t_i.$$

A total of twenty-four can be obtained only if t_{1234}, all the t_{hij}, and all but two of the t_{ij} are unity.

We can now try to reconstruct a set of generators with these values of t. One letter, say A, must be in all four words; another letter, say B, must be in the first three words but not the fourth, and so on. There are only two cases to be considered: $t_{12} = t_{34} = 0$ and $t_{12} = t_{13} = 0$. In the first case each of the generators must have six letters; in the second case $w(2) = w(3) = 6$. Thus an arrangement with $w(1) = w(2) = 7$ and $w(3) = w(4) = 5$ is impossible. Burton and Connor demonstrated the nonexistence differently, using some other results from their paper.

7.9 Double Confounding and Quasi-Latin Squares

In a Latin square design p treatments or treatment combinations are applied to p^2 plots arranged in a square array. The plots are assigned to the treatments in such a way that each treatment or treatment combination appears exactly once in each row and in each column. In consequence, the treatment differences are not confounded with either the row or the column effects. Yates (1937) introduced the term quasi-Latin square to denote a design in which more than p treatment combinations are used in a $p \times p$ square array. There is double confounding. One set of effects is confounded with rows and another set is confounded with columns.

In a 4×4 square we may have one complete replication of a 2^4 design or two replicates of a 2^3 factorial. In dividing a 2^4 design into four blocks of four runs each, we may confound the four-factor interaction and two of the two-factor interactions, or else two of the three-factor interactions and their product, which would be a two-factor interaction. A design in which $ABCD$, AB, and CD are confounded with columns, and ABC, ACD, and BD are confounded with rows is

(1)	*bcd*	*abd*	*ac*
ab	*acd*	*d*	*bc*
cd	*b*	*abc*	*ad*
abcd	*a*	*c*	*bd*

With this design we can estimate all four main effects and three of the six two-factor interactions, but we have only 2 d.f. that can be used for error. These are the 2 d.f. corresponding to ABD and BCD.

When two replicates of the 2^3 factorial are incorporated in one 4×4 square, we might choose to have ABC and AB confounded with rows and AC and BC confounded with columns. There are two ways of achieving this:

	BC		AC			BC		AC	
ABC	*abc*	*c*	*b*	*a*		*a*	*b*	*abc*	*c*
	bc	*ac*	(1)	*ab*	or	(1)	*ab*	*ac*	*bc*
AB	(1)	*ab*	*abc*	*c*		*abc*	*c*	(1)	*ab*
	a	*b*	*ac*	*bc*		*bc*	*ac*	*b*	*a*

In either case the first two rows and the last two rows are separate replicates. The first two columns and the last two columns also constitute replicates. If, however, the rows and columns are randomized, the design will not in practice have this nice pattern with runs in the same replicate being in adjacent rows and columns. The analysis is similar to that in the example of partial confounding given earlier. Each of the partially confounded effects (the four interactions) is estimated from the two rows (or columns) in which it is not confounded.

Yates gives a design for 2^6 in an 8×8 square confounding ACE, ADF, BDE, BCF, $ABCD$, $ABEF$, and $CDEF$ between the rows and ABF, ADE, BCD, CEF,

ABCE, *ACDF*, and *BDEF* between the columns. A problem with this design was noted by Yates (1948) and was discussed in detail by Grundy and Healy (1950). The difficulty lies in the fact that, when the columns and rows are randomized, there is a considerable probability of obtaining an arrangement in which the contrast for one of the main effects turns out to be the contrast between diagonally opposite quarters of the design. This is illustrated in the arrangement that we wrote down for the 2^4 factorial in the 4×4 square; the *ABD* contrast pits the four points in the top left quadrant and the four in the bottom right quadrant against the points in the other two quadrants.

Yates presents other designs involving the use of several squares. He also gives quasi-Latin squares for 3^n designs and for two replicates of a design with two factors at three levels and one at two levels in a 6×6 square. He discusses the use of quasi-Latin squares in split-plot designs, choosing his terminology from the weaving of tartans. A design in which main effects are confounded with both rows and columns is called a plaid square. If main effects are confounded in only one direction, we have a half-plaid square.

An example of a half-plaid square for a 2^3 factorial repeated twice, with *A* confounded with rows and *BC* and *ABC* confounded with columns is

BC		*ABC*	
abc	*ac*	*a*	*ab*
(1)	*b*	*c*	*bc*
a	*ab*	*abc*	*ac*
bc	*c*	*b*	(1).

Although we have introduced double confounding with the quasi-Latin square, it is not essential that the designs have the same number of rows and columns. The following Latin rectangle designs are given by Healy (1951) using four rows and eight columns. The first design consists of four replicates of a 2^3 factorial. Each row consists of a complete replicate, and the treatment combinations are arranged within the rows so that only *ABC* is confounded with columns. As the design is written, columns 1 and 2 form a replicate confounding *ABC*. So do columns 3 and 4, columns 5 and 6, and columns 7 and 8.

(1)	*a*	*b*	*ab*	*c*	*ac*	*bc*	*abc*
ab	*b*	*a*	(1)	*abc*	*bc*	*ac*	*c*
ac	*c*	*abc*	*bc*	*a*	(1)	*ab*	*b*
bc	*abc*	*c*	*ac*	*b*	*ab*	(1)	*a*

In the second design there are two replicates of a 2^4 factorial. Rows 1 and 2, and rows 3 and 4 each comprise a complete replicate of the 2^4. The first four columns are a complete replicate confounding AB, ACD, and BCD. In the last four columns, CD, ABC, and ABD are confounded.

(1)	*bd*	*ac*	*abcd*	*ab*	*ad*	*bc*	*cd*
abc	*a*	*bcd*	*d*	*acd*	*c*	*abd*	*b*
cd	*bc*	*ad*	*ab*	(1)	*bd*	*ac*	*abcd*
abd	*acd*	*b*	*c*	*bcd*	*abc*	*d*	*a*

He also gives some designs for the 2^5 factorial in a 4×8 rectangle. One of these confounds ABC, ADE, and $BCDE$ with rows and AC, DE, ABD, and so on, with columns. The other makes use of fractional factorials, which will be discussed later.

7.10 Confounding in 3^n Designs

We have already mentioned how in the 3^3 experiment the sums of squares for each of the two-factor interactions and the three-factor interaction can be subdivided into components with two degrees of freedom. The same procedure leads to a method of confounding for dividing the twenty-seven points into three blocks of nine points each. If, for example, AB^2C^2 is to be confounded with blocks, we place in the first block the nine points for which $x_1 + 2x_2 + 2x_3 \equiv 0 \pmod 3$; the second block contains the points with $x_1 + 2x_2 + 2x_3 \equiv 1 \pmod 3$; the third block has $x_1 + 2x_2 + 2x_3 \equiv 2 \pmod 3$. The three blocks are

(I) 000, 012, 021, 101, 110, 122, 202, 211, 220;

(II) 002, 011, 020, 100, 112, 121, 201, 210, 222;

(III) 001, 010, 022, 102, 111, 120, 200, 212, 221,

where, for example, $a_2b_1c_1$ is written as 211.

Confounding two of the components of the three-factor interaction to obtain nine blocks of three points each also confounds a main effect. If we take, for example, the three points for which $x_1 + x_2 + x_3 \equiv i \pmod 3$ and $x_1 + 2x_2 + 2x_3 \equiv j \pmod 3$, we also have $2x_1 \equiv i + j \pmod 3$ or $x_1 \equiv 2(i + j) \pmod 3$. We can avoid confounding main effects by confounding components of two-factor interactions such as AB, AC, and BC^2.

Partial confounding can also be used for 3^n designs with several replicates. We may take a 3^3 with four replicates and confound each of the four pairs of degrees of freedom for the three-factor interaction in one of the replicates. The analysis follows the same general pattern as the analysis of the 2^3 example that was discussed earlier.

7.11 Confounding with Factors at More Than Three Levels

The procedure that we have used in the previous section for the 3^n series is a special case of a more general procedure, due to Bose and Kishen (1940) and Bose (1947*a*). They consider confounding in an experiment with n factors each at s levels where s is a prime or a power of a prime.

They take a geometric approach to the problem. The treatment combinations become points in n-dimensional Euclidean space, and the blocks are sets, or intersections of sets, of parallel hyperplanes, or flats. We shall not go into the details of this work. Reference to it will be made again in a later chapter on the construction of incomplete block designs.

Let $s = p^m$ where p is a prime. We denote each of the treatment combinations by a set of n coordinates $(z_1, z_2, \ldots, z_n)$. These coordinates are elements of the Galois field of p^n elements, $GF(p^n)$. If $n = 1$, as is the case when $s = 3$ or 5, the elements can be taken as the integers $0, 1, \ldots, s - 1$, and arithmetic is carried out mod s. If $s = 4 = 2^2$, the elements are $0, 1, x, 1 + x$ where $x^2 = x + 1$, and arithmetic is mod 2.

Consider the set of points for which $\sum a_i z_i = c$ where a_i and c are elements of the field. The equation defines a hyperplane, or a "flat," which contains s^{m-1} points. Letting c take in turn each value in the field gives us a set of s blocks, each with s^{n-1} points. If $\sum b_i z_i = c$ denotes another set of hyperplanes, the intersections of pairs of hyperplanes, one from each set, form a set of s^2 blocks, each of size s^{n-2}. In the first case, the $s - 1$ d.f. corresponding to $\sum a_i y_i$, or the $A^{a_1}B^{a_2}\ldots$ interaction, are confounded with blocks. In the second case, we also confound $A^{b_1}B^{b_2}\ldots$ and its generalized interactions with $A^{a_1}B^{a_2}$.

7.12 Confounding in the 4^2 Design

We illustrate now two approaches to the problem of dividing a 4^2 design with factors A and B into four blocks of four runs each.

(i) We denote each treatment combination by a pair of elements from $GF(2^2)$. For convenience we write y for $x + 1$. Then $x^2 = y$ and $xy = 1$. The sixteen points in the design are

$$\begin{array}{cccc} 00 & 10 & x0 & y0 \\ 01 & 11 & x1 & y1 \\ 0x & 1x & xx & yx \\ 0y & 1y & xy & yy. \end{array}$$

There are 9 degrees of freedom for AB interaction. We elect to confound what might be called the AB^x component by taking as blocks the flats $z_1 + xz_2 = c$. The blocks are

$z_1 + xz_2 =$	0	1	x	y
	00	$0y$	01	$0x$
	$1y$	10	$1x$	11
	$x1$	xx	$x0$	xy
	yx	$y1$	yy	$y0$.

(ii) We may replace each four-level factor by two pseudofactors, each at two levels, A by P and Q, B by R and S. Let the coordinate one denote a pseudofactor at its high level and zero denote it at its low level. The points are

0000	1000	0100	1100
0010	1010	0110	1101
0001	1001	0101	1110
0011	1011	0111	1111

This really amounts to replacing 0, 1, x, and y in the previous representation by 00, 10, 01, and 11, respectively.

The three degrees of freedom for A correspond to the main effects P and Q and the interaction PQ. The main effects of B are R, S, and RS. The other nine interactions between the pseudofactors correspond to the interaction AB. We can now elect to split the 2^4 design into four blocks confounding PQR, QRS, and PS. This method will give the same blocks as the previous method.

7.13 Confounding in Other Factorial Experiments

The methods described in this chapter can be applied, although not so easily, to factorial experiments in which the factors do not all have the same number of levels. If some of the factors have two levels and the others have four levels, the confounding can be achieved by the use of pseudofactors, which has just been described. Yates (1937) discusses confounding in designs with some three-level factors and some four-level factors. Binet et al. (1955) consider dividing a single replicate of a design with factors at several levels into blocks.

Exercises

1. Apply Yates' algorithm to the following set of data for a 2^4 experiment:

(1)	3.96	c	3.71	d	4.53	cd	4.52
a	4.05	ac	3.87	ad	4.69	acd	4.72
b	2.71	bc	2.67	bd	3.01	bcd	3.32
ab	2.94	abc	3.02	abd	3.37	$abcd$	3.53

2. Use the algorithm to evaluate the contrasts and obtain the analysis of variance table for the 3^3 design given as an example in Chapter 5.
3. Devise an algorithm in the style of Yates for calculating the contrasts in a $2^2 \times 3$ design. Use it to carry out the analysis of variance for the following design which has two levels of P, two levels of N, and three levels of K.

	K_0		K_1		K_2	
	N_0	N_1	N_0	N_1	N_0	N_1
P_0	23	36	25	46	25	31
P_1	34	28	19	54	19	38

4. The data in exercise 1 were actually taken from a $2^2 \times 4$ experiment [Raese and Decker (1966)]. Factors C and D correspond to the factor with four levels. Use the results of exercise 1 to obtain the analysis of variance table for the $2^2 \times 4$ design.
5. In the section on partial confounding we considered the example of a 2^3 design with four replicates, each divided into two blocks of four points each. Prove that the estimates we obtained are actually the least squares estimates.
6. Prove the necessity of the four conditions on the words in the defining contrast subgroup given at the beginning of the section on the composition of that group.
7. Find the principal block when the 2^8 design is split into blocks of sixteen runs each, using the defining contrast subgroup generated by $ABCD$, $ABEF$, $ABGH$, $ACEG$.
8. Derive the quasifactorial design for a 2^6 design in an 8×8 square by Yates, which was mentioned in the text.
9. Derive a design for a 2^5 factorial in a 4×8 rectangle, confounding ABC, ADE, $BCDE$ with rows and AC, DE, ABD with columns.
10. Obtain the partially confounded design for a 3^3 factorial with four replicates, each in three blocks of nine points each. Confound ABC with blocks in the first replicate, ABC^2 in the second, AB^2C in the third, and AB^2C^2 in the fourth.
11. A reasonable procedure when a plot is missing in a 2^n design is to replace it by a value which makes the highest order interaction zero. Apply this technique to the data of exercise 1, assuming bc to be missing. What is the variance of the estimate of an effect when a missing plot is included?
12. (Continuation). In the 2^n design the highest order interaction has only one contrast. However, in a 3^n design the highest order interaction has 2^n degrees

of freedom; thus, there are 2^n orthogonal contrasts, each of which might be equated to zero to calculate a missing value. A less ambiguous procedure would be to omit all the highest order interaction terms from the model and use the procedure of the earlier chapters. Investigate this for the 3^3 experiment of exercise 2, assuming that the point 012 is missing.

CHAPTER 8

Fractions of 2^n Factorial Designs

As the number of factors in a 2^n design increases, the number of points soon outgrows the facilities and budgets of most experimenters. The full 2^8 experiment calls for 256 runs. Only eight of the degrees of freedom correspond to main effects, and twenty-eight to two-factor interactions. The remaining 219 degrees of freedom correspond to interactions involving three or more factors. If these higher order interactions are all negligible, the experimenter has 219 degrees of freedom for error, which is more than he really needs for an adequate estimate. On the other hand, he has to make 256 runs to estimate only 8 + 28 effects, and only 14 per cent of the degrees of freedom are used for estimation.

We have already seen how Latin squares can be used to estimate the main effects in a p^3 factorial experiment with only p^2 runs. Graeco-Latin squares and hyper-Graeco-Latin squares result in even greater savings, but at the cost of assuming that some of the interactions do not exist.

This chapter is devoted to an investigation of fractions of 2^n designs. One of the fractions that will be discussed can be derived from a Latin square of side 2, either

$$\begin{matrix} 1 & 2 \\ 2 & 1 \end{matrix} \quad \text{or} \quad \begin{matrix} 2 & 1 \\ 1 & 2. \end{matrix}$$

If we let rows represent the levels of A, columns the levels of B, and 1 and 2 the levels of C, we have two four-point fractions of the 2^3 design: (1), bc, ab, ac, and b, c, a, abc. These are half replicates of the 2^3 design, or 2^{3-1} designs, from which the main effects A, B, and C are estimable, if we neglect their interactions. The 2^{n-k} fractions, or $1/2^k$ fractions of 2^n designs were introduced by Finney (1945).

They will be the topic of the first part of the chapter, after which some other fractions will be discussed.

Fractional factorials are commonly used in industrial experimentation where runs are expensive and where an acceptable prior estimate of the variance is available. They may also be used for screening experiments. These are experiments performed in the early stages of an investigation in which many factors are considered, most of which may turn out to have little or no effect upon the response under investigation. The purpose of the screening experiment is to screen several factors with the purpose of spotting those factors, if any, that have appreciable effects. The factors that seem to be interesting are then investigated more closely in subsequent experiments. The emphasis is not so much upon estimating the effects with small variance, but upon finding out which of the factors merit further study. We shall assume throughout most of this chapter that the experimenter is interested in estimating the main effects, and perhaps the two-factor interactions; that he is prepared to assume before carrying out the experiment that the higher order interactions are negligible; and that he is not particularly concerned with using the actual experiment to provide himself with an estimate of the variance. The emphasis will therefore be on finding small experiments in which a high percentage of the degrees of freedom are used for estimation. In these small experiments there will be few degrees of freedom for error even if higher order interactions are used, and it will be assumed that the experimenter has some prior estimate of the error variance. Daniel (1959) has introduced a graphical method for deciding which effects in a 2^n factorial are significant when there is no estimate of error. If there is no prior estimate available such graphical procedures seem to be the only recourse.

8.1 The 2^{n-k} Fractional Factorials

When in the previous chapter we divided the 2^5 design into four blocks of eight runs, each block constituted a quarter replicate, or a 2^{5-2} fraction. The term 2^{n-k} fraction or $(1/2^k)$th replicate is used to describe only those sets of 2^{n-k} treatment combinations that could occur in the same block when the complete factorial is divided into 2^k blocks confounding exactly $2^k - 1$ effects in the manner derived in the previous chapter. Thus we may define a 2^{n-k} fraction as a set of 2^{n-k} different points which constitute a subgroup of the group of treatment combinations, or a coset of such a subgroup. This definition throws very little light upon the nature of fractional factorials and their importance. Henceforth we shall regard a subgroup as a trivial coset of itself, and use the term *coset* to embrace both the subgroup itself and the other cosets.

The first fraction listed (the principal block) was the fraction defined by $I = -ABD = -ACE = +BCDE$. The other fractions in order were defined by $I = +ABD = -ACE = -BCDE$, $I = -ABD = +ACE = -BCDE$, and $I = +ABD = +ACE = +BCDE$. The four fractions are said to belong to the same family. The term *principal fraction* has been used in the literature in two different senses and we shall refrain from employing it here. Some authors use it to denote the fraction containing (1) by analogy with the term principal block. Box and Hunter (1961) use it to denote the fraction for which the defining contrasts all have positive signs; in our example that would be the fraction containing *de*, *a*, *be*, *abd*, *cd*, *ace*, *bc*, and *abcde*. We now consider that fraction in more detail.

For each treatment combination in the fraction the relationship $1 = x_1x_2x_4 = x_1x_3x_5 = x_2x_3x_4x_5$ holds. Multiplying by x_1 and recalling that $x_1^2 = 1$, we have

$$x_1 = x_2x_4 = x_3x_5 = x_1x_2x_3x_4x_5$$

at each point so that A, BD, CE, and $ABCDE$ are indistinguishable. The four effects A, BD, CE, and $ABCDE$ are said to constitute an alias set and to be aliased (or confounded) with each other. The alias sets are the cosets of the defining contrast subgroup (which is sometimes called the alias subgroup in this context). There are 2^{n-k} alias sets (including the subgroup itself) and each set has 2^k members. Each effect is aliased with the effects in the same alias set but with no others.

If the elements of the vector of effects $\boldsymbol{\beta}$ are written by order of alias sets so that $\boldsymbol{\beta}' = (\beta_0, \beta_{124}, \beta_{135}, \beta_{2345}, \beta_1, \beta_{24}, \beta_{35}, \beta_{12345}, \ldots)$ the cross-product matrix $\mathbf{X'X}$ in the normal equations consists of square matrices $8\mathbf{J}_4$ along the main diagonal and zeros elsewhere. The rank of $\mathbf{X'X}$ is 8. A matrix of full rank is obtained by choosing one effect from each of the alias sets and striking out the remaining rows and columns. If, therefore, we were to assume a restricted model which included the mean, the five main effects, either BC or DE, and either CD or BE, but no other terms, the corresponding least squares equations would have a unique set of solutions.

The A contrast from this fraction is

$$(a + abd + ace + abcde - de - be - cd - bc).$$

Its expectation is $8(\beta_1 + \beta_{24} + \beta_{35} + \beta_{12345})$, so that when we use the contrast to estimate A we are actually estimating $A + BD + CE + ABCDE$. The other fractions give chains $A \pm BD \pm CE \pm ABCDE$, with the signs corresponding to the signs of the defining contrasts. The full set of alias chains is

$$\begin{array}{ll}
I + ABD + ACE + BCDE & A + BD + CE + ABCDE \\
B + AD + ABCE + CDE & AB + D + BCE + ACDE \\
C + ABCD + AE + BDE & AC + BCD + E + ABDE \\
BC + ACD + ABE + DE & ABC + CD + BE + ADE
\end{array}$$

For A to be estimated we must assume that BD, CE, and $ABCDE$ are all zero. If we plan to use this fraction with the intention of estimating A we must decide a priori that we are going to ignore the other three interactions. The prior decision to ignore the five-factor interaction will cause little concern, but it may not be reasonable to assume without further investigation that the two-factor interactions BD and CE are negligible. If these two particular interactions are the only ones that concern the experimenter, he can avoid the difficulty by choosing a different fraction such as one of the 2^{5-2} defined by $I = \pm ABC = \pm ADE = \pm BCDE$. The two interactions will then be aliased with each other, but not with any of the main effects. There is not a quarter replicate from which the mean, the five main effects, BD and CE can all be estimated. If, on the other hand, the experimenter is not willing to declare prior to carrying out his experiment that any of the two-factor interactions can be ignored, he cannot estimate without bias the five main effects from any 2^{5-2} fraction. In that case he is going to have to perform a larger experiment or to omit one of the factors.

8.2 The Classification of Fractions

We have seen that in the use of fractional replicates it is necessary to assume a priori that certain effects are zero. It is a common practice in industrial experimentation to make the blanket assumption in the absence of prior information to the contrary that all interactions that involve three or more factors can be ignored. We shall make that assumption in this chapter. An interaction which is taken to be zero in this way will be said to be suppressed. We shall sometimes suppress the two-factor interactions also, and consider estimating only the main effects. It will be convenient to use the abbreviation 2 f.i. for the expression *two-factor interaction.*

An effect will be said to be estimable if, and only if, there is a contrast in the data which has for its expectation the particular effect biased only by other effects which we choose to suppress. The effect being estimated is said to be aliased with the effects that were suppressed. Designs in which main effects are aliased with each other are of no interest. The remaining designs are divided into several classes.

One method of classification has three classes:

(i) Main effects only designs. These are fractions in which some of the main effects are aliased with 2 f.i. and are estimable only when those 2 f.i. are suppressed;

(ii) Main effects clear plans. These are fractions in which all the main effects are estimable whether the 2 f.i. are suppressed or not, but some of the 2 f.i. are aliased with each other;

(iii) 2 f.i. clear plans. These are fractions in which all the main effects and all the 2 f.i. are estimable.

These classes were originally called *types A*, *B*, and *B′* by Box and Wilson (1951). Box and Hunter used the terms *resolution III*, *resolution IV*, and *resolution V*. Daniel (1962) called them *three-*, *four-*, *and five-letter plans*. The last two terminologies are in common use. The justification for Daniel's nomenclature becomes clear when the alias sets are considered. If one of the defining contrasts is a three-letter word, say ABC, then A is aliased with BC, B with AC, and C with AB; the three main effects are estimable only if the three 2 f.i. are suppressed. With $ABCD$ as a defining contrast, A is aliased with BCD, but AB is aliased with CD. With a five-letter word $ABCDE$, the main effects and 2 f.i. are all aliased only with higher order interactions.

We shall use the following definition of the resolution of a fractional factorial, which is applicable to other fractions as well as the 2^{n-k} series. An interaction involving t factors will be called a tth order effect; main effects are first-order effects. Let t be any integer.

A design is said to be of resolution $(2t + 1)$ if it satisfies the condition that all effects of order t or less are estimable whenever all effects of order higher than t are suppressed.

A design is said to be of resolution $2t$ if all effects of order $(t - 1)$ or less are estimable whenever all effects of order $(t + 1)$ or higher are suppressed.

8.3 Some 2^{n-k} Fractions

The 2^{3-1} Design

The smallest usable 2^{n-k} fraction is the 2^{3-1} consisting of four points. Two such designs are useful: one, defined by $I = ABC$, contains the four points, a, b, c, and abc; the other, defined by $I = -ABC$, consists of (1), ab, ac, and bc. Both these designs are three-letter plans. Each main effect is aliased with a two-factor interaction. In the first of these half replicates the estimates of the main effects are

$$\hat{\beta}_1 = (abc + a - b - c)/4 \quad \text{with expectation} \quad A + BC;$$

$$\hat{\beta}_2 = (abc + b - a - c)/4 \quad \text{with expectation} \quad B + AC;$$

$$\hat{\beta}_3 = (abc + c - a - b)/4 \quad \text{with expectation} \quad C + AB.$$

The variance of each estimate is $\sigma^2/4$.

In the traditional one-at-a-time experiment, which appeals to some experimenters, each factor is varied in turn while the other factors are maintained at their low levels. With three factors this design consists of the four points (1), a, b, and c. It is interesting to compare this design with the half replicates. Each uses four points to estimate the mean and the three main effects. However, the estimate of the main effect A, for example, from the one-at-a-time experiment is $\beta_1 = (a - (1))/2$ with expectation $A - AB - AC + ABC$ and variance $\sigma^2/2$. The 2^{3-1} fraction thus provides an estimate which has a smaller variance and which is biased by fewer interactions.

The one-at-a-time experiment should not, however, be dismissed completely because of its relative inefficiency. It is attractive to some experimenters in laboratories and pilot plants, especially when they are not quite sure at the beginning of a series of experiments how far apart to take the levels of the quantitative factors; if a ten-degree change of temperature is tried and does not appear to produce an adequate change in response, perhaps the high and low levels of the temperature factor should be twenty degrees apart. A useful compromise is to begin with the one-at-a-time experiment and then to add a fifth point, *abc*. The five points then comprise the half replicate defined by $I = +ABC$ plus the base point (1). With the fifth point included in the experiment, one can retain the same model with mean and main effects only, or else add to the model a term for one of the interactions. It will be found that when (1) is added to the half replicate and β_{123} is added to the model, the least squares estimates of the main effects A, B, and C remain the same as the estimates from the half replicate alone. On the other hand, if β_{12} were the interaction added to the model, the estimates of the mean, A and B, would be unchanged but that of C would be altered.

8.4 Designs with Eight Points

The two half replicates of the 2^4 factorial defined by $I = \pm ABCD$ are four-letter plans with eight points. Each of the four main effects is aliased with a three-factor interaction. The 2 f.i. are aliased in pairs; AB is aliased with CD, AC with BD, and AD with BC. If it is known beforehand that one of the four factors does not interact with any of the others, the interactions involving this factor can be suppressed and then the other three 2 f.i. will be estimable.

To illustrate these designs we consider a complete 2^4 design which formed part of a gasoline experiment. Each of the data points is the octane requirement of a car (from which eighty has been subtracted). The four factors were all qualitative.

(1) 7, *a* 7, *b* 9, *ab* 1, *c* 11, *ac* 12, *bc* 14, *abc* 7,

d 10, *ad* 8, *bd* 12, *abd* 5, *cd* 6, *acd* 6, *bcd* 10, *abcd* 6.

The corresponding contrasts, each measuring 16β, are

T 131, A -27, B -3, AB -25,

C $+13$, AC $+7$, BC $+7$, ABC $+1$,

D -5, AD $+1$, BD $+9$, ABD -7,

CD -27, ACD $+3$, BCD $+3$, $ABCD$ $+1$.

The procedure for the use of Yates' algorithm is to regard each fraction as a complete 2^3 design in A, B, and C, to which the fourth factor D has been added by putting $x_4 = +x_1x_2x_3$, or $x_4 = -x_1x_2x_3$. We give the calculations for the two half replicates. The fourth column of figures in each section is the column of contrasts; these figures, which estimate 8β, should be compared with one half of the values of the contrasts for the complete design.

$$I = +ABCD$$

(1)	7	15	28	66	
ad	8	13	38	−12	A
bd	12	18	−10	0	B
ab	1	20	−2	−26	$AB + CD$
cd	6	1	−2	10	C
ac	12	−11	2	8	$AC + BD$
bc	14	6	−12	4	$BC + AD$
abcd	6	−8	−14	−2	D

$$I = -ABCD$$

d	10	17	31	65	
a	7	14	34	−15	A
b	9	17	−7	−3	B
abd	5	17	−8	1	$AB - CD$
c	11	−3	−3	3	C
acd	6	−4	0	−1	$AC - BD$
bcd	10	−5	−1	3	$BC - AD$
abc	7	−3	2	3	$-D$

The last contrast in each table is the ABC contrast. In the first fraction this contrast estimates $8(ABC + D)$; suppressing ABC we have $8\hat{\beta}_4 = -2$. In the second half replicate the contrast estimates $8(ABC - D)$, so that we have $-8\hat{\beta}_4 = 3$, or $8\hat{\beta}_4 = -3$.

It should also be noted that adding the values of the contrasts in the two half replicates gives the values for the complete design. Thus the A contrast for the complete factorial is $-27 = (-12) + (-15)$; the AB contrast is $-26 + 1 = -25$, and the CD contrast is $-26 - 1 = -27$. We are reminded in the second half replicate that it is not wise to conclude, just because the estimate of $AB - CD$ is -1, that both AB and CD are small. Although there is no mathematical justification for it, some scientists would as a practical matter be prepared, as Daniel (1962) has pointed out, to assume for example in the first half replicate that, since the A and C effects are large and the B and D effects are small, the relatively large size of the $AC + BD$ contrast can be attributed to the AC interaction rather than the BD interaction. This procedure, lacking as it does a firm mathematical basis, can lead to pitfalls unless the scientist has a considerable amount of prior information about the probable relationships between the factors.

If we add to the first of these half replicates a fifth factor E by putting $x_5 = x_1x_2$, we have the 2^{5-2} design defined by $I = ABCD = ABE = CDE$.

This resolution III design consists of the points *e*, *ad*, *bd*, *abe*, *cde*, *ac*, *bc*, *abcde*, and is similar to the 2^{5-2} designs that were mentioned earlier.

Adding *F* by putting $x_6 = x_1x_3$ gives the 2^{6-3} design defined by $I = ABCD = ABE = CDE = ACF = BDF = BCEF = ADEF$, which consists of the points *ef*, *ad*, *bdf*, *abe*, *cde*, *acf*, *bc*, *abcdef*.

Finally, we may add a seventh factor by letting $x_7 = x_2x_3$ and obtain *efg*, *adg*, *bdf*, *abe*, *cde*, *acf*, *bcg*, *abcdefg*. The defining contrast subgroup for this fraction is generated by *ABCD*, *ABE*, *ACF*, and *BCG*. The alias chains for main effects and 2 f.i. are

$$A + BE + CF + DG$$

$$B + AE + CG + DF$$

$$C + AF + BG + DE$$

$$D + AG + BF + CE$$

$$E + AB + CD + FG$$

$$F + AC + BD + EG$$

$$G + AD + BC + EF$$

A three-letter plan for $2^n - 1$ factors in 2^n runs can always be constructed in this way by starting with a basic 2^n design and then adding new factors by equating them to interactions.

8.5 Designs with Sixteen Points

The half replicates of the 2^5 design defined by $I = +ABCDE$ or by $I = -ABCDE$ are designs of resolution V. Each main effect is aliased with a four-factor interaction; each 2 f.i. is aliased with a three-factor interaction. There are five main effects and ten 2 f.i. Thus there are as many effects to be estimated as there are degrees of freedom. A fraction in which each degree of freedom is used to estimate an effect is said to be saturated. Daniel (1956) introduced the terms *degree-of-freedom efficiency* and *variance efficiency* for a fractional factorial. The variance efficiency of a fraction is obtained by comparing the average of the variances of the estimates of the effects to the smallest variance that could be hoped for with factors at the levels ± 1. For a main effect such as *A*, the minimum variance of the estimate $\hat{\beta}_1$ is obtained when half the points are at $x_1 = +1$ and half are at $x_1 = -1$; the variance is then $V(\hat{\beta}_1) = \sigma^2/N$. The 2^{n-k} fractions all have 100 per cent variance efficiency. We shall discuss later some non-orthogonal fractions which have lower efficiencies. The degree-of-freedom efficiency is the ratio of the number of effects estimated to the number of degrees of freedom in the design. The 2^{5-1} fractions have 100 per cent degree-of-freedom efficiency.

The half replicate defined by $I = -ABCDE$ consists of the sixteen points

$$(1), ae, be, ab, ce, ac, bc, abce, de, ad, bd, abde, cd, acde, bcde, abcd.$$

The complementary half replicate, defined by $I = +ABCDE$, contains all treatment combinations with an odd number of letters.

Four-letter plans for six, seven, or eight factors can be obtained by starting with a basic 2^4 design and adding the other factors by putting $E = ABC$, $F = ABD$, $G = ACD$, and $H = BCD$. The design for all eight factors consists of the points

$$(1), aefg, befh, abgh, cegh, acfh, bcfg, abce,$$

$$dfgh, adeh, bdeg, abdf, cdef, acdg, bcdh, abcdefgh.$$

The defining contrast subgroup is generated by $ABCE$, $ABDF$, $ACDG$, and $BCDH$; each defining contrast contains an even number of letters.

Designs like this can also be obtained by the important method of folding over. Suppose that we begin with a resolution III design, the 2^{7-4} design with eight points presented in the previous section. We can regard it as a 2^{8-5} design in which all the runs were made at the low level of the factor H, and add the generator $-H$ to the defining contrast subgroup. The folding-over process consists of repeating each run with the levels of every factor changed. Thus, folding over *efg* gives *abcdh*. The design is

$$efg, adg, bdf, abe, cde, acf, bcg, abcdefg,$$

$$abcdh, bcefh, acegh, cdfgh, abfgh, bdegh, adefh, h.$$

For the first eight points we have $x_1x_2x_5 = x_1x_3x_6 = x_2x_3x_7 = -x_8 = +1$; for the second eight points we have $x_1x_2x_5 = x_1x_3x_6 = x_2x_3x_7 = -x_8 = -1$. For the whole set of sixteen points we have $(x_1x_2x_5)(x_1x_3x_6) = (x_1x_2x_5)(x_2x_3x_7) = (x_1x_2x_5)(-x_8) = (x_1x_3x_6)(x_2x_3x_7) = (x_1x_3x_6)(-x_8) = (x_2x_3x_7)(-x_8) = +1$, and also $x_1x_2x_3x_4 = 1$.

Thus $x_2x_3x_5x_6 = x_1x_3x_5x_7 = -x_1x_2x_5x_8 = \cdots = +1$ and the generators of the fraction are $BCFG$, $ABEG$, $-ABEH$, $ABCD$.

In general, folding over will convert any design of resolution III into a design of resolution IV or higher. The defining contrasts for the combined design are those defining contrasts of the original fraction that contain an even number of letters, and, since there are no two-factor interactions in that set, they all contain four letters or more. The original 2^{n-k} fraction has $2^{k-1} - 1$ defining contrasts with an even number of letters and 2^{k-1} with an odd number. In the complementary fraction obtained by folding over, the signs of each of the odd defining contrasts are changed, and the signs of each of the even contrasts are unaltered. It also follows that folding over a resolution V design produces a design of resolution VI or higher.

Folding over can also be considered from the point of view of breaking chains of aliases. Since the three-letter defining contrasts change signs in the second fraction, the 2 f.i. change signs in those alias chains which include main effects. Thus, in our example, the alias chain for A in the first eight points (considering only main effects and 2 f.i.) is $A + BE + CF + DG - AH$, while the corresponding alias chain in the second set is $A - BE - CF - DG + AH$. The A contrast from the first eight points gives an estimate of $\beta_1 + \beta_{25} + \beta_{36} + \beta_{47} - \beta_{12}$. From the second eight points we have an estimate of $\beta_1 - \beta_{25} - \beta_{36} - \beta_{47} + \beta_{12}$. Adding them gives an estimate of $2\beta_1$.

Another possibility is to take a design of resolution III and repeat it, changing only the level of A without adding the extra factor H. This gives a design in which A and all the 2 f.i. which contain A are estimable, biased only by interactions involving three or more factors. In our example the design would be

$$efg,\ adg,\ bdf,\ abe,\ cde,\ acf,\ bcg,\ abcdefg,$$
$$aefg,\ dg,\ abdf,\ be,\ acde,\ cf,\ abcg,\ bcdefg.$$

The set of defining contrasts for the combined design consists of all the defining contrasts of the original design which do not contain A. The alias chains containing A are $A + BE + CF + DG$ and $A - BE - CF - DG$. The alias chains containing AB are $AB + E + FG + CD$ and $AB - E - FG - CD$.

8.6 Other 2^{n-k} Fractions of Resolution V

For six factors the smallest 2^{n-k} fraction of resolution V is a half replicate. This can be defined either by the six-factor interaction, $I = \pm ABCDEF$, or by any one of the five-factor interactions. We would usually prefer one of the halves defined by the six-factor interaction, because the main effects would be aliased with five-factor interactions and the 2 f.i. with four-factor interactions.

For seven or eight factors, sixty-four points are needed. For seven factors this requires a half replicate. With eight factors we can obtain a suitable quarter replicate with two five-letter words and a six-letter word in the defining contrasts, for example, $I = ABCDE = ABFGH = CDEFGH$.

2^{n-k} fractions of resolution V for nine, ten, or eleven factors call for 128 points each. Examples of suitable sets of defining contrasts are

$$2^{9-2}:\quad I = ABCDEF = ABCGHJ = DEFGHJ;$$
$$2^{10-3}:\quad I = ABCDEF = ABCGHJ = ABDEGHK;$$
$$2^{11-4}:\quad I = ABCDEF = ABFJK = AEFGKL = ACEHL.$$

The first is a design of resolution VI. In the last two fractions only sets of generators for the defining contrast subgroups are given. For twelve or more factors at least 256 runs are needed. A set of generators for a 2^{12-4} design of resolution V is $ABCDJK$, $ABEFJL$, $ADEGLM$, $ABGHKL$.

8.7 Blocking Fractional Factorials

Blocking in fractional factorial designs is facilitated by the use of blocking factors. These are dummy factors which are supposed to interact with each other but not with the real factors. An example will illustrate the procedure. Suppose that we wish to divide the 2^{4-1} design of resolution IV into four blocks of two runs each. We introduce two blocking factors, $X = -AC$ and $Y = -AB$. This gives us eight treatment combinations: (1), *adxy*, *bdy*, *abx*, *cdx*, *acy*, *bcxy*, and *abcd*. We place into one block the points at the high level of X and the high level of Y, into the second block the points at low X and high Y, and so on. The blocks in the design are

(1), *abcd*; *ab*, *cd*; *ac*, *bd*; *ad*, *bc*.

This particular example illustrates another point. The design is a fold-over design, and any fold-over design of resolution IV can be divided into blocks of size two without losing its resolution IV property. Each block consists of a pair of complementary treatment combinations. This blocking procedure corresponds to taking the two-factor interactions as blocking factors.

We have already discussed the saturated designs of resolution III for $2^n - 1$ factors in 2^n runs in which only the main effects are estimated. If we use any one of the factors as a blocking factor, we have a design of resolution III for $2^n - 2$ factors in two blocks of size 2^{n-2}. Choosing two of the factors and their interaction, which will correspond to a third factor, gives a design for $2^n - 4$ factors in four blocks, and so on.

As an example we may construct a design of resolution III for twelve factors in four blocks of four.

We take the basic 2^4 design in the factors A, B, C, and D, and assign $X = AD$ and $Y = BC$ as blocking factors, giving the sixteen points *xy*, *ay*, *bx*, *ab*, *cx*, *ac*, *bcxy*, *abcy*, *d*, *axy*, *bd*, *abdxy*, *cd*, *acdx*, *bcdy*, *abcdxy*. Then, upon adding new factors $E = AB, F = AC, G = BD, H = CD, J = ABC, K = ABD, L = ACD$, and $M = BCD$, the blocks are the columns of the array

1	2	3	4
abehlm	*bfhjkm*	*aghjkl*	*efgh*
acfgkm	*cegjlm*	*abcefj*	*bckl*
bdfgjl	*abdegk*	*defklm*	*adjm*
cdehjk	*acdfhl*	*bcdghm*	*abcdefghjklm*.

An interesting example is found in the National Bureau of Standards list (p. 8). A 2^{7-1} design of resolution V is laid out in an 8×8 square. All main effects and 2 f.i. are estimable with both row and column confounding. The fraction is defined by $I = +ABCDEFG$. The blocking factors for the columns are $X = ACF$, $Y = ABE$, $Z = EFG$. For the row confounding we have $U = ACD$, $V = ABF$, and $W = BCE$.

(1)	*acef*	*adfg*	*cdeg*	*befg*	*abcg*	*abde*	*bcdf*
adef	*cd*	*eg*	*acfg*	*abdg*	*bcdefg*	*bf*	*abce*
bcef	*ab*	*abcdeg*	*bdfg*	*cg*	*aefg*	*acdf*	*de*
abcd	*bdef*	*bcfg*	*abeg*	*acdefg*	*dg*	*ce*	*af*
cdfg	*adeg*	*ac*	*ef*	*bcde*	*abdf*	*abcefg*	*bg*
aceg	*fg*	*cdef*	*ad*	*abcf*	*be*	*bcdg*	*abdefg*
bdeg	*abcdfg*	*abef*	*bc*	*df*	*acde*	*ag*	*cefg*
abfg	*bceg*	*bd*	*abcdef*	*ae*	*cf*	*defg*	*acdg*

8.8 Analyzing 2^{n-k} Fractions

In the example of the 2^{4-1} design defined by $I = +ABCD$, we proceeded as if the design were a 2^3 design in A, B, and C to which D had been added by putting $x_4 = x_1x_2x_3$. We applied Yates' algorithm to the basic 2^3 design and ignored the factor D until we wrote down the effects in the last column of the table. We could just as well have applied the algorithm to the 2^3 in A, B, and D, or in A, C, and D, or in B, C, and D. The general technique is to consider the 2^{n-k} fraction as a complete factorial in a set of $(n - k)$ of the factors, apply Yates' algorithm, and then incorporate the other factors in the last column.

It is not always the case that any subset of $(n - k)$ factors will provide a basic set. Consider for example the 2^{6-2} defined by $I = ABCD = ABEF = CDEF$. We cannot analyze this as a 2^4 in A, B, C, and D. Nor can we take as the basic set of four factors A, B, E, and F, or C, D, E, and F. We can take A, B, C, and E and put $x_4 = x_1x_2x_3$, $x_6 = x_1x_2x_5$, or else A, B, C, and F and put $x_5 = x_1x_2x_6$, but we cannot have both E and F out of the basic set. There is no defining contrast that contains E but not F, or F but not E, and it is impossible to express E as a product of factors that does not involve F. Indeed, any basic set of four factors in this example will have to contain at least one member from each of the pairs A, B; C, D; E, F.

In theory the problem of assigning the remaining effects and interactions to the alias sets with the correct signs presents no difficulties; all one has to do is to write down all the 2^{n-k} cosets of the defining contrast subgroup and cross out those interactions that are to be suppressed. Unfortunately, that procedure rapidly becomes a burden as the number of factors increases. We have considered in detail a 2^{7-4} design; even in a 2^{15-11} there are only sixteen points but there are 32,767 effects, including fifteen main effects and 105 two-factor interactions. Before continuing we digress to consider another problem.

Given a set of 2^{n-k} points which form a fractional factorial, how do we find the set of defining contrasts? This question has been known to arise in practice when, by the time the experiment has been completed, the information on how the fraction was constructed has been lost. It has been discussed by Nelder (1963). The method of solving the problems presented in this section is due to Margolin (1967).

We assume that we have already found by inspection a basic set of factors or letters. Margolin gives a simple method of doing this and calls the letters in the basic set live letters and the others dead. It will be helpful to consider an example. Consider the eight-point fraction *bc*, *acde*, *abdf*, *ef*, *ag*, *bdeg*, *cdfg*, *abcefg*. E, F, and G are a set of live letters. The first step is to find a defining contrast that contains A and some of the live letters, but no other dead letters. We shall then do the same for B, C, and D, and the four contrasts will be a set of generators for the defining-contrast subgroup. These contrasts are called the special defining contrasts for the dead letters. Margolin gives two rules.

To determine the sign of the special contrast for any dead letter in the defining relationship of the fraction, we examine the treatment combination in which all $(n - k)$ of the live factors are at their high levels. If, at this point, the particular dead factor is also at its high level, then the special contrast for that factor has a positive sign. If the factor is at its low level, the contrast has a negative sign. In the example the special contrasts for A, B, and C have positive signs; that for D has a negative sign.

We now find which live factors appear in the special contrast of the dead factor being considered. We consider each live factor in turn and look at two points, the point at which all the live factors are at their low levels, and the point at which the particular live factor is at its high level and the others are at their low levels. If the dead factor is at the same level in both points, then the live factor does not appear in the special contrast; if the levels are different, the live factor is present.

In the example we consider first the special contrast for A. We have already established that it has a positive sign. The point at which E, F, and G are all at their low levels is *bc*. In each of the points *acde*, *abdf*, *ag*, A is at its high level and therefore all three letters, E, F, and G, appear in the special contrast for A, which is thus $AEFG$. The other special contrasts are BEG, CFG, and $-DEF$.

We return now to the problem of the aliases. We have $A = EFG$, $B = EG$, $C = FG$, and $D = -EF$. We next write down the seven sets of three effects in the live 2^3 that have the identity for their generalized interaction: $(E, F, EF = -D)$, $(E, G, EG = B)$, $(E, FG = C, EFG = A)$, (F, FG, G), (F, EG, EFG), (G, EF, EFG), (EF, EG, FG). When the interactions are expressed as dead factors, the products give the set of three letter defining contrasts,

$$-DEF,\ BEG,\ ACE,\ CFG,\ ABF,\ -ADG,\ -BCD,$$

and the alias chains for the main effects and 2 f.i. follow immediately, $A + CE + BF - DG$, $B + EG + AF - CD$, and so on.

8.9 Designs Obtained by Omitting 2^{n-k} Fractions

We begin by considering an example. In the 2^{4-k} series there is nothing between the half replicate of resolution IV, from which the main effects can be estimated, and the complete factorial with sixteen points. The main effects and the six 2 f.i. call for only ten degrees of freedom. Can we estimate all of them with only twelve runs? John (1961*a*) showed that there were two ways in which this could be done by omitting a quarter replicate from the complete design. The problem arose in a practical situation: an experiment had to be planned with four factors at two levels each and there was only enough raw material available to make twelve runs. How might those twelve runs best be chosen?

Consider the family of quarter replicates defined by

(i)	$I = +AB = +ACD = +BCD$	*d, ab, c, abcd*;
(ii)	$I = -AB = +ACD = -BCD$	*bd, a, bc, acd*;
(iii)	$I = +AB = -ACD = -BCD$	(1), *abd, cd, abc*;
(iv)	$I = -AB = -ACD = +BCD$	*b, ad, bcd, ac*;

and suppose that we omit the first of them. The remaining fractions form three overlapping half replicates

(ii) + (iii) $I = -BCD$, (ii) + (iv) $I = -AB$, (iii) + (iv) $I = -ACD$.

We can estimate A from (ii) + (iii) since there A is aliased with $ABCD$; A is not, however, estimable from either of the other two half replicates. B is estimable from (iii) + (iv); C, D, and CD from (ii) + (iv); AC and AD from (ii) + (iii); BC and BD from (iii) + (iv). We shall see later that these are the least squares estimates. AB is estimable from both (ii) + (iii) and (iii) + (iv); the least squares estimate is the average of these two estimates.

We recall the model

$$E(y) = \beta_0 + \sum_i \beta_i x_i + \sum_{\substack{i\\ i<j}} \sum_j \beta_{ij} x_i x_j + \cdots$$

that we presented before, and again consider the normal equations $\mathbf{X'Y} = \mathbf{X'X}\hat{\boldsymbol{\beta}}$. The alias sets for the quarter replicates are *I, AB, ACD, BCD*; *A, B, CD, ABCD*; *C, ABC, AD, BD*; and *AC, BC, D, ABD*. We write the elements of the parameter vector $\boldsymbol{\beta}$ in that order: $\boldsymbol{\beta}' = (\beta_0, \beta_{12}, \beta_{134}, \beta_{234}, \beta_1, \beta_2, \beta_{34}, \beta_{1234}, \ldots)$.

The cross-product matrix, $\mathbf{X'X}$, for the complete design is $2^4\mathbf{I}_4$. The cross-product matrix for the omitted fraction (i) consists of submatrices $4\mathbf{J}_4$ along the main diagonal and zeros elsewhere. The cross-product matrix for the twelve-point design is the difference between these two matrices and consists of submatrices $16\mathbf{I}_4 - 4\mathbf{J}_4$ along the diagonal. Each of these submatrices is of rank 3. When we suppress from the model the higher order interactions and strike out the corresponding rows and columns of $\mathbf{X}'\mathbf{X}$, the first submatrix is reduced to $16\mathbf{I}_2 - 4\mathbf{J}_2$ with inverse $(2\mathbf{I}_2 + \mathbf{J}_2)/32$ and the other three submatrices become $16\mathbf{I}_3 - 4\mathbf{J}_3$ with inverses $(\mathbf{I}_3 + \mathbf{J}_3)/16$.

The inverse of the cross-product matrix is the variance-covariance matrix of the least squares estimates of the parameters. The estimate of A, for example, has variance $V(\hat{\beta}_1) = (1 + 1)\sigma^2/16 = \sigma^2/8$. This is the variance of the estimate from the half replicate (ii) + (iii), and thus, since the least squares estimates are the unique minimum variance estimates, the estimate from the half replicate is indeed the least squares estimate of A. Similarly, the variance of the average of the estimates of AB from the two intersecting half replicates (ii) + (iii) and (iii) + (iv) is $3\sigma^2/32$, which is the same as $(2 + 1)\sigma^2/32$, the variance of the least squares estimate.

The calculations in the previous example were simplified because we chose to omit the fraction for which the defining contrasts each had positive signs. The general case is only slightly more complicated. Suppose that we have omitted from a 2^n factorial any 2^{n-k} fraction. We can again write the elements of the parameter vector in order of alias sets. Then the cross-product matrix for the omitted fraction will consist of identical square submatrices $2^{n-k}\,\mathbf{P}$ along the diagonal and zeros elsewhere. The elements of $\mathbf{P}$ may be positive or negative, depending on the signs of the defining contrasts. They do, however, satisfy the following three conditions:

$$p_{ij} = p_{ji}; \qquad p_{ij} = \pm 1; \qquad p_{hj}p_{ij} = p_{hi}.$$

For any matrix $\mathbf{P}_m$ of order m with elements satisfying these conditions it follows that $\mathbf{P}_m^2 = m\mathbf{P}_m$; $\mathbf{J}_m$ is a special case of the class of $\mathbf{P}$ matrices. This is true in particular of the matrix obtained by striking out any $(2^k - m)$ columns of $\mathbf{P}$ and the corresponding rows.

Proceeding as before we see that the cross-product matrix for the actual design consists of singular matrices $2^n\mathbf{I} - 2^{n-k}\mathbf{P}$ along the main diagonal and zeros elsewhere. If we can find at least one effect in each alias set that can be suppressed, we can strike out that row and column and obtain a nonsingular matrix. We can readily show that when $a\mathbf{I}_m + b\mathbf{P}_m$ is not singular, its inverse is $c\mathbf{I}_m + d\mathbf{P}_m$ ($\mathbf{P}_m$ being the same in both matrices), where $ac = 1$ and $d = -bc/(a + mb)$. In this instance we need the inverse of $2^n\mathbf{I}_m - 2^{n-k}\mathbf{P}_m$, which is $c\mathbf{I}_m + d\mathbf{P}_m$ where $c = 2^{-n}$ and $d = 2^{-n}(2^k - m)^{-1}$.

We have now shown that if none of the effects in an alias set is suppressed, none of them is estimable, but, if at least one effect is suppressed, the others are estimable. Furthermore, if m of the effects in an alias set are to be estimated, their least squares estimates will all have the same variance $(c + d)\sigma^2$ or

$$V(\hat{\beta}) = 2^{-n}(2^k - m + 1)\sigma^2/(2^k - m).$$

The covariance of estimates of parameters in the same alias set is $d\sigma^2$ or $2^{-n}\sigma^2/(2^k - m)$. The covariance of estimates of effects in different alias sets is zero.

We now turn to the problem of finding estimates that have these variances. If we can do so, they will be the unique least squares estimates, and the covariances will look after themselves. Now that we have established the variance-covariance matrix of the estimates, there will be no loss of generality if we simplify matters by assuming that each of the defining contrasts in the omitted fraction has a positive sign. Suppose that it is defined by $I = P = Q = PQ = \cdots$. The actual design then consists of $2^k - 1$ overlapping half replicates defined by $I = -P$, $I = -Q$, $I = -PQ$, and so on. Each pair of these half replicates has a quarter in common: for example, the half replicates defined by $I = -P$ and by $I = -Q$ both share the points for which $I = -P = -Q = +PQ$.

Let S be an effect that is to be estimated. In the half replicate defined by $I = -P$, S is aliased with PS, and if PS is suppressed, S can be estimated from that half replicate. If $2^k - m$ of the effects in the same alias set as S are suppressed, S is estimable from $2^k - m$ overlapping half replicates. Each of these estimates has variance $2^{-(n-1)}\sigma^2$; their covariance is $2^{-n}\sigma^2$. Let $\hat{\beta}$ denote the average of these estimates. Then

$$V(\hat{\beta}) = (2^k - m)^{-2}\{(2^k - m)2^{-(n-1)} + (2^k - m)(2^k - m - 1)2^{-n}\}\sigma^2$$
$$= 2^{-n}(2^k - m + 1)\sigma^2/(2^k - m),$$

which is the same as the variance of the least squares estimate.

We have thus established the following method of calculating the least squares estimates of effects. To obtain the least squares estimate of an effect S, take the estimates of S from each of the half replicates in which it is estimable and average them.

8.10 Three-Quarter Replicates

The simplest designs of this type are the three-quarter replicates or $3(2^{n-2})$ designs [John (1962)]. These fractions are obtained by omitting one of the quarters from a full factorial and are said to be defined by the missing quarter. Thus the twelve-point design just presented is the $3(2^{4-2})$ design defined by $I = AB = ACD = BCD$. Had the set of points b, ad, bcd, ac been omitted instead, we should have had the $3(2^{4-2})$ defined by $I = -AB = -ACD = BCD$. We can add other factors to $3(2^{n-2})$ designs and obtain $3(2^{n-k})$ designs.

The smallest of these designs has six points. We can omit from the 2^3 factorial any one of the quarters defined by $I = \pm AB = \pm AC = \pm BC$. These are (1), abc; a, bc; b, ac; ab, c. The alias sets are I, AB, AC, BC and A, B, C, ABC. Each of the main effects is estimated from the half replicate in which it is aliased with ABC. The 2 f.i. and the mean are hopelessly confounded with each other. The design is a resolution IV design. The variance efficiency of the estimates of the main effects is $8/12 = 2/3$.

This design can be made into a $3(2^{5-3})$ fraction of resolution III by adding two more factors, D and E. They are equated to any two of the interactions

$\pm AB$, $\pm AC$, $\pm BC$. Suppose that we let $D = -AB$ and $E = -BC$. Then D and E are estimated from the half replicates in which they are aliased with AC. If we omit the quarter defined by $I = -AB = -AC = +BC$, the six points are (1), *abc*, *bde*, *acde*, *abe*, *ce*.

A and D are estimated from the half replicate defined by $I = -BC$. The A contrast is $abe + acde - ce - bde$. The D contrast is $acde + bde - abe - ce$. B and the general mean are estimated from the half replicate defined by $I = +AC$. The B contrast is $abc + bde - (1) - acde$; $4\hat{\beta}_0 = (1) + abc + bde + acde$. C and E are estimated from the half replicate defined by $I = +AB$. The C contrast is $abc + ce - (1) - abe$; the E contrast is $abe + ce - (1) - abc$.

In general, any $3(2^{n-2})$ fraction can be made into a design of resolution III for $3(2^{n-2}) - 1$ factors. New factors are assigned to interactions in such a way that three of the effects in each alias set correspond to main effects and one is left over. Each main effect is then estimated from the half replicate in which it is aliased with the suppressed effect that is left over. In this way we can convert the $3(2^{4-2})$ fraction mentioned earlier into a design of resolution III for eleven factors. We can let $E = AB$, $F = ACD$, $G = CD$, $H = ABC$, $J = AD$, $K = AC$, $M = BC$ with BCD, $ABCD$, BD, and ABD left over. The design would then be

(ii)	*bdfhk*	*afghm*	*bcfjm*	*acdfgjk*
(iii)	*egjkm*	*abdej*	*cdegh*	*abcehkm*
(iv)	*bghjk*	*adhjm*	*bcdgm*	*ack*,

where (ii), (iii), and (iv) refer to the original quarters. The main effects B, E, H, and M are estimated from (iii) + (iv); D, F, G, and J are estimated from (ii) and (iv); the mean, A, C, and K are estimated from (ii) + (iii).

The design that we gave earlier is not the only $3(2^{4-2})$ of resolution III. We could also have omitted one of the quarters defined by $I = \pm A = \pm BCD = \pm ABCD$. Some experimenters might prefer to use this fraction because in it the one effect that is estimated from two half replicates is the main effect A rather than a 2 f.i.

The $3(2^{4-2})$ defined by $I = AB = AC = BC$ and the $3(2^{4-2})$ defined by $I = AB = CD = ABCD$ are also of interest. Neither is of resolution V, the latter having an alias set, AC, BC, AD, BD, which contains only two-factor interactions. Both, however, are of resolution IV and can be made to accommodate six factors. Consider in particular the first design. The alias sets are I, AB, AC, BC; AD, BD, CD, $ABCD$; A, B, C, ABC; and D, ABD, ACD, BCD. We can add two more factors, E and F, by setting them equal to two of the interactions $\pm ABD$, $\pm ACD$, $\pm BCD$. The main effects will be clear of 2 f.i.; the 2 f.i. will be hopelessly confounded. If we let $E = +ABD$ and $F = +ACD$, the twelve points will be

aef	*bcef*	*be*	*ace*	*abf*	*cf*
bcd	*ad*	*acdf*	*bdf*	*cde*	*abde*.

It should be noted that this is a fold-over design and that it can be divided into six blocks of two runs each without losing its resolution IV property. The blocks are the columns of the array in which we have written the points.

This illustrates a general procedure for obtaining a resolution IV design for $3(2^{n-3})$ factors in $3(2^{n-2})$ runs. We take a $3(2^{n-2})$ fraction defined by omitting a quarter replicate in which every defining contrast contains an even number of letters. The alias sets then consist either of four effects each with an even number of letters, or else of four effects each with an odd number of letters. The new factors are added by equating them to odd interactions, leaving one over in each alias set. The 2 f.i. will then all have an even number of letters, and so they will not appear in the same alias sets as the main effects.

8.11 The $3(2^{5-2})$ Designs

We have already seen that the 2^{5-1} fractions defined by $I = \pm ABCDE$ are saturated fractions of resolution V with sixteen points. It is interesting to investigate the gains, if any, that can be made by using twenty-four points. To what extent can we, by a suitable choice of twenty-four points, obtain better estimates, i.e., estimates with smaller variances? It is not immediately obvious that any improvement can be made, apart from the fact that we shall have an estimate of error with 8 d.f. from the actual data; after all, in a three-quarter replicate some of the effects are only estimable from a single half replicate, which is what we have in the 2^{5-1} designs.

Apart from permutations of the letters and choices of sign in the defining contrasts of the omitted fraction, there are only six $3(2^{5-2})$ fractions of resolution V. The defining contrasts and alias sets for them are given in the table on page 166.

Designs 4 and 5 have the property that each alias set contains exactly two of the effects that are to be estimated. In these designs we have $V(\hat{\beta}_h) = V(\hat{\beta}_{hi}) = 3\sigma^2/64$; the half replicate gives $V(\hat{\beta}_h) = V(\hat{\beta}_{hi}) = \sigma^2/16$. Designs 1, 2, and 3 each have three alias sets with only one suppressed effect. In each case these alias sets contain main effects, which means that in these designs several of the main effects are only estimated from single half replicates. Taking the extra points has not improved the precision of the estimates of those main effects.

When we consider blocking, the differences between the designs become more apparent. The choice of the blocking variables in one of these designs affects not only the number of blocks but also their sizes [John (1964*b*)]. When one of the defining contrasts is used as a blocking factor in a $3(2^{n-2})$ design, we have two blocks; one of these is a half replicate and the other is the remaining quarter. If one of the other effects is used, we have two blocks of equal size. If two effects in the same alias set are used for blocking, their product, which is a defining contrast, is also a blocking factor. Thus, if two of the defining contrasts are blocking factors, so is the third, and the blocks are the quarters; if one of the other alias sets contains two blocking factors, there are four blocks, two

1				2				3			
I	*A*	*BCD*	*ABCD*	*I*	*AB*	*ACD*	*BCD*	*I*	*AB*	*ACDE*	*BCDE*
B	*AB*	*CD*	*ACD*	*A*	*B*	*CD*	*ABCD*	*A*	*B*	*CDE*	*ABCDE*
C	*AC*	*BD*	*ABD*	*C*	*ABC*	*AD*	*BD*	*C*	*ABC*	*ADE*	*BDE*
BC	*ABC*	*D*	*AD*	*AC*	*BC*	*D*	*ABD*	*AC*	*BC*	*DE*	*ABDE*
E	*AE*	*BCDE*	*ABCDE*	*E*	*ABE*	*ACDE*	*BCDE*	*E*	*ABE*	*ACD*	*BCD*
BE	*ABE*	*CDE*	*ACDE*	*AE*	*BE*	*CDE*	*ABCDE*	*AE*	*BE*	*CD*	*ABCD*
CE	*ACE*	*BDE*	*ABDE*	*CE*	*ABCE*	*ADE*	*BDE*	*CE*	*ABCE*	*AD*	*BD*
BCE	*ABCE*	*DE*	*ADE*	*ACE*	*BCE*	*DE*	*ABDE*	*ACE*	*BCE*	*D*	*ABD*

4				5				6			
I	*A*	*BCDE*	*ABCDE*	*I*	*AB*	*CDE*	*ABCDE*	*I*	*ABC*	*ADE*	*BCDE*
B	*AB*	*CDE*	*ACDE*	*B*	*A*	*BCDE*	*ACDE*	*A*	*BC*	*DE*	*ABCDE*
C	*AC*	*BDE*	*ABDE*	*C*	*ABC*	*DE*	*ABDE*	*B*	*AC*	*ABDE*	*CDE*
BC	*ABC*	*DE*	*ADE*	*BC*	*AC*	*BDE*	*ADE*	*AB*	*C*	*BDE*	*ACDE*
D	*AD*	*BCE*	*ABCE*	*D*	*ABD*	*CE*	*ABCE*	*D*	*ABCD*	*AE*	*BCE*
BD	*ABD*	*CE*	*ACE*	*BD*	*AD*	*BCE*	*ACE*	*AD*	*BCD*	*E*	*ABCE*
CD	*ACD*	*BE*	*ABE*	*CD*	*ABCD*	*E*	*ABE*	*BD*	*ACD*	*ABE*	*CE*
BCD	*ABCD*	*E*	*AE*	*BCD*	*ACD*	*BE*	*AE*	*ABD*	*CD*	*BE*	*ACE*
	$R = ABE$				$R = ACD$				$R = ACE$		
	$S = ABCD$				$S = BCE$				$S = BCD$		
	$RS = CDE$				$RS = ABDE$				$RS = ABDE$		

with 2^{n-2} points each and two with 2^{n-3} points each. Finally, if two effects in different alias sets, neither of which is a defining contrast, are used for blocking, their product is in a third alias set and we have four blocks of equal size. In all cases, for an effect to be estimable there must be a suppressed effect in the alias set in addition to any blocking factors.

Designs 4, 5, and 6 can each be blocked into four blocks of six points each by using as blocking factors the three factors R, S, and RS, which are given in each case. Design 6 can be run in separate quarters. Designs 1, 2, and 3 may be split into blocks of sizes 8, 8, 4, and 4. A particularly interesting property of designs 4 and 5 is that they may be blocked, using $ABCDE$, into the half replicate defined by $I = \pm ABCDE$ and the remaining quarter. An implication of this is that we may elect to complete either of these designs if, after the half replicate has been run, we find ourselves able to make an additional eight runs. Alternatively, we may choose to run the half replicate first anyway, as a precaution against finding ourselves in a situation in which we are not able to complete the twenty-four point sequence.

8.12 Some $3(2^{n-k})$ Designs of Resolution V

The smallest 2^{8-k} design of resolution V is a quarter replicate with sixty-four points. The following $3(2^{8-4})$ fraction has resolution V in forty-eight points. Omitting one of the factors gives a design of resolution V for seven factors in forty-eight points. We omit from the complete 2^6 factorial the quarter defined by $I = ABE = CDF = ABCDEF$, and add new factors by putting $G = ABCD$ and $H = ACEF$. The three quarters of the design are

(i) *fg*, *aef*, *befh*, *abfgh*, *c*, *aceg*, *bcegh*, *abch*, *dh*, *adegh*, *bdeg*, *abd*, *cdfgh*, *acdefh*, *bcdef*, and *abcdfg*;
(ii) *eg*, *a*, *bh*, *abegh*, *cef*, *acfg*, *bcfgh*, *abcefh*, *defh*, *adfgh*, *bdfg*, *abdef*, *cdegh*, *acdh*, *bcd*, and *abcdeg*;
(iii) *gh*, *aeh*, *be*, *abg*, *cfh*, *acefgh*, *bcefg*, *abcf*, *df*, *adefg*, *bdefgh*, *abdfh*, *cdg*, *acde*, *bcdeh*, and *abcdgh*.

The three quarters may be used as blocks.

With ninety-six points we can obtain a design of resolution V for eleven factors. This is achieved by omitting from the 2^7 factorial the quarter defined by $I = AB = CDEFG = ABCDEFG$ and adding new factors $H = ABDEF$, $J = ABCEG$, $K = AEFG$, and $L = BCDE$. The alias sets are presented in John (1969).

8.13 Adding Fractions

The three-quarter replicate can just as well be obtained by adding a quarter replicate to a half replicate. The addition of other fractions to a half replicate does not, however, give such neat results. There are some useful designs that can be obtained in this way and in them, too, the least squares estimates of the estimable effects can be pieced together from the estimates from the constituent fractions. We shall illustrate the procedure by an example and state the results for the general case.

Suppose that we take a 2^{n-1} design defined by $I = -P$ where P is some effect, and add to it the 2^{n-3} fraction defined by $I = P = Q = PQ = R = PR = QR = PQR$. Let S be some other effect. The alias set, which contains S, consists of S, PS, QS, PQS, RS, PRS, QRS, and $PQRS$. This alias set is composed of four pairs of effects, such as S, PS, which are themselves aliased in the half replicate. If one of the effects in each pair is suppressed, the other four effects can be estimated from the half replicate without the need for any additional points. The extra points are needed only to break chains of aliases, such as $S-$PS, when neither effect is suppressed.

As an example, suppose that in a 2^4 factorial we take the half replicate defined by $I = -ABCD$ and add the 2^{4-3} defined by $I = ABCD = A = BCD = B = ACD = AB = CD$. There are two alias sets:

$$M, ABCD; \quad A, BCD; \quad B, ACD; \quad AB, CD;$$
$$C, ABD; \quad AC, BD; \quad BC, AD; \quad D, ABC.$$

As usual, we suppress the interactions with three or four factors.

We shall call a pair of effects such as S and PS a nonnegligible pair if neither S nor PS is suppressed. The first alias set contains one nonnegligible pair AB, CD; the second set has two such pairs AC, BD and AD, BC. The points in the design are

$$abc, abd, a, acd, b, bcd, c, d, ab, abcd.$$

We can take the extra eighth replicate (ab, $abcd$) and combine it in turn with each of the eighths in the original half replicate that have the same defining contrasts. This will give us four overlapping 2^{4-2} fractions:

$$I = A = B = AB, \qquad abc, abd, ab, abcd;$$
$$I = A = CD = ACD, \qquad a, acd, ab, abcd;$$
$$I = B = CD = BCD, \qquad b, bcd, ab, abcd;$$
$$I = AB = ACD = BCD, \qquad c, d, ab, abcd.$$

From the first of these quarters we have an estimate of CD: $\widehat{CD} = (ab + abcd - abc - abd)/4$. In the half replicate the CD contrast gives an estimate of $CD - AB$: $(a + acd + b + bcd - abc - abd - c - d)/8$. If we subtract this from the estimate CD, we obtain an estimate of AB:

$$\widehat{AB} = (2ab + 2abcd - abc - abd - a - acd - b - bcd + c + d)/8.$$

Both $\widehat{CD}$ and $\widehat{AB}$ have variances $\sigma^2/4$. Also the mean, A, and B are estimable from the original half replicate. Thus all the effects in the first alias set that are not suppressed are estimable.

We find in considering the submatrix of the cross-product matrix that if we delete the suppressed effects and write the remaining effects in the order M, A, B, AB, and CD the submatrix is

$$\mathbf{U}_1 = 2\begin{pmatrix} 4\mathbf{I}_3 + \mathbf{J}_3 & \mathbf{E}_{3,2} \\ \mathbf{E}_{2,3} & 8\mathbf{I}_2 - 3\mathbf{J}_2 \end{pmatrix}$$

where $\mathbf{E}_{m,n}$ is a matrix of m rows and n columns with every element unity. The corresponding covariance submatrix is

$$\mathbf{U}_1^{-1} = 2^{-4}\begin{pmatrix} 2\mathbf{I}_3 & -\mathbf{E} \\ -\mathbf{E} & \mathbf{I}_2 + 3\mathbf{J}_2 \end{pmatrix}.$$

This shows that $V(\hat{\beta}_0) = V(\hat{\beta}_1) = V(\hat{\beta}_2) = \sigma^2/8$ and $V(\hat{\beta}_{12}) = V(\hat{\beta}_{34}) = \sigma^2/4$, and so the estimates that we have given for M, A, B, AB, and CD are the least squares estimates.

On the other hand, the cross-product submatrix for the other alias set is

$$\mathbf{U}_2 = 2\begin{pmatrix} 4\mathbf{I}_2 + \mathbf{J}_2 & \mathbf{J}_2 & \mathbf{J}_2 \\ \mathbf{J}_2 & 8\mathbf{I}_2 - 3\mathbf{J}_2 & \mathbf{J}_2 \\ \mathbf{J}_2 & \mathbf{J}_2 & 8\mathbf{I}_2 - 3\mathbf{J}_2 \end{pmatrix},$$

where the first two columns correspond to C and D, the next two to the nonnegligible pair AC and BD, and the last two to the nonnegligible pair AD and BC. The matrix $\mathbf{U}_2$ is singular. The main effects C and D are estimable from the original half replicate, but the addition of the two extra points does not break either of the chains $AC - BD$ or $AD - BC$.

The results in the general case where a 2^{n-k} fraction is added to a disjoint half replicate are as follows:

(i) Any effects that are estimable in the half replicate are estimable in the augmented design (the worst that can happen is that the estimates will be the same);

(ii) If an alias set contains more than one nonnegligible pair, the only estimable effects that can be estimated are those that are estimated in the half replicate, and the estimates from the half replicate are the least squares estimates of the effects from all $2^{n-1} + 2^{n-k}$ points;

(iii) If an alias set contains only one nonnegligible pair and, also, m pairs with one effect suppressed and one to be estimated ($m \leq 2^{k-1} - 1$), all the effects are estimable.

In the last case, when $m < 2^{k-1} - 1$, and in the case in which an alias set has no nonnegligible pairs, the least squares estimates are not simply the estimates from the half replicate. They may be found by following the procedures derived in John (1966*b*) or by using the results obtained by Plackett (1950).

This technique of adding fractions may be used to obtain a design of resolution V for seven factors in thirty-six runs. The smallest such 2^{7-k} fraction has sixty-four runs; any of the half replicates having as the defining contrast an interaction involving five, six, or seven factors will suffice. A design with forty-eight runs is obtained by using a $3(2^{7-3})$ fraction; one can, for example, omit the quarter defined by $I = ABC = DEF = ABCDEF$ from the complete 2^6 design and add a new factor by putting $G = ABCD$. The thirty-six point design is a subset of the forty-eight point fraction. It is obtained as follows. We add to the 2^{6-1} defined by $I = -ABCDEF$ the four-point fraction generated by $I = +ABCDEF = +ABDEF = -ABC = +BCDEF = \ldots$; the seventh factor is again added by putting $G = ABCD$.

Also of interest is an eighty-point design of resolution V for nine factors. This is obtained by adding to the 2^{7-1} defined by $I = -AB$ the 2^{7-3} defined by $I = AB = CDEFG = ABCDEFG = ACF = BCF = ADEG = BDEG$ and then putting $H = AEFG$ and $J = BCDE$.

8.14 Nested Fractions

We have already mentioned that one of the differences between industrial and agricultural experimentation is that in the former the runs can often be made in sequence. Advantage can be taken of this in planning an experimental strategy. There is also a disadvantage because experiments may by misfortune be terminated before all the runs have been made. The designer would be prudent to anticipate such risks.

We expect the size of an experiment to be determined in the planning stage by budgetary considerations or, perhaps on some occasions, by calculations involving the power of the t or F tests. Very often, however, time plays an important part because the scientist is allotted a pilot plant for only a limited time. For example, he might have the use of a pilot plant for two weeks, or for ten working days. If he can make two runs a day, and if he allows the first day to set up his equipment, that gives him eighteen runs. A 2^4 design followed by two check points on the last day might be appropriate.

But what are the consequences if something goes wrong with the machinery and three days are lost while a pump is being replaced? Such misfortunes are not uncommon. The loss of three days in the experiment mentioned in the previous paragraph would mean that instead of eighteen runs the experimenter could only carry out the first twelve of the sixteen points in the 2^4 design. Suppose that the last four runs that had been scheduled (but never made) had been (1), *abd*, *cd*, *abc*. Then he would have been lucky because the twelve points that were actually carried out would have formed a $3(2^{4-2})$ of resolution V.

On the other hand, if he had lost (1), *bd*, *bcd*, *ab* he would have been left with a twelve-point fraction that can, at best, be called messy. The only half replicate that it contains is the one defined by $I = +ABD$.

This suggests that the experimenter should modify complete randomization by arranging his runs so that if the experiment ends prematurely he still has a useful fraction to analyze. The experimental strategy should be directed toward designs in which fractions are nested within the overall design so that a partially completed design has a high salvage content.

An obvious possibility is to choose as the first eight points one of the fractions defined by $I = +ABCD$ or by $I = -ABCD$ so that even if he finishes only half his assigned runs he does at least have a resolution IV design. He might then follow that by a set of four points at the same level of one of the factors, and thereby obtain one of the $3(2^{n-2})$ replicates of the type $I = A = BCD = ABCD$.

We have already noted that the $3(2^{5-2})$ fraction defined by $I = E = ABCD = ABCDE$ can be split into a half replicate and a quarter replicate by using any of the defining contrasts as a blocking factor. This indicates two strategies. An experimenter who starts on the twenty-four point sequence but doubts his ability to finish the program can make use of the fact that the 2^{5-1} design of resolution V defined by $I = \pm ABCDE$ is included in his set of points and he may decide to run that half replicate first. On the other hand, an experimenter who had carried out a complete 2^4 design in A, B, C, and D (holding E fixed) could, if he found himself with time for another eight runs, proceed to carry out one of the 2^{4-1} fractions defined by $I = \pm ABCD$ at the other level of E, thus converting his 2^4 design into a $3(2^{5-2})$ design of resolution V.

Indeed, if after he has carried out the $3(2^{5-2})$ fraction defined by $I = E = ABCDE = ABCD$ which consists of all points except the omitted fraction

$$e, ade, bde, abe, cde, ace, bce, abcde,$$

he wishes to go further and add the set

$$e, ade, bde, abe,$$

he would then have a twenty-eight point $7(2^{5-3})$ design with alias sets

$$I, E, ABCDE, ABCD, ABD, ABDE, CE, C;$$

$$A, AE, BCDE, BCD, BD, BDE, ACE, AC;$$

$$B, BE, ACDE, ACD, AD, ADE, BCE, BC;$$

$$AB, ABE, CDE, CD, D, DE, ABCE, ABC.$$

Each alias set contains at least one four-factor interaction, and so the design is of resolution VII.

8.15 Resolution III Designs

A resolution III design is defined as a design in which all the main effects are estimable when all the interactions are suppressed. It can be shown, by following the arguments that will be used in the next section for resolution IV designs, that in a resolution III design the mean β_0 must also be estimable. Hence, a resolution III design for n factors must contain at least $n + 1$ points. We shall use, henceforth, the notation $2^n/\ /N$ to denote an N-point fraction of a 2^n factorial.

We have already seen examples of designs in which this lower bound is attained. These are the saturated designs for $2^n - 1$ factors in 2^n runs, and for $3(2^{n-2}) - 1$ factors in $3(2^{n-2})$ runs. The former series of designs have variance efficiency one; in the $3(2^{n-k})$ series of saturated designs the efficiency is only two-thirds and, if $n + 1$ is divisible by four, it may be preferable to use the Plackett and Burman designs (1946). These are main effects designs for $2^{n-1}/\ /N$ with $N = 4k$ where k is an integer. They have variance efficiency one. The Plackett and Burman designs will be discussed with the main effects plans of Addelman in the next chapter.

Other main effects plans have been considered under the heading of weighing designs by Hotelling (1944*a*), Kishen (1945), Mood (1946), Raghavarao (1959), and Yang (1966) and (1968). A discussion of them is found in the paper by Margolin (1969*a*). We shall merely point out here that a $2^n/\ /n + 1$ design of resolution III can always be obtained in the following way. The ith run, $1 \leq i \leq n$, has the ith factor at its low level, and all the other factors at their high levels; the $(n + 1)$th point has all the factors at their low levels.

Taking the model

$$E(Y) = \beta_0 + \sum_i \beta_i x_i,$$

we have normal equations $\mathbf{X'Y} = \mathbf{X'X}$ where $\mathbf{X}$ is the design matrix and includes a vector $\mathbf{1}$ for β_0. The moment matrix $\mathbf{X'X}$ and its inverse, the covariance matrix, are

$$\mathbf{X'X} = 4\mathbf{I}_{n+1} + (n - 3)\mathbf{J}_{n+1}, \qquad (\mathbf{X'X})^{-1} = \frac{\mathbf{I}}{4} - \frac{(n - 3)\mathbf{J}}{4(n - 1)^2},$$

so that $V(\hat{\beta}_i) = [(n^2 - 3n + 4)/4(n - 1)^2]\sigma^2$ and $\lim_{n\to\infty} V(\hat{\beta}_i) = \sigma^2/4$.

We can do considerably better by using Yang's results. We can also do considerably worse by replacing the last run with the point at which all the factors take their high levels. We should then have the classical one-at-a-time design. This is a resolution III $2^n/\ /n + 1$ design with $V(\hat{\beta}_i) = \sigma^2/2$ for each i and for all n.

8.16 Designs of Resolution IV

Webb (1968), noting that folding over converted a 2^{n-k} design of resolution III into a 2^{n+1-k} fraction of resolution IV, showed that folding over will convert any resolution III design for n factors into a resolution IV design for $n + 1$ factors. He also showed that any resolution IV design for n factors must contain at least $2n$ points. He called such a design with only $2n$ points a minimal design and conjectured that any minimal resolution IV design must be a fold-over design. This conjecture was proved by Margolin (1969*c*); Margolin had proved it earlier (1969*b*) in the special case in which each factor appeared exactly n times at its high level and n times at its low level.

A design is of resolution IV if, and only if, the main effects are estimable when all interactions with three or more factors are suppressed. The mean may, or may not, be estimable. We have already mentioned the resolution IV designs for $2^3//6$, $3(2^{3-2})$. In these designs the two alias sets are I, AB, AC, BC, and A, B, C, ABC, and the mean is not estimable unless at least one of the 2 f.i. is suppressed.

We consider a model that contains only the mean, the main effects, and the 2 f.i.:

$$E(y) = \beta_0 + \sum_h \beta_h x_h + \sum_h \sum_i \beta_{hi} x_h x_i, \qquad 1 \le h < i \le n,$$

and a design with N points.

The normal equations are

$$\mathbf{X}'\mathbf{Y} = \mathbf{X}'\mathbf{X},$$

where $\mathbf{X}$ is a matrix of $1 + n + n(n-1)/2$ columns and N rows.

We denote by $\mathbf{X}_h$ the vector with elements x_{hj} where x_{hj} is the value of x_h for the jth datum point; similarly $\mathbf{X}_{hi}$ is the vector with elements $x_{hj}x_{ij}$; $\mathbf{X}_{hi}$ is said to be the Hademard product of $\mathbf{X}_h$ and $\mathbf{X}_i$. The vector $\mathbf{X}_0 = \mathbf{1}$ corresponds to the mean; $\mathbf{X}_0, \ldots, \mathbf{X}_h, \ldots, \mathbf{X}_{hi}, \ldots$ are the columns of $\mathbf{X}$.

If all the main effects β_h, $h > 0$, are to be estimable, then the vectors $\mathbf{X}_h$, $h > 0$, and $\mathbf{X}_0$ must be linearly independent. Furthermore, the main effects vectors $\mathbf{X}_h$ must be linearly independent of the interaction vectors $\mathbf{X}_{hi}$. Suppose that this were not so and that, for example, $\mathbf{X}_1 = \mathbf{Xc}$ where $\mathbf{c}$ is a vector having $c_1 = 0$ and at least one element other than c_1 not zero. We are given that β_1 is estimable. Let $\hat{\beta}_1 = \mathbf{a}'\mathbf{Y}$. Then

$$E(\hat{\beta}_1) = \beta_1 = \mathbf{a}'E(\mathbf{Y}) = \mathbf{a}'\left\{\mathbf{X}_0\beta_0 + \sum_h \mathbf{X}_h\beta_h + \sum_h \sum_i \mathbf{X}_{hi}\beta_{hi}\right\},$$

and, since this is an identity, $\mathbf{a}'\mathbf{X}_1 = 1$ and $\mathbf{a}'\mathbf{X}_0 = \mathbf{a}'\mathbf{X}_h = \mathbf{a}'\mathbf{X}_{hi} = 0$, $h > 1$, which contradicts the assumption that $\mathbf{X}_1$ is a linear combination of the other vectors. The converse, that if all the main effects vectors are linearly independent of the mean and interaction vectors, then the main effects are estimable, is easily shown.

Consider now the set of $2n$ vectors $\mathbf{X}_1, \mathbf{X}_2, \ldots, \mathbf{X}_0, \mathbf{X}_{12}, \mathbf{X}_{13}, \ldots$. The elements of $\mathbf{X}_{1h}$ are obtained by multiplying each element x_{hj} of $\mathbf{X}_h$ by x_{1j}, which is equal to ± 1; the elements of $\mathbf{X}_0$ are the squares of the elements of $\mathbf{X}_1$. The vectors $\mathbf{X}_h$ are linearly independent, i.e., there exists no set of scalars c_h such that $\sum_h c_h x_{hj} = 0$ for all j. In particular, it is not true that either $+1(\sum x_{hj})$ or $-1(\sum x_{hj}) = 0$ for each j. Hence, the set of vectors $\mathbf{X}_0, \mathbf{X}_{12}, \ldots$ are linearly independent, and it has already been shown that $\mathbf{X}_1, \mathbf{X}_2, \ldots$ are independent of them. It follows that $\mathbf{X}_1, \mathbf{X}_2, \ldots, \mathbf{X}_0, \mathbf{X}_{12}, \ldots$, is a linearly independent set of $2n$ vectors. Thus $r(\mathbf{X}) = r(\mathbf{X}'\mathbf{X}) \geq 2n$. But $r(\mathbf{X}'\mathbf{X}) \leq N$ and so $N \geq 2n$, which proves that every resolution IV design for n factors must contain at least $2n$ points. Since we have already seen several examples of resolution IV $2^n//2n$ designs, we know that this lower bound is attainable.

We now show that when a resolution III design for n factors is folded over, the resulting design is of resolution IV and can indeed be made into a resolution IV design for $n + 1$ factors. We consider the $(n + 1)$th factor to be held at its low level throughout the resolution III design (if we choose to ignore this factor there will be no difficulty). We write the $\mathbf{X}$ matrix for the resolution III design as

$$\mathbf{X} = (\mathbf{1} \quad \mathbf{U} \quad \mathbf{V}),$$

where $\mathbf{1}$ corresponds to the mean, $\mathbf{U} = (\mathbf{X}_1, \mathbf{X}_2, \ldots, \mathbf{X}_{n+1})$ and $\mathbf{V}$ consists of the vectors $\mathbf{X}_{hi}$. The rank of $\mathbf{U}$ is $n + 1$. For the folded-over portion, with the $(n + 1)$th factor now at its high level,

$$\mathbf{X} = (\mathbf{1} \quad -\mathbf{U} \quad \mathbf{V}).$$

For the combined design

$$\mathbf{X}'\mathbf{X} = \begin{pmatrix} \mathbf{1} & \mathbf{U} & \mathbf{V} \\ \mathbf{1} & -\mathbf{U} & \mathbf{V} \end{pmatrix}' \begin{pmatrix} \mathbf{1} & \mathbf{U} & \mathbf{V} \\ \mathbf{1} & -\mathbf{U} & \mathbf{V} \end{pmatrix} = \begin{pmatrix} 2N & \mathbf{0} & 2\mathbf{1}'\mathbf{V} \\ \mathbf{0} & 2\mathbf{U}'\mathbf{U} & \mathbf{0} \\ 2\mathbf{V}'\mathbf{1} & \mathbf{0} & 2\mathbf{V}'\mathbf{V} \end{pmatrix}.$$

The main effect submatrix $2\mathbf{U}'\mathbf{U}$ is nonsingular. The estimable main effects are orthogonal to the mean and the interactions.

We have already seen that there are elementary weighing designs for $n - 1$ factors in n runs with resolution III. Folding them over gives minimal resolution IV designs $2^n//2n$ for any n. These are not necessarily the most efficient $2^n//2n$ designs, and some more efficient designs are given in the papers by Webb (1968) and Margolin (1969*a*) that have already been cited. We shall not give Margolin's proofs of Webb's conjecture that all minimal resolution IV designs are fold-over designs. The reader is referred to Margolin's original papers.

Exercises

1. Analyze the fraction obtained by omitting the quarter defined by $I = A = -BCDE = -ABCDE$.

2. (Continuation) Analyze the $3(2^{5-2})$ obtained by omitting the quarter defined by $I = -AB = -CDE = ABCDE$.

3. Find the defining contrasts for the following 2^{6-3} fraction: *c*, *ae*, *bd*, *adf*, *bef*, *abcf*, *cdef*, and *abcde*.

4. The point (1) is added to the 2^{3-1} fraction, *a*, *b*, *c*, and *abc*. Find the least squares estimates of the main effects in the following cases:
(i) All interactions are suppressed;
(ii) All interactions except AB are suppressed;
(iii) All interactions except ABC are suppressed.

5. Obtain a $7(2^{7-6})$ fraction of resolution IV and divide it into seven blocks of two runs each without losing the resolution IV property.

6. An experimenter runs two half replicates of a 2^4 factorial. They are defined by $I = ABC$ and $I = ABD$ respectively. There are sixteen observations altogether since the four overlapping points are duplicated. How would you analyze the data, assuming that all higher order interactions are suppressed? How is the analysis modified if the two half replicates are in separate blocks?

7. Healy (1951) gives a plan for a single replicate of a 2^5 factorial in a Latin rectangle, i.e., a rectangular array of thirty-two plots arranged in four rows and eight columns. The interactions ABC, $ABDE$, and CDE are confounded with the rows. The first four columns consist of the 2^{5-1} defined by $I = -BCE$, and AB, CD, $ABCD$ are confounded with columns. The second four columns contain the remaining sixteen points with AC, DE, $ACDE$ confounded. Reconstruct his design. Assuming that all higher order interactions are suppressed, confirm that all the main effects and two-factor interactions are estimable and show how they are estimated.

8. Finney (1946) describes a potato growing experiment in which four organic fertilizers were applied in a Latin square design. Each of the sixteen plots was split into two subplots to which treatment combinations from a 2^2 factorial in N (nitrogen) and K (potash) were applied. Each whole plot either had (1) on one subplot and *nk* on the other, or *n* on one and *k* on the other. The following data are yields of potatoes.

D	*A*	*B*	*C*
n 61	(1) 48	*n* 80	(1) 85
k 52	*nk* 67	*k* 68	*nk* 87
B	*C*	*D*	*A*
(1) 51	*n* 68	(1) 72	*n* 75
nk 57	*k* 50	*nk* 94	*k* 62
A	*D*	*C*	*B*
(1) 32	*n* 93	(1) 84	*n* 85
nk 61	*k* 70	*nk* 104	*k* 85
C	*B*	*A*	*D*
n 85	(1) 96	*n* 91	(1) 89
k 69	*nk* 91	*k* 66	*nk* 83

Analyze the data. Is it necessary to suppress the *NK* interaction? How would the analysis be modified if *A*, *B*, *C*, and *D* were the four-treatment combinations in a 2^2 factorial? How would it be modified if they were a 2^{3-1} of resolution III? Finney always confounded *NK* between whole plots. Could he perhaps have confounded the main effects *N* and *K* sometimes? Would he have gained anything by it?

CHAPTER 9

Fractional Factorials with More Than Two Levels

The fractions of the 2^n designs that were discussed in the previous chapter are of considerable practical use as well as of academic interest. They enable us to handle many factors without having to make an unreasonable number of observations, and it seems only natural that experimenters should wish to seek the advantages of fractionation with factors having more than two levels. New complications then arise; not only do factors at more than two levels call for more data points in the full design, but they also have more effects to be measured.

In a 2^n experiment with all interactions involving three or more factors suppressed, there are $1 + n + n(n - 1)/2 = (n^2 + n + 2)/2$ parameters to be estimated. In a 3^n experiment each main effect has two degrees of freedom, corresponding to linear and quadratic effects; each two-factor interaction has four degrees of freedom, giving a total of $2n^2 + 1$ degrees of freedom. In a 5^n design the estimation of two-factor interactions soon gets out of hand; with only three factors there are already 12 d.f. for main effects and forty-eight for two-factor interactions.

The interest in fractionation of 3^n, 4^n, and 5^n designs is not merely for mathematical amusement. The model that was used for 2^n designs contained no terms in x_h^2, and so, if the experimenter has cause to believe that for some factor there is a quadratic component to be estimated, he must take three or more levels. (It would be wrong to think that we always pay attention to estimating the quadratic component, even when we suspect that it is there. In response-surface fitting, which will be discussed in the next chapter, we often elect to ignore the quadratic effects in the earlier stages of experimentation and to make decisions about future experiments on the basis of first-order models.) The consideration of curvature thus leads us to 3^n factorials, and we shall begin this chapter by recalling some of the properties of 3^n designs described in Chapter 7.

We shall continue to use the concept of the resolution of a design that was introduced in the previous chapter. In view of the fact that the number of two-factor interactions increases so quickly with n, resolution V designs are of little interest and we shall not say much about them. The major use of these fractional factorials is in screening experiments, and so most of this chapter will be concerned with resolution III designs, which will often be called main effects plans. In the next section we shall recall the simplest of the main effects plans, the Graeco-Latin squares, and follow them with a discussion of 3^{n-k} factorials.

We shall then turn to the series of orthogonal resolution III designs, which are of considerable practical interest, developed by Plackett and Burman (1946) and by Addelman (1962). In that section we shall find some fractions of $2^m 3^n$ designs, i.e., designs with $m + n$ factors, of which m appear at two levels and the other n at three levels. The discussion of $2^m 3^n$ designs will be developed further in later sections, and the chapter will conclude with a few examples of nonorthogonal fractions.

We have purposely restricted our discussion in this chapter to these topics. The reader who is particularly interested may wish to read some of the earlier papers on the subject of fractionation of s^n designs where $s = p^m$ and p is a prime. Among the papers of interest are some on orthogonal arrays. We note in particular papers by Bose (1947*a*), Rao (1946) and (1947*b*), Bose and Bush (1952), and Finney (1945).

An orthogonal array (N, k, s, d) is a set of N treatment combinations with k factors each at s levels, which has the property that for any subset of d out of the k factors all s^d of the treatment combinations occur the same number of times; d is called the strength of the array. An orthogonal array of strength 2 is an orthogonal main effects plan. The levels of the factors will usually be denoted by $0, 1, 2, \ldots, s - 1$. If s is prime, these symbols will actually be elements of the Galois field $GF(s)$. We shall sometimes write the levels of a two-level factor symbolically as zero and one; this will be purely symbolic and they will be treated as the two elements of $GF(2)$ with $1 + 1 = 0$. In practice, the levels will still be ± 1. The 2^{3-1} defined by $I = -ABC$ may be written 000, 110, 101, and 011. Consider the subset of factors A, B. We see that the set of treatment combinations 00, 01, 10, and 11 occurs once each. The same thing is true for the treatment combinations of the subsets A, C and B, C. The design is thus an orthogonal array (4, 3, 2, 2).

9.1 3^n Designs

The 3^2 and 3^3 experiments were discussed in Chapter 7. At that time we took a 3^2 example and divided the four degrees of freedom for $A \times B$ interaction into four components: lin A lin B, lin A quad B, quad A lin B, and quad A quad B. We shall no longer use that method of subdivision of the interaction sum of squares because it does not lead to reasonable procedures for finding fractions. Instead we shall adopt a procedure used previously for confounding.

The three levels of each factor are represented by 0, 1, and 2, and arithmetic is carried out mod 3. The twenty-seven points of the 3^3 design form a finite geometry in three-dimensional space. The sum of squares for A may be divided into two single degrees of freedom, corresponding to linear and quadratic effects. The linear contrast is the difference between the sum of the observations at $x_1 = 2$ and the sum at $x_1 = 0$, which we write as $\{x_1 = 2\} - \{x_1 = 0\}$. The quadratic contrast is $\{x_1 = 0\} - 2\{x_1 = 1\} + \{x_1 = 2\}$. The equations $x_1 = 0$, $x_1 = 1$, and $x_1 = 2$ determine hyperplanes (in three dimensions they are just planes). The degrees of freedom correspond to the two orthogonal contrasts between the totals of observations on the parallel planes.

For the $A \times B$ interaction we take the two sets of parallel planes $x_1 + x_2 \equiv 0, 1, 2 \pmod 3$ and $x_1 + 2x_2 \equiv 0, 1, 2 \pmod 3$. For each set of three planes there are two orthogonal contrasts, and hence 2 degrees of freedom. We call the first pair of degrees of freedom the AB interaction and the second pair the AB^2 interaction. Nothing further is gained by considering the planes $2x_1 + x_2 \equiv 0, 1, 2$ because the set of points with $2x_1 + x_2 \equiv c$ has $2(2x_1 + x_2) \equiv 2c$, or, reducing mod 3, $x_1 + 2x_2 = 2c$, and this again gives the AB^2 interaction.

The three-factor interaction $A \times B \times C$ has 8 degrees of freedom and four components, ABC, AB^2C, ABC^2, and AB^2C^2, corresponding to sets of hyperplanes, $x_1 + x_2 + x_3 \equiv 0, 1, 2$, $x_1 + 2x_2 + x_3 \equiv 0, 1, 2$, $x_1 + x_2 + 2x_3 \equiv 0, 1, 2$, and $x_1 + 2x_2 + 2x_3 \equiv 0, 1, 2$, respectively. It is conventional to write the components of interactions with the exponent of the first factor unity. We gave an example of confounding the 3^3 factorial in three blocks of nine points each. Each of the blocks is a one-third replicate of the full factorial, which we may write as a 3^{3-1} fraction or, extending the notation that we introduced in the last sections of the previous chapter, as a $3^3//9$ design. Indeed, each of the blocks is a resolution III design, which leads us into the next section.

9.2 Latin Squares as Main Effects Plans

We originally presented the $s \times s$ Latin squares as designs for eliminating plot differences in two directions: rows and columns. There is nothing in our model or in the derivation of the analysis to prevent us from letting the rows represent the s levels of one factor and the columns represent the s levels of a second factor. Thus we see the Latin square of side s as an orthogonal main effects plan for three factors, each at s levels in s^2 runs. The Graeco-Latin square is a main effects plan for four factors, and so on.

When we divided the 3^3 design into three blocks of nine points, the first block consisted of the points for which $x_1 + 2x_2 + 2x_3 = 0$. If we let x_1 denote the row and x_2 the column, this gives the Latin square

000	012	021
101	110	122
202	211	220.

There are eight degrees of freedom in a 3×3 Latin square, and so there is room for a fourth factor. We put $x_4 = x_1 + x_2$ and obtain a saturated main effects plan or $3^4//9$:

0000	0121	0212
1011	1102	1220
2022	2110	2201.

In the more general case, where $s = p^m$, the $s \times s$ Latin square can be made to accomodate $s + 1$ factors, each at s levels, in an orthogonal main effects plan. An interesting pioneering example of a 5^{5-3}, due to L. H. C. Tippett, is reported by Fisher (1947). The factors involved were four different components of a spindle used in a cotton mill, each at five levels, and time, observations having been made at five different periods.

The case $s = 4 = 2^2$ was also considered in Chapter 7. Here we have two possible methods of attack. The finite geometry approach that we have adopted so far implies the use of finite fields. In the case of $s = 3$ or 5, the elements of the field are the residue classes mod s. For $s = 4$ we can denote the levels of each factor by 0, 1, x, and y $(=1 + x)$ with arithmetic mod 2. When we blocked the 4^2 experiment into four blocks of four, we took as the blocking factor C: $x_3 = yx_1 + yx_2$. This also gives us a Latin square $4^3//16$ main effects plan:

000	$10y$	$x01$	$y0x$
$01y$	110	$x1x$	$y11$
$0x1$	$1xx$	$xx0$	yxy
$0yx$	$1y1$	xyy	$yy0$.

We also mentioned at that time the second method of attack, which involves replacing each four-level factor by two two-level factors.

The remarks that were made earlier about the bias resulting from interaction in Latin square designs are again relevant. In the last design A is aliased with $B \times C$, B with $A \times C$, and C with $A \times B$.

In addition to the Graeco-Latin squares, there are other larger 3^{n-k} fractions which are main effects plans. The principle of their construction is the same as that for 2^{n-k} designs of resolution III; we take a basic 3^n design and add other factors by equating them to interactions, or rather to components of interactions. The following example will illustrate the procedure. We shall obtain a twenty-seven point resolution III 3^{13-10} design and show how it becomes a 3^{9-6} in nine blocks of three points each.

We begin by taking a 3^3 design in the three factors A, B, and C and then add new factors by setting

$$x_4 = x_1 + 2x_3, \qquad x_5 = 2x_1 + x_2 + x_3, \qquad x_6 = x_2 + 2x_3,$$

$$x_7 = 2x_1 + x_2 + 2x_3, \qquad x_8 = x_1 + x_2 + x_3, \qquad x_9 = x_1 + x_2,$$

$$x_{10} = x_1 + x_3, \qquad x_{11} = x_1 + x_2 + 2x_3, \qquad x_{12} = x_1 + 2x_2,$$

$$x_{13} = x_2 + x_3.$$

This is equivalent to taking as defining contrasts AC^2D^2, AB^2C^2E, BC^2F^2, AB^2CG, $ABCH^2$, ABJ^2, ACK^2, ABC^2L^2, AB^2M^2, and BCN^2. The twenty-seven points in the design follow:

0000000000000	0221100122011	0112200211022
1122221120000	1010021212011	1201121001022
2211112210000	2102212002011	2020012121022
1001202111110	1222002200121	1110102022102
2120120201110	2011220020121	2202020112102
0212011021110	0100111110121	0021211202102
2002101222220	2220201011201	2111001100212
0121022012220	0012122101201	0200222220212
1210210102220	1101010221201	1022110010212.

Suppose now that we decide to block the design by using the last two factors, M and N, as blocking factors, i.e., by assigning to the same block those points that have the same value for x_{12} and the same value for x_{13}. There will be nine blocks corresponding to the pairs of coordinates 00, 01, 02, . . ., and in the arrangement in which we have written the points they are the first three points, the middle three points, and the last three points of each column. There are eight degrees of freedom between blocks. Four of these are the 2 degrees of freedom each for M and N. The remaining 4 degrees of freedom are for the generalized interactions of M and N, namely K and L, which also become blocking factors since we have

$$x_{12} + x_{13} = x_1 + x_3 = x_{10}, \qquad x_{12} + 2x_{13} = x_1 + x_2 + 2x_3 = x_{11}.$$

The blocked design is thus a resolution III 3^{9-6} design in the first nine factors.

9.3 3^{n-k} Fractions with Interactions

There is an extensive list of 3^{n-k} fractions in the pamphlet prepared by Connor and Zelen (1959) for the National Bureau of Standards. In it they consider designs of resolution IV and resolution V with blocking for $4 \le n \le 10$.

The smallest resolution IV design for $n = 4$ is the one-third replicate, of which an example will be given in the next section. For $5 \le n \le 9$, they have resolution IV designs with eighty-one points. Among the resolution V designs, eighty-one points are necessary for $n = 4$ or 5. For $6 \le n \le 9$, resolution V designs are listed with 243 points.

The $3^5/ /81$ design can be obtained by using any of the five-factor interactions as a defining contrast. The resolution IV $3^9/ /81$ design is obtained by setting

$$x_5 = x_1 + 2x_3 + x_4, \qquad x_6 = 2x_1 + x_2 + x_3 + 2x_4, \qquad x_7 = x_2 + 2x_3 + x_4,$$
$$x_8 = 2x_1 + x_2 + 2x_3, \qquad x_9 = x_1 + x_2 + x_3 + x_4.$$

They obtain their $3^9/ /243$ design by adding new factors to the basic 3^5 design as follows:

$$x_6 = x_1 + x_2 + 2x_3 + x_5, \qquad x_7 = 2x_1 + x_2 + 2x_4 + x_5,$$
$$x_8 = x_2 + x_3 + 2x_4 + 2x_5, \qquad x_9 = x_1 + x_2 + x_3 + x_4.$$

EXAMPLE OF A 3^{4-1} DESIGN. Vance (1962) reports on a series of experiments in the course of which a 3^{4-1} factorial was used. There were four major operating variables in the treatment of lube oil at a refinery, and these were taken as the factors in the experiment. The objective of the experimental program was to find a set of operating conditions that would optimize a measure of quality in the lube oil, which will be referred to as the response. Their previous experience in the use of the plant had convinced Vance and his colleagues that the quadratic effects should be estimated, and so at least three levels of each factor should be included.

In this preliminary experiment they were prepared to allow two-factor interactions to be aliased with each other. Their main concern was to obtain preliminary estimates of the linear and quadratic effects, aliased only by higher order interactions, and to use this information to provide a basis for the design of subsequent experiments. This approach is similar to the procedure used in fitting response surfaces, which will be discussed in Chapter 10.

They had a choice of several possible fractions. In the 2^4 factorial there are only two half replicates that have $ABCD$ as the defining contrast. There are, however, no fewer than twenty-four 3^{4-1} fractions in which the defining contrast is some form of the four-factor interaction. We can take as the points of the fraction the set of points on any of the hyperplanes defined by

$$x_1 + a_2x_2 + a_3x_3 + a_4x_4 \equiv b \pmod 3,$$

where $a_i = 1$ or 2 and $b = 0, 1,$ or 2.

There is usually no particular reason for choosing any one of these fractions over the others. Vance elected to use the fraction defined by

$$x_1 + x_2 + x_3 + x_4 = 0.$$

The experimental points and the reported responses are

0000	4.2	0012	5.9	0021	8.2
0102	13.1	0111	16.4	0120	30.7
0201	9.5	0210	22.2	0222	31.0
1002	7.7	1011	16.5	1020	14.3
1101	11.0	1110	29.0	1122	55.0
1200	8.5	1212	37.4	1221	66.3
2001	11.4	2010	21.1	2022	57.9
2100	13.5	2112	51.6	2121	76.5
2202	31.0	2211	74.5	2220	85.1.

We can now carry out the analysis of variance procedure for the basic 3^3 factorial in the usual way and obtain sums of squares for the main effects A, B, C, and the interactions. These sums of squares are given in Table 9.1. The sum of squares for D is calculated just as the sum of squares for the other main effects, by adding the squares of the totals at the several levels of x_4, dividing by nine, and subtracting the correction for the mean.

The two-factor interactions are hopelessly jumbled. If we work in terms of the symbolic components of interaction we can untangle the web to a certain extent, but it is hardly worthwhile. Consider for example the $A \times D$ interaction, which has four degrees of freedom. The AD interaction involves contrasts between sums of points with $x_1 + x_4 \equiv$ const (mod 3). Since we also have the defining relationship $x_1 + x_2 + x_3 + x_4 \equiv 0$, it follows that for these points $(x_1 + x_2 + x_3 + x_4) + 2(x_1 + x_4) \equiv$ const, i.e., $x_2 + x_3 \equiv$ const, and so the AD contrasts are also BC contrasts. Similarly, we have $x_1 + 2x_2 + 2x_3 + x_4 \equiv$ const, and so the AD contrasts are also AB^2C^2D contrasts.

An equivalent approach is to argue that the defining contrasts are I, $ABCD$, $A^2B^2C^2D^2$. Multiplying by AD and letting $A^3 = B^3 = C^3 = D^3 = I$ gives an alias set AD, A^2BCD^2, B^2C^2. Finally, squaring the last two members to make the exponent of the first letter unity gives the set AD, AB^2C^2D, BC. We can proceed in this way to obtain the aliases of each component (2 d.f.) in the 3^3 experiment. If we ignore interactions involving three or more factors, the alias sets are

$$A;\quad B;\quad C;\quad AB, CD;\quad AB^2;\quad AC, BD;\quad AC^2;\quad BC, AD;$$

$$BC^2;\quad (ABC), D;\quad (AB^2C), BD^2;\quad (ABC^2), CD^2;\quad (AB^2C^2), AD^2.$$

The crucial difference between this and the 2^{4-1} fractional factorial defined by $I = ABCD$ is that in the latter the $A \times D$ interaction is completely aliased, or confounded, with the $B \times C$ interaction. In the present case two of the degrees of freedom of the $A \times D$ interaction are aliased with BC, and the other two degrees of freedom come from the $A \times B \times C$ interaction sum of squares. A more meaningful aliasing would have been one where, for example, lin $A \times$ lin B was aliased with quad $C \times$ quad D, but we cannot obtain such relationships using this method of obtaining fractions.

TABLE 9.1

ANALYSIS OF VARIANCE

Source	SS	d.f.
Total	15544.5	26
A	4496.3	2
(lin A	4399.2	1)
(quad A	97.1	1)
B	2768.7	2
(lin B	2647.5	1)
(quad B	121.2	1)
C	5519.8	2
(lin C	5516.0	1)
(quad C	3.8	1)
D	283.4	2
(lin D	213.6	1)
(quad D	69.8	1)
(Total for main effects	13068.2	8)
$A \times B$	310.8	4
$A \times C$	1232.9	4
$B \times C$	669.8	4
$A \times B \times C$ (excluding D)	262.8	6

9.4 Orthogonal Main Effects Plans

We have already discussed saturated 2^{n-k} and 3^{n-k} fractions, in which the basic design is "loaded" with additional factors by equating them to interactions. This procedure has given us orthogonal main effects plans for $2^n - 1$ two-level factors in 2^n runs and for $(3^n - 1)/2$ three-level factors in 3^n runs. Plackett and Burman (1946) discovered a series of orthogonal saturated resolution III designs for $N - 1$ factors in N runs when $N = 4k$ where k is an integer. Their designs were mentioned in passing in Chapter 8, and a discussion of them has been postponed until now. They listed designs for all values of $N = 4k$ up to $N = 100$, with the exception of $N = 92$. The case $N = 92$ was solved by Baumert, Golomb, and Hall (1962). The other major contribution to orthogonal main effects plans to be discussed at this time is due to Addelman (1962), who showed how new plans could be obtained by collapsing and replacing factors in the basic series of designs by other factors which have different numbers of levels.

9.5 The Plackett and Burman Designs

These are designs for factors at two levels which we denote by + and −, standing for $x_i = +1$ and $x_i = -1$, respectively. Let **X** denote the design matrix. If there are N points and n factors, **X** is a matrix with N rows and n columns in which x_{hi} is the value (+ or −) of x_i at the hth point. We have the main effects model

$$E(y_h) = \beta_0 + \sum_i \beta_i x_{hi},$$

or $E(\mathbf{Y}) = \mathbf{D}\boldsymbol{\beta}$ where $\mathbf{D} = (\mathbf{1}, \mathbf{X})$ and $\boldsymbol{\beta}' = \beta_0, \beta_1, \beta_2, \ldots, \beta_n$.

We wish to find a design such that the matrix $\mathbf{D}'\mathbf{D}$ is diagonal. Since $x_{ki}^2 = 1$ for all h and all i, the diagonal elements of $\mathbf{D}'\mathbf{D}$ are each equal to N, and so we require $\mathbf{D}'\mathbf{D} = N\mathbf{I}$, i.e., that the column vectors of **X** be orthogonal to each other and to **1**.

We denote by $\mathbf{X}_i$ the ith column of **X**, consisting of the values taken by x_i at each design point. If $\mathbf{1}'\mathbf{X}_i = c_i$, Margolin (1967) defines $\mathbf{X}_i$ to be a column vector of value c_i. If $c_i = 0$, $\mathbf{X}_i$ is called a zero-sum column vector. If $c_i \neq 0$, $\mathbf{X}_i$ is called a nonzero-sum column vector. If $\mathbf{D}'\mathbf{D}$ is to be diagonal, each $\mathbf{X}_i$ must be a zero-sum column vector, which means that half the entries must be plus and half minus. Similarly, if we let $(\mathbf{X}_i, \mathbf{X}_j)$ denote the inner product $\mathbf{X}_i'\mathbf{X}_j$, we need to have $(\mathbf{X}_i, \mathbf{X}_j) = 0$ for all pairs $i, j (i \neq j)$. If $n = N - 1$, which is the largest number of factors at two levels that we can hope to accommodate, the problem of finding a suitable design becomes one of finding a matrix **D** of N rows and N columns with the following properties:

(i) Each element is ± 1;
(ii) One column of **D** is **1**;
(iii) $N^{-1/2}\mathbf{D}$ is orthogonal. Such a matrix is called a Hademard matrix.

We shall now show how Plackett and Burman obtained their designs. The reader who does not wish to follow the derivation, which involves the use of finite fields, can safely skip to the end of this section, where we list the design for $N = 12$. This method of constructing a Hademard matrix is due to Paley (1933).

Let z be any nonzero element of the finite field $GF(p^m)$; z is said to be a quadratic residue of the field if there is an element w in the field such that $w^2 = z$; otherwise, z is a nonquadratic residue. If $p \geq 3$, it can be shown that the number of nonzero elements that are quadratic residues is equal to the number that are nonquadratic residues. We introduce the Legendre symbol, $\chi(z)$, defined as follows (it will be used in a slightly different fashion in Chapter 13):

$$\chi(0) = 0;$$
$$\chi(z) = +1, \quad \text{if } z \text{ is a quadratic residue;}$$
$$\chi(z) = -1, \quad \text{if } z \text{ is a nonquadratic residue.}$$

It follows that $\sum \chi(z) = 0$, summing over all z in $GF(p^m)$. Three other results about the Legendre symbols are needed, and we state them without proof.

(i) Let z' be another element of $GF(p^m)$. Then $\chi(z)\chi(z') = \chi(zz')$, in particular $\chi(-z) = \chi(-1)\chi(z)$;

(ii) $\chi(-1) = +1$, if $p^m = 4k + 1$, $\chi(-1) = -1$, if $p^m = 4k - 1$ where k is an integer;

(iii) Let u and v be two elements of the field $u \neq v$, $p > 2$. Then, summing over all z in the field $\sum \chi(z - u)\chi(z - v) = -1$.

Using these results, we can construct a Hademard matrix of order N where $N = p^m + 1 = 4k$.

We number the rows and columns of a matrix $\mathbf{X}$, h, and i where h and i are elements of $GF(p^m)$, and add one extra row, which has every element minus; $\mathbf{X}$ is a matrix of $p^m + 1$ rows and p^m columns. The elements of the first p^m rows are assigned as $x_{hh} = +1$, $x_{hi} = -\chi(h - i)$ $(h \neq i)$.

Consider the matrix $\mathbf{D} = (\mathbf{1}, \mathbf{X})$. For this matrix we have

$$\mathbf{1}'\mathbf{X}_i = -1 + 1 - \sum_h \chi(h - i) = 0,$$

$$\begin{aligned}(\mathbf{X}_i, \mathbf{X}_j) &= 1 - \chi(i - j) - \chi(j - i) + \sum_h \chi(h - i)\chi(h - j) \\ &= 1 - \chi(i - j) - \chi(j - i) - 1.\end{aligned}$$

But with $p^m = 4k - 1$ we have $\chi(j - i) = \chi(-1)\chi(i - j) = -\chi(i - j)$, so that $(\mathbf{X}_i, \mathbf{X}_j) = 0$. The matrix $\mathbf{D}$ is thus a Hademard matrix.

We also note that if $\mathbf{D}$ is any Hademard matrix of order N,

$$\mathbf{D}^* = \begin{pmatrix} \mathbf{D} & -\mathbf{D} \\ \mathbf{D} & \mathbf{D} \end{pmatrix}$$

is a Hademard matrix of order $2N$, and so a design for $2N - 1$ factors in $2N$ runs can be constructed from a design for $N - 1$ factors in N runs.

If $p^m = 4k + 1$, Plackett and Burman show that the foregoing construction procedure can be modified to obtain a Hademard matrix of order $2(p^m + 1)$, and the reader is referred to their paper for the details. They used this method to obtain designs for $N = 52$, 76, and 100. Their design for $N = 36$ was obtained by trial.

It should also be noted that if $N = p + 1$, i.e., $m = 1$, then, apart from the last row which represents the base point (1), $\mathbf{X}$ is a circulant matrix. We have $x_{h+1,i+1} = x_{h,i}$, and so in their listing the authors give only one line of each matrix for those values of N.

As an example we now derive the Plackett and Burman design for $N = 12$, $p = 11$. The elements of $GF(11)$ are 0, 1, 2, . . ., 10 with arithmetic being carried out mod 11. The quadratic residues are $1 = 1^2 = 10^2$, $4 = 2^2 = 9^2$, $9 = 3^2 =$

$8^2, 5 = 4^2 = 7^2, 3 = 5^2 = 6^2$. Then $\chi(z) = +1$ if $z = 1, 3, 4, 5, 9$ and $\chi(z) = -1$ if $z = 2, 6, 7, 8, 10$. The design matrix $\mathbf{X}$ follows. This is a better design for eleven factors than the $3(2^{11-9})$ because each main effect is estimated from all twelve points with variance efficiency one. If there are only ten factors, we can use the extra factor for blocking and obtain two blocks of six. We cannot, however, obtain more than two blocks because in the design there are no factors for which, writing α, β, and γ for subscripts, $x_{h\alpha} = x_{h\beta}x_{h\gamma}$ for all h.

The Plackett and Burman Design for $N = 12$

A	B	C	D	E	F	G	H	J	K	L	
+	+	−	+	+	+	−	−	−	+	−	*abdefk*
−	+	+	−	+	+	+	−	−	−	+	*bcefgl*
+	−	+	+	−	+	+	+	−	−	−	*acdfgh*
−	+	−	+	+	−	+	+	+	−	−	*bdeghj*
−	−	+	−	+	+	−	+	+	+	−	*cefhjk*
−	−	−	+	−	+	+	−	+	+	+	*dfgjkl*
+	−	−	−	+	−	+	+	−	+	+	*aeghkl*
+	+	−	−	−	+	−	+	+	−	+	*abfhjl*
+	+	+	−	−	−	+	−	+	+	−	*abcgjk*
−	+	+	+	−	−	−	+	−	+	+	*bcdhkl*
+	−	+	+	+	−	−	−	+	−	+	*acdejl*
−	−	−	−	−	−	−	−	−	−	−	(1)

9.6 Addelman's Designs

During the discussion of the two-way layout in Chapter 4 the condition of proportional frequencies was derived. Suppose that in a two-way layout with factors A and B there are n_{ij} observations at the ith level of A and the jth level of B, and that $\sum_j n_{ij} = n_{i.}$, $\sum_i n_{ij} = n_{.j}$. We denote the sum of all the observations at the ith level of A by $T_{i.}$, and the sum of the observations at the jth level of B by $T_{.j}$. It was shown that a necessary and sufficient condition for the two factors to be orthogonal, i.e., for any comparison between the averages of the observations at two levels of A, e.g., $n_{i'.}^{-1}T_{i'.} - n_{i.}^{-1}T_{i.}$, to be orthogonal to any B comparison, such as $n_{.j'}^{-1}T_{.j'} - n_{.j}^{-1}T_{.j}$, is that the cell frequencies be proportional to the marginal frequencies. This condition is that

$$n_{ij} = n_{i.}n_{.j}/N$$

for all i and all j; $N = \sum_i n_{i.} = \sum_j n_{.j}$ is the total number of observations. Addelman (1962) makes use of this property to develop new orthogonal main effects plans [see also Addelman and Kempthorne (1961)].

In an earlier section we presented a 3^{4-2} orthogonal main effects plan: four factors in nine points. Suppose that we replace one of the three-level factors, say D, by a two-level factor, E. We may let E take its low level whenever $x_4 = 0$ or 2, and its high level whenever $x_4 = 1$. Then the condition of proportional frequencies will be satisfied and we shall have an orthogonal main effects fraction for a 2×3^3 design. The correspondence between the two designs is

0000		0000
0121		0121
0212		0210
1011		1011
1102	$\longrightarrow$	1100
1220		1220
2022		2020
2110		2110
2201		2201.

The important thing is that two levels of D are assigned to one level of E, and the third level of D is assigned to the other level of E. The condition of proportional frequencies will be satisfied whether the two levels of D go to high E or to low E, or whichever level of D is chosen to be the odd one. The latter choice does, however, have a bearing upon the efficiency in larger designs where the new factor has more than two levels. We could replace two, three, or all four of the three-level factors by two-level factors. Changing all four factors to two levels would not have much merit in practice because with three points at one level of a factor and six points at the other level, we have $V(2\hat{\beta}) = \sigma^2(1/3 + 1/6) = \sigma^2/2$. This is the same variance we obtain with a 2^{4-1}, and that design is of resolution IV and has one fewer point. The other three designs, $2^1 3^3 / /9$, $2^2 3^2 / /9$, and $2^3 3 / /9$, are all useful in practical applications.

Suppose now that we go a step further and in a 4^{4-2} Graeco-Latin square design replace one of the four-level factors, D, by a three-level factor, E. There are now two possibilities of interest, both of which give us proportional frequencies and hence orthogonal main effects plans. We make the correspondence $x_4 = 3 \rightarrow x_0 = 0$, $x_4 = 1$ or $2 \rightarrow x_5 = 1$, $x_4 = 2 \rightarrow x_5 = 2$. This is a symmetric assignment. There will be four points each at $x_5 = 0$ and $x_5 = 2$, and eight at $x_5 = 1$. To obtain a measure of relative efficiency in this choice, we can compare the variances of the linear and quadratic contrasts with the variances that would have been obtained had we been able to take one third of the points at each level of E. The relative efficiencies are then linear 3/4 and quadratic 18/16. Alternatively, we could assign two levels of D to $x_5 = 0$, and one each to $x_5 = 1$ and $x_5 = 2$. In that case the relative efficiency of the linear contrast would be one, but the relative efficiency of the quadratic contrast would be 18/22. The difference is hardly worth getting excited about. We might sum it up by saying that if we are more interested in the linear contrast we should take more observations at the end points. If we want more emphasis on the quadratic term we should take more observations at the middle value of x_5. In either case this idea of going from four levels to three is a useful one. This procedure of replacing a factor at s levels by another factor at t levels ($t < s$) is called collapsing the factor.

Collapsing is not confined to replacing a factor by one other factor with fewer levels. We have already mentioned while discussing confounding that a four-level factor can be considered as a pair of two-level factors. If, however, we are interested only in main effects, we can go further and replace a single four-level factor by three two-level factors. This is achieved by the correspondence $x = 0 \to 000$, $x = 1 \to 011$, $x = 2 \to 101$, $x = 3 \to 110$. Indeed, if a factor has s levels, where $s = t^m$, t being a prime, or a power of a prime, it may be replaced by $(s - 1)/(t - 1)$ factors each at t levels. The 4^{5-3} hyper-Graeco-Latin square design can be converted in this way into $4^n 2^{15-3n}//16$, $n = 0, 1, 2, 3, 4$.

Another technique that Addelman uses is the addition of fractions to increase the number of levels of a factor. Suppose, for example, that we wish to construct a 6×3^3 design. We may take a 3^{4-2} fraction and run it twice, letting x_1 take the values 0, 1, and 2 in the first replicate and 3, 4, and 5 in the second replicate. If a 5×3^3 were required, we could then collapse the six-level factor, and have the desired design.

Addelman uses as basic designs from which to build his plans several of the standard saturated fractional factorials 2^{7-4}, 3^{4-2}, 2^{15-11}, 5^{6-4}, 3^{13-10}, 2^{31-27}. In addition, he uses the following interesting basic plan for $3^7//18$. The derivation of this design is given by Addelman and Kempthorne (1961). The eighteen points fall into two nine-point groups:

0000000	0021011
0112111	0100122
0221222	0212200
1011120	1002221
1120201	1111002
1202012	1220110
2022102	2010212
2101210	2122020
2210021	2201101.

In the first half the defining relationships are $x_3 = x_1 + x_2$, $x_4 = x_1 + 2x_2$, $x_5 = x_1^2 + x_2$, $x_6 = x_1^2 + x_1 + x_2$, $x_7 = x_1^2 + 2x_1 + x_2$. In the second half $x_3 = x_1 + x_2 + 2$, $x_4 = x_1 + 2x_2 + 1$, $x_5 = 2x_1^2 + x_2$, $x_6 = 2x_1^2 + 2x_1 + x_2 + 1$, $x_7 = 2x_1^2 + x_1 + x_2 + 1$.

With these basic plans the tools are now available to generate a large number of orthogonal main effects plans, especially in the $2^m 3^n$ series.

The Plackett and Burman designs do not lend themselves to modification, either by collapsing or by replacing factors. The only choice they allow is to leave out one or more factors, and doing that does not improve the precision of the estimates of the main effects of the factors that remain. With the 2^{n-k} fractions it is also possible under suitable circumstances to reverse the replacement procedure by exchanging three two-level factors for a four-level factor. This requires, however, that the triple of coordinates of the three factors replaced takes only the set 111, 100, 010, 001, or else the set 011, 101, 110, 000, giving $C = \pm AB$.

Consider changing the 2^{7-4} design, *(1)*, *adef*, *befg*, *abdg*, *cdeg*, *acfg*, *bcdf*, *abce* into a 3×2^4 orthogonal main effects plan. The generators of the defining contrast subgroups are I, $-ACD$, $ABCE$, $-ABF$, $-BCG$. At each point $x_1 + x_3 + x_4 \equiv 0 \pmod 2$. We can therefore replace A, C, and D by a four-level factor, letting $000 \to 0$, $011 \to 1$, $101 \to 2$, $110 \to 3$. We then introduce the three-level factor by collapsing the four-level factor that we have just acquired; we collapse levels 2 and 3 to $x_8 = 2$. The final design follows, together with the original eight points:

	BEFGH
(1)	00000
adef	01102
befg	11110
abdg	10012
cdeg	01011
acfg	00112
bcdf	10101
abce	11002.

9.7 Fractions of 2^m3^n Designs of Resolution V

A listing of fractions of 2^m3^n designs was prepared by Connor and Young (1961) in the National Bureau of Standards series. More recently Margolin (1967) has investigated such fractions. In this section we shall give a brief sketch of Margolin's results. We shall use A to denote a factor with two levels and R, S, T, U to denote factors with three levels. When we talk of a resolution V design in this context we shall mean a design in which all such interactions as AR_L, AR_Q, R_LS_L, R_QS_L, R_LS_Q, R_QS_Q are estimable. Interactions with three or more factors will be suppressed.

In a complete 3^m factorial the linear contrast for R is given by $\{R = 2\} - \{R = 0\}$, i.e., by the difference between the totals at the highest level of R and the lowest level of R. The R_LS_L contrast is $\{R = 2, S = 2\} + \{R = 0, S = 0\} - \{R = 0, S = 2\} - \{R = 2, S = 0\}$. (This is not the kind of contrast between hyperplanes that we considered earlier.) In a 3^{n-p} fraction of resolution IV the main effects will be clear, but there will be some linear dependencies between the interaction terms. As an example, Margolin points out that if we take the 3^{4-1} fraction defined by $x_r + x_s + x_t + x_u \equiv 1 \pmod 3$, there is a linear dependency between R_LS_L, R_QS_Q, T_LU_L, and T_QU_Q. Suppose now that we collapse the factor U into a two-level factor A by letting $x_u = 0$, $x_u = 2 \to x_a = 0$, and $x_u = 1 \to x_a = 1$. Then the linear U contrast and the interactions involving U_L disappear. Contrasts involving levels of A correspond to U_Q. The disappearance of U_L removes the dependency between the other three interactions. The AT_Q interaction, for example, corresponds to the former T_QU_Q.

Using this idea Margolin obtains several fractions of 2^m3^n designs by collapsing factors in resolution IV 3^{n-p} designs. The example that we have just mentioned is a design of resolution V for $2^13^3/\,/27$. Collapsing a second factor would give $2^23^2/\,/27$. He also gives a $2^23^4/\,/81$ fraction.

A second method of obtaining fractions is to run different 3^{n-p} fractions at each level of a two-level factor. An obvious example of a $2^13^5/\,/2^13^4$ would be to run a resolution V 3^{5-1} fraction at low A and another 3^{5-1} (or the same one) at high A. Similarly, a $2^53^1/\,/96$ fraction may be formed by running one of the resolution V 2^{5-1} fractions at each level of the three-level factor. Margolin gives two designs formed in this way. A $2^43^1/\,/24$ design for four factors A, B, C, and D at two levels each and E at three levels is obtained by running the half replicate defined by $I = -ABCD$ at $x_5 = 0$ and $x_5 = 1$ and running the half replicate defined by $I = +ABCD$ at $x_5 = 2$.

9.8 Two Nonorthogonal Fractions of the 4^3 Design

We mentioned earlier that a four-level factor may be replaced by a pair of two-level factors. The 4^3 factorial with factors P, Q, and R can thus be considered as a 2^6 factorial where A and B correspond to P, C and D to Q, and E and F to R. The three degrees of freedom for the main effect of P correspond to A, B, and AB. The nine degrees of freedom for PQ correspond to AC, AD, ACD, BC, BD, BCD, ABC, ABD, and $ABCD$.

The first fraction that we shall consider is a main effects only plan with twelve points. This will be a $3(2^{6-4})$ fraction with all six main effects and the interactions AB, CD, and EF estimable. All other interactions are suppressed. Consider the family of 2^{4-2} fractions of the factorial in A, B, C, and D defined by $I = \pm AD = \pm BC = \pm ABCD$ and add new factors by putting $E = +ACD$, $F = +ABC$. The four quarters are

(1), adf, bce, abcdef;
cef, acde, bf, abd;
de, aef, bcd, abcf;
cdf, ac, bdef, abe.

Any three of them constitute a $3(2^{6-4})$ fraction for which the alias sets are I, AD, BC, $ABCD$; C, $ACD(=E)$, B, ABD; D, A, BCD, $ABC(=F)$; CD, AC, $BD(=EF)$, AB. This is a main effects only plan. Furthermore, because the defining contrasts AD, BC, and $ABCD$ are all suppressed, the design may be run in three blocks of four runs each, the blocks being the three "quarters" chosen.

This design is actually a three-quarter fraction of the Latin square. Suppose that we let columns correspond to the levels *(1)*, a, b, ab of P and rows to the levels *(1)*, c, d, cd of Q, and that we denote the levels *(1)*, e, f, and ef of R by 0, 1, 2, and 3, respectively. Furthermore, let us denote points in the four "quarters" by α, β, γ, δ, respectively. When this is done we have a Graeco-Latin square:

(*1*)	*aef*	*bf*	*abe*	0α	3γ	2β	1δ
cef	*ac*	*bce*	*abcf*	3β	0δ	1α	2γ
de	*adf*	*bdef*	*abd*	1γ	2α	3δ	0β
cdf	*acde*	*bcd*	*abcdef*	2δ	1β	0γ	3α.

Thus our design actually amounts to using the fourth factor (Greek letters) as a blocking factor and taking any three of the four blocks.

The second design is a resolution V fraction with forty-eight points. This design is the $3(2^{6-2})$ fraction defined by $I = ACE = BDF = ABCDEF$. The alias sets follow. The effects that correspond to degrees of freedom in the $P \times Q \times R$ interaction, and are therefore suppressed, have been underlined. This design may be run in three blocks of sixteen runs each. An example is given by John (1970).

$I, \underline{ACE}, \underline{BDF}, \underline{ABCDEF};$ $\quad A, CE, \underline{ABDF}, \underline{BCDEF};$

$B, \underline{ABCE}, DF, \underline{ACDEF};$ $\quad AB, \underline{BCE}, \underline{ADF}, CDEF;$

$C, AE, \underline{BCDF}, \underline{ABDEF};$ $\quad AC, E, \underline{ABCDF}, BDEF;$

$BC, ABE, CDF, \underline{ADEF};$ $\quad ABC, BE, \underline{ACDF}, DEF;$

$D, \underline{ACDE}, BF, \underline{ABCEF};$ $\quad AD, CDE, ABF, \underline{BCEF};$

$BD, \underline{ABCDE}, F, \underline{ACEF};$ $\quad ABD, \underline{BCDE}, AF, CEF;$

$CD, \underline{ADE}, \underline{BCF}, ABEF;$ $\quad ACD, DE, \underline{ABCF}, BEF;$

$BCD, \underline{ABDE}, CF, AEF;$ $\quad ABCD, \underline{BDE}, \underline{ACF}, EF.$

CHAPTER 10

Response Surfaces

Most of the experimental designs that have been considered so far have had their origins in agricultural experimentation. The development of 2^{n-k} fractional factorials has taken place in industrial experimentation, but they were introduced in an agricultural context by Finney (1945). The concept of response surfaces and designs for their exploration began in the chemical industry. The early work was done by statisticians and chemical engineers in the Imperial Chemicals Industries in Great Britain. The first major paper in the field was published by Box and Wilson (1951). Its title, "On the Experimental Attainment of Optimum Conditions," is indicative of its philosophy.

Suppose for simplicity that we wish to investigate the production of some chemical product in a pilot plant, and that there are two factors, temperature and pressure, that are to be considered. The experimental procedure will be to make several runs on the pilot plant at different temperatures and pressures and to observe on each run a response, y, such as the yield of the product, or the purity of the product, or the amount of raw material converted, and so on. We have a two-dimensional factor space with two quantitative factors. We can conceive, therefore, of the expected response $E(y)$ as a function of the levels of the factors. If x_1 denotes the temperature and x_2 the pressure we have

$$y = f(x_1, x_2) + e,$$

with the usual assumptions about the random error e. The function $f(x_1, x_2)$ is single valued and, writing η for $E(y)$, the surface represented by $\eta = f(x_1, x_2)$ is called the response surface. If the x_1 and x_2 axes are taken in the usual way and the y axis is taken perpendicular to the plane of the paper, the response surface can be represented on the paper, just as altitude is represented on a map, by drawing the lines of equal response as contours.

It may be that prior knowledge of the chemical process involved will enable us to know beforehand the form of $f(x_1, x_2)$. The discussion of such models is outside the scope of this book. We shall consider the situation in which $f(x_1, x_2)$ is approximated by a polynomial of degree d, in which case the model is called a dth-order model. The set of points (x_1, x_2) at which runs can be made is called the factor space.

One possibility would be for us to take several levels of temperature and several levels of pressure to straddle the factor space, carry out a complete, or fractional, factorial experiment, and fit the model to the data as we have done in earlier chapters. This ignores three important features of this kind of experiment.

1. The usual object of the experiment is not to investigate $f(x_1, x_2)$ over the whole factor space but to locate the region in the factor space where the response is at its highest and to map it. If we regard the response surface as a mountain, the object is to find the peak and to explore the area in which the peak is found. The industrialist is rarely interested in regions where the yield of his process will be relatively low.

2. A polynomial model will usually be quite unrealistic over the whole space, but its use in a relatively small region around the peak is acceptable as a Taylor series expansion approximating a more complicated function.

3. Experimental runs on pilot plants, or actual commercial reactors, usually have relatively high precision and can be conducted sequentially.

Our problem can thus be summarized: How do we find the region of the factor space that interests us and, having found it, how do we design experiments to map $f(x_1, x_2)$ over that region? We shall discuss the second question first.

The remainder of the chapter can be divided into two areas: the design and analysis of experiments for fitting first- and second-order models, and the method of steepest ascent and *EVOP*.

The two most important papers on the topic of response surfaces are the one by Box and Wilson (1951) mentioned before, and the paper by Box and J. S. Hunter (1957). Box and a group of workers at Imperial Chemical Industries prepared a book, *Design and Analysis of Industrial Experiments*, edited by Davies (1954). The chapter on the determination of optimum conditions contains several examples and is an excellent practical exposition. The reader who wishes to pursue this subject further will find numerous references in the review paper on response surface methodology by Hill and W. G. Hunter (1966).

10.1 First-Order Designs

These are designs for fitting a first-degree equation:

$$y = \beta_0 + \sum \beta_i x_i + e_i, \qquad i = 1, 2, \ldots, k.$$

A simple example of such a design is the main effects only 2^{n-k} fractional factorial where the coordinates of x_i take the values ± 1. Before considering optimal designs it should be recalled that the coefficients β_i are affected by the choice of the scale of x_i. If x_i denotes the length of some physical component measured in feet, we can triple the value of β_i by measuring x_i in units of yards instead. It is desirable, therefore, to scale the factors uniformly, and we shall do this by imposing the following restrictions upon the coordinates x_{ji} where x_{ji} is the value of x_i at the jth experimental point:

$$\sum_j x_{ji} = 0, \qquad \sum_j x_{ji}^2 = b,$$

where b is a constant. The first of these requirements is a shift in the origin so that $\bar{x}_i = 0$ for all i; the second amounts to a change of scale to provide uniform scatter among the x_i.

Let $\mathbf{X}$ denote the design matrix (x_{ji}). The normal equations may be written

$$\begin{pmatrix} \mathbf{1}' \\ \mathbf{X}' \end{pmatrix} \mathbf{Y} = \mathbf{S}\hat{\boldsymbol{\beta}},$$

where

$$\mathbf{S} = (\mathbf{1}, \mathbf{X})'(\mathbf{1}, \mathbf{X}) = \begin{bmatrix} N & \mathbf{0} \\ \mathbf{0} & \mathbf{X}'\mathbf{X} \end{bmatrix},$$

and $\boldsymbol{\beta}' = (\beta_0, \beta_1, \ldots, \beta_k)$. In comparing designs we shall take as a criterion the average variance of the regression coefficients, or

$$\bar{v} = (k + 1)^{-1}\left\{V(\hat{\beta}_0) + \sum_i V(\hat{\beta}_i)\right\} = (k + 1)^{-1} \operatorname{tr}(\mathbf{S}^{-1}).$$

For an optimal design we seek, therefore, to minimize

$$\operatorname{tr}(\mathbf{S}^{-1}) = \sum_{h=0}^{k} \lambda_h^{-1},$$

where λ_h are the latent roots of $\mathbf{S}$ and $\lambda_0 = N$. The trace is minimized subject to the condition $\sum \lambda_h = \operatorname{tr} \mathbf{S} = N + kb$ or $\sum_{h=1}^{k} \lambda_h = kb$, in which case,

$$\lambda_h = b$$

for all h where $h = 1, 2, \ldots, k$. This requires that $\mathbf{S}$ be a diagonal matrix, or, equivalently, that the columns of $\mathbf{X}$ be mutually orthogonal. This condition for $\mathbf{X}$ is both necessary and sufficient for a design to be optimal and for such a design we have $v(\hat{\beta}_i) = b^{-1}\sigma^2$, $i > 0$.

Some examples of optimal designs are the 2^{n-p} fractions and the Plackett and Burman designs for n factors. For $k = 3$ we can use the vertices of the regular tetrahedron with the center at the origin and any orientation. Also, for $k = 2$ the vertices of the regular polygon inscribed in a circle with the center at the origin form an optimal design.

10.2 Second-Order Designs

These are designs for fitting the second-degree surface (for $k = 3$, the general quadric surface):

$$E(y_j) = \beta_0 + \sum_i \beta_i x_{ji} + \sum_i \beta_{ii} x_{ji}^2 + \sum_{h}\sum_{\substack{i \\ h<i}} \beta_{hi} x_{jh} x_{ji}.$$

It is convenient to write the elements of $\boldsymbol{\beta}$ in the order

$$\boldsymbol{\beta}' = (\beta_0, \beta_{11}, \beta_{22}, \ldots, \beta_{kk}, \beta_1, \beta_2, \ldots, \beta_k, \beta_{12}, \ldots).$$

Writing the normal equations as $\mathbf{TY} = \mathbf{S}\hat{\boldsymbol{\beta}}$, we have

$$(\mathbf{TY})' = \sum y, \sum x_1^2 y, \ldots, \sum x_1 y, \ldots, \sum x_1 x_2 y, \ldots.$$

We shall confine our investigation to designs that satisfy conditions of symmetry and scaling, which are similar to those for first-order designs. The following conditions, with $h \neq i$, are imposed upon the set of points in the design:

$$\sum_j x_{jh} = \sum_j x_{jh}^3 = \sum_j x_{jh} x_{ji} = \sum_j x_{jh}^2 x_{ji} = \sum x_{jh}^3 x_{ji} = 0,$$

$$\sum_j x_{jh}^2 = b, \qquad \sum_j x_{jh}^2 x_{ji}^2 = c, \qquad \sum_j x_{ji}^4 = c + d.$$

The matrix $\mathbf{S}$ now takes a convenient form. In its top left-hand corner is the principal minor, $\mathbf{U}$, with $(k + 1)$ rows and columns which will now be discussed. The remaining rows and columns consist of zero elements everywhere except on the main diagonal. The diagonal element corresponding to $\hat{\beta}_i$ is b, so that $\hat{\beta}_i = \sum x_{ji} y_j / b$ and $V(\hat{\beta}_i) = \sigma^2/b$. Similarly, $\hat{\beta}_{hi} = \sum x_{jh} x_{ji} y_j / c$ and $v(\hat{\beta}_{hi}) = \sigma^2/c$.

The submatrix $\mathbf{U}$ may be written, with $a = N$, as

$$\mathbf{U} = \begin{bmatrix} a & b\mathbf{1}_k' \\ b\mathbf{1}_k & c\mathbf{J}_k + d\mathbf{I}_k \end{bmatrix}.$$

The inverse, when it exists, is a similarly patterned matrix,

$$\mathbf{U}^{-1} = \begin{bmatrix} p & q\mathbf{1}_k' \\ q\mathbf{1}_k & s\mathbf{J}_k + t\mathbf{I}_k \end{bmatrix}.$$

Taking the product $\mathbf{U}\mathbf{U}^{-1} = \mathbf{I}_{k+1}$ gives a set of five consistent equations in the four unknowns, p, q, s, and t.

$$ap + kbq = 1 \tag{10.1}$$

$$aq + b(ks + t) = 0 \tag{10.2}$$

$$bp + q(kc + d) = 0 \tag{10.3}$$

$$bq + ds + ct + kcs = 0 \tag{10.4}$$

$$dt = 1. \tag{10.5}$$

The last equation gives $t = 1/d$. Eliminating t from Equations 10.2 and 10.4 gives $s = -(b^2 - ac)q/bd$, and solving Equations 10.1 and 10.3 gives $bp = -(kc + d)q$ and $q = -b/[a(kc + d) - kb^2]$.

If we now write $[d\{a(kc + d) - kb^2\}]^{-1} = A$, these solutions become $p = d(kc + d)A$, $q = -bdA$, $s = (b^2 - ac)A$, and $t = 1/d$. If $kb^2 = a(kc + d)$ the design is singular.

It would be convenient if we could choose the design points in such a way that $\operatorname{cov}(\hat{\beta}_0, \hat{\beta}_{hh}) = 0$ for all h, and $\operatorname{cov}(\hat{\beta}_{hh}, \hat{\beta}_{ii}) = 0$ for all $h, i, h \neq i$. The first condition cannot be satisfied, for we have

$$\operatorname{cov}(\hat{\beta}_0, \hat{\beta}_{hh}) = q\sigma^2 = -bdA\sigma^2,$$

and $b > 0$, $d > 0$, $A \neq 0$. Hence, $q \neq 0$ and the covariance cannot be made to vanish. On the other hand, $\operatorname{cov}(\hat{\beta}_{hh}, \hat{\beta}_{ii}) = s\sigma^2$, and a necessary and sufficient condition for this to vanish is that $b^2 - ac = 0$. It is customary to call the property $\operatorname{cov}(\hat{\beta}_{hh}, \hat{\beta}_{ii}) = 0$ the orthogonality property for response surface designs, and $b^2 = ac$, the orthogonality condition.

Box and Hunter (1957) write $\sum_j x_{jh}^2 = N\lambda_2$; $\sum_j x_{jh}^2 x_{ji}^2 = N\lambda_4$. The orthogonality condition then becomes $\lambda_2^2 = \lambda_4$.

Another interesting property for a response surface design is rotatability [Box and Hunter (1957)]. A design is defined to be rotatable if its variance contours are concentric circles about the origin. Equivalently, if $\mathbf{x}_0 = (x_{01}, x_{02}, \ldots, x_{0k})$ denotes a point in the factor space and if $\hat{y}_0$ is the estimated response at $\mathbf{x}_0$, the design is said to be rotatable if, and only if, $V(\hat{y})$ is a function only of the distance $\rho = (x_{01}^2 + x_{02}^2 + \cdots + x_{0k}^2)^{1/2}$ of $\mathbf{x}_0$ from the origin.

Box and Hunter (1957) derive the necessary and sufficient conditions for a design of any order to be rotatable. We shall confine ourselves to considering second-order designs with the restrictions that were listed at the beginning of this section. Then

$$\begin{aligned}\sigma^{-2}V(\hat{y}_0) = {} & V(\hat{\beta}_0) + \sum_h V(\hat{\beta}_h)x_{0h}^2 + \sum_h V(\hat{\beta}_{hh})x_{0h}^4 + \sum_{\substack{h \\ h<i}}\sum_i V(\hat{\beta}_{hi})x_{0h}^2 x_{0i}^2 \\ & + 2\sum_h \operatorname{cov}(\hat{\beta}_0, \hat{\beta}_{hh})x_{0h}^2 + 2\sum_{\substack{h \\ h<i}}\sum_i \operatorname{cov}(\hat{\beta}_{hh}, \hat{\beta}_{ii})x_{0h}^2 x_{0i}^2.\end{aligned}$$

This is to be a function of ρ^2. The first term, $V(\hat{\beta}_0)$, is a constant. The second-degree terms are $\rho^2\{V(\hat{\beta}_h) + 2\operatorname{cov}(\hat{\beta}_0, \hat{\beta}_{hh})\}$. The fourth-degree terms must reduce to a constant multiple of ρ^4 so that the coefficient of x_{0h}^4 has to be half the coefficient of $x_{0h}^2 x_{0i}^2$, or

$$2V(\hat{\beta}_{hh}) = V(\hat{\beta}_{hi}) + 2\,\text{cov}\,(\hat{\beta}_{hh}, \hat{\beta}_{ii}).$$

In terms of the elements of the matrix $\mathbf{U}^{-1}$, this gives

$$2(s + t) = c^{-1} + 2s,$$

or

$$2ct = 1.$$

But $t = 1/d$, and it follows that a necessary and sufficient condition for rotatability is $d = 2c$, or since $d = \sum_j x_{hj}^4 - \sum_j x_{hj}^2 x_{ij}^2$,

$$\sum_j x_{hj}^4 = 3 \sum_j x_{hj}^2 x_{ij}^2.$$

For a rotatable design, $A^{-1} = 2c\{a(k + 2)c - kb^2\}$ and $\sigma^{-2}V(\hat{y}_0) = p + (b^{-1} + 2q)\rho^2 + (s + t)\rho^4$. For a design that is both orthogonal and rotatable, $b^2 = ac$ and $A^{-1} = 4ac^2$, so that in the notation of Box and Hunter $p = (k + 2)/(2N)$, $q = -1/(2N\lambda_2)$, $s = 0$, and $t = 1/(2N\lambda_2^2)$.

10.3 3^k Designs and Some Fractions

In this and the following sections we shall consider a few examples of designs for fitting second-order models. The 3^k factorials are obvious candidates for consideration because each factor is at three levels and the model can certainly be fitted. For the complete factorial we have, when x_h takes the levels $-1, 0, +1$,

$$a = 3^k, \qquad b = 2 \cdot 3^{k-1}, \qquad c = 4 \cdot 3^{k-2}, \qquad d = 2 \cdot 3^{k-2}.$$

We see that $b^2 - ac = 0$ and $c = 2d$. It follows that the complete 3^k factorials are orthogonal designs for k factors, but they are not rotatable. We have $A^{-1} = 4 \cdot 3^{3k-4}$ so that

$$p = \frac{2k + 1}{3^k}, \qquad q = -\frac{1}{3^{k-1}}, \qquad s = 0, \qquad t = \frac{1}{2 \cdot 3^{k-2}},$$

and the variance of an estimated response at the point $\mathbf{x}_0$ is given by

$$\frac{NV}{\sigma^2} = (2k + 1) - \frac{9}{2}\sum_h x_{0h}^2 + \frac{9}{2}\sum_h x_{0h}^4 + \frac{9}{4}\sum_{\substack{h \\ h<i}}\sum_i x_{0h}^2 x_{0i}^2.$$

The smallest of these factorials is the 3^2 design which we considered in Chapter 4. For this design the adjusted variance $NV\sigma^{-2}$ is given by

$$NV\sigma^{-2} = 5 - 9(x_{01}^2 + x_{02}^2)/2 + 9(x_{01}^4 + x_{02}^4)/2 + 9x_{01}^2 x_{02}^2/4.$$

At the origin we have $V\sigma^{-2} = 5/9$. At $(\pm 1, 0)$ and $(0, \pm 1)$, $V\sigma^{-2} = 5/9$ again and at the corner points $(\pm 1, \pm 1)$, $V\sigma^{-2} = 7.25/9$. There is a minimum of

3.2/9 at $(\pm 0.4, \pm 0.4)$. Box and Hunter (1957) give diagrams for the variance contours for three designs for $k = 2$. The first of these is the 3^2 design, in which they have scaled the factors so that $\sum x_1^2 = \sum x_2^2 = 9$, i.e., the levels of x_h are $-\sqrt{(3/2)}, 0, +\sqrt{(3/2)}$. The third design is the 3^2 factorial rotated through 45°, and the second is one of the polygonal designs that will be considered in the next section.

For the 3^3 design we shall write $v = NV\sigma^{-2}$. At the origin we have $v = 7$; at the center points of the faces of the cube, $v = 7$; at the midpoints of the edges, $v = 9.25$; and at the corners, $v = 13.75$. The minimum value of v occurs when $x_1^2 = x_2^2 = x_3^2 = 1/3$ and is 4.75. For any radius ρ the minimum variance occurs on the main diagonals of the cube.

We are again faced with the fact that there are twenty-seven points in the 3^3 design and only ten coefficients in the second-order equation. Debaun (1959) presents some fractions of the 3^3 design. These are quite different from the fractions based upon hyperplanes that we considered in the previous chapter; those fractions are of little help to us in the present situation. Other designs based on fractions of 3^k factorials are given by Box and Behnken (1960).

Debaun considers five subsets of the complete factorial, each of which is symmetrical about the origin. They are

(i) The complete factorial;
(ii) The cube with points $(\pm 1, \pm 1, \pm 1)$;
(iii) The center point (0, 0, 0);
(iv) The octahedron with points $(\pm 1, 0, 0)$, $(0, \pm 1, 0)$, and $(0, 0, \pm 1)$ at the centers of the faces of the cube;
(v) The cuboctahedron, whose twelve points are the midpoints of the edges of the cube given by $(\pm 1, \pm 1, 0)$, $(\pm 1, 0, \pm 1)$, and $(0, \pm 1, \pm 1)$.

He goes on to consider and compare designs made by combining these fractions. The designs that he considers are

(i) The complete factorial;
(ii) Cube plus octahedron plus n center points;
(iii) Cube plus 2 octahedra plus n center points;
(iv) Cuboctahedron plus n center points;
(v) Cube plus cuboctahedron plus n center points;
(vi) Cuboctahedron plus octahedron plus n center points.

In comparing the designs he uses as his criterion the information per point, which is v^{-1}. He shows that the complete factorial is a relatively inefficient design by this criterion and recommends designs (iii) and (iv). We shall now consider design (iv) and give an example.

The Cuboctahedron

For the cuboctahedron with n center points we have $N = a + 12 + n$, $b = 8$, $c = 4$, $d = 4$, and $A^{-1} = 64n$.

If there are no center points the design is singular. If, however, $n > 0$, we have

$$p = \frac{1}{n}, \qquad q = -\frac{1}{2n}, \qquad s = \frac{4 - n}{16n}, \qquad t = \frac{1}{4},$$

and the design is orthogonal if, and only if, $n = 4$.

In that case we have

$$v = 4 - 2\sum x_h^2 + 4\sum x_h^4 + 4(x_1^2x_2^2 + x_1^2x_3^2 + x_2^2x_3^2).$$

Schneider and Stockett (1963) give an example of a cuboctahedron with three center points. The experiment was designed to determine the effects of column temperature (z_1), gas/liquid ratio (z_2), and packing height (z_3) in reducing the unpleasant odor of a chemical product that was being sold for household use. The coded factors are

$$x_1 = \frac{z_1 - 80}{40}, \qquad x_2 = \frac{z_2 - 0.5}{0.2}, \qquad x_3 = \frac{z_3 - 4}{2}.$$

The second-order model

$$y = \beta_0 + \beta_1x_1 + \beta_2x_2 + \beta_3x_3 + \beta_{11}x_1^2 + \beta_{22}x_2^2 + \beta_{33}x_3^2 + \beta_{12}x_1x_2 + \beta_{13}x_1x_3 + \beta_{23}x_2x_3$$

was fitted to the data. Afterwards the authors replaced x_2 by x_4, which is the ratio of liquid rate to packing height, and obtained a somewhat better fit (although it was still far from ideal). The three center points provide an unbiased estimate of error with two degrees of freedom. The data and the analysis of variance table follow.

y	x_1	x_2	x_3	x_4
66	−1	−1	0	11.75
39	+1	−1	0	11.75
43	−1	+1	0	3.75
49	+1	+1	0	3.75
58	−1	0	−1	11.00
17	+1	0	−1	11.00
−5	−1	0	+1	3.67
−40	+1	0	+1	3.67
65	0	−1	−1	23.50
7	0	+1	−1	7.50
43	0	−1	+1	7.80
−22	0	+1	+1	2.50
−31	0	0	0	5.50
−35	0	0	0	5.50
−26	0	0	0	5.50

TABLE 10.1
ANALYSIS OF VARIANCE

Source	SS	d.f.	MS	F
Raw Total	24,874.0	15		
Mean	129.1	1		
Corrected total	24,744.9	14		
First-order terms	17,940.2	3	5,980	295
Second-order terms	5,253.3	6	876	44
Lack of fit	1,510.7	3	504	25
Error	40.7	2	20	

10.4 Polygonal Designs for $k = 2$

A useful first-order design for two factors consists of the vertices of a regular polygon with n sides inscribed in the unit circle. Each point in the design may be expressed by $x_1 = \cos\theta$, $x_2 = \sin\theta$, where θ is the angle between the radius vector and the x_1 axis. Taking the first point to be $x_1 = 1$, $x_2 = 0$, the other points correspond to $\theta = 2j\pi/n, j = 1, 2, \ldots, n - 1$; $x_1 + ix_2 = \cos\theta + i\sin\theta$ are the roots of the equation $x^n - 1 = 0$ with $(1, 0)$ corresponding to $j = n$.

The sum of the roots of $x^n - 1 = 0$ is zero. Hence, $\sum_{j=1}^{2n} \cos\theta + i\sin\theta = 0$ so that $\sum\cos\theta = \sum\sin\theta = 0$ and $\sum x_{j1} = \sum x_{j2} = 0$. Furthermore, the sum of the squares of the roots is also zero. Hence,

$$\sum(\cos\theta + i\sin\theta)^2 = \sum\cos^2\theta - \sum\sin^2\theta + 2i\sum\sin\theta\cos\theta = 0.$$

Also, $\sum\cos^2\theta + \sum\sin^2\theta = n$. We thus have

$$\sum x_{j1}^2 = \sum x_{j2}^2 = \sum\cos^2\theta = \sum\sin^2\theta = n/2;$$

$$\sum x_{j1}x_{j2} = \sum\sin\theta\cos\theta = 0.$$

The first-order design is orthogonal and satisfies the conditions that we imposed on first-order designs in the previous sections. It is easily shown that the design retains these properties if it is rotated through any angle α, i.e., if $x_{j1} = \cos(\alpha + 2j\pi/n)$ and $x_{j2} = \sin(\alpha + 2j\pi/n)$. A larger design may be made up by using several polygons and letting the hth polygon have n_h vertices and radius ρ_h; in particular, we may add n_0 center points at the origin.

When we turn to second-order designs we note that for the single polygon $x_{j1}^2 + x_{j2}^2 \equiv 1$ at each point, and the design is singular. Suppose that we add n_0 center points and let $N = n + n_0$. It is necessary that $n \geq 5$.

The sum of the cubes of the roots of $x^n = 1$ is zero. Hence,

$$\sum (\cos^3 \theta - 3 \sin^2 \theta \cos \theta) + i \sum (3 \sin \theta \cos^2 \theta - \cos^3 \theta) = 0.$$

But $\sum (\cos^3 \theta - 3 \sin^2 \theta \cos \theta) = 4 \sum \cos^3 \theta - 3 \sum \cos \theta$, and we have

$$\sum \cos^3 \theta = \sum \sin^3 \theta = \sum \cos^2 \theta \sin \theta = \sum \cos \theta \sin^2 \theta = 0.$$

The sum of the fourth powers also vanishes, giving $\sum \cos^3 \theta \sin \theta = \sum \cos \theta \sin^3 \theta = 0$, and also $\sum (\cos^4 \theta + \sin^4 \theta - 6 \cos^2 \theta \sin^2 \theta) = 0$. But

$$\sum (\cos^2 \theta + \sin^2 \theta)^2 = \sum (\cos^4 \theta + \sin^4 \theta + 2 \cos^2 \theta \sin^2 \theta) = n;$$

hence,

$$\sum \sin^2 \theta \cos^2 \theta = n/8.$$

Furthermore,

$$\sum \cos^4 \theta = \sum \cos^2 \theta (1 - \sin^2 \theta) = n/2 - n/8 = 3n/8 = \sum \sin^4 \theta.$$

In the notation of the previous section, we now have

$$a = N = n + n_0, \qquad b = n/2, \qquad c = n/8, \qquad d = n/4.$$

It is not surprising that the rotatability condition $d = 2c$ is satisfied. The orthogonality condition $b^2 = ac$ leads to

$$n^2/4 = (n + n_0)n/8, \qquad \text{or} \qquad n = n_0,$$

which calls for more center points than many experimenters might regard as practicable. The elements of the covariance matrix are

$$p = n_0^{-1}, \qquad q = -n_0^{-1}, \qquad s = n_0^{-1} - n^{-1}, \qquad t = 4n^{-1}.$$

The variance of the estimate of the response at a point a distance ρ from the origin is given by

$$\sigma^{-2} V(\hat{y}) = n_0^{-1} + (2n^{-1} - 2n_0^{-1})\rho^2 + (n_0^{-1} + 3n^{-1})\rho^4.$$

If the design is orthogonal, this becomes

$$\sigma^{-2} V(\hat{y}) = n^{-1}(1 + 4\rho^4),$$

and the minimum variance occurs at the origin.

It is interesting to consider the case in which the variance is the same at the center of the circle ($\rho = 0$) and on the circumference ($\rho = 1$). For this to occur we must have

$$(2n^{-1} - 2n_0^{-1}) + (n_0^{-1} + 3n^{-1}) = 0,$$

or $n = 5n_0$, a particular example of which is given by the pentagon with a single center point. The minimum variance now occurs at $\rho^2 = 1/2$, and is given by $3\sigma^2/n$; the variance at $\rho = 0$ and $\rho = 1$ is $5\sigma^2/n$.

EXAMPLE. The following example is given by J. S. Hunter (1960). There are two factors, reaction time and temperature, for a batch process, and the levels are given by

$$x_1 = \frac{\text{time} - 90}{10}, \qquad x_2 = \frac{\text{temp} - 205}{10}.$$

This experiment is actually part of a larger experiment and we shall refer to it again later. The design was an octagon with four center points. The fitted second-order equation was

$$y = 87.3750 - 1.3838x_1 + 0.3625x_2 - 2.1438x_1^2 - 3.0938x_2^2 - 4.8750x_1x_2.$$

(Hunter retained four decimal places in the calculated coefficients for use later in reducing this equation to canonical form.) The data and the analysis of variance table follow.

y	x_1	x_2	y	x_1	x_2
88.0	0	0	89.7	0	0
86.8	0	0	85.0	0	0
78.8	−1	−1	81.2	$\sqrt{2}$	0
84.5	+1	−1	83.3	$-\sqrt{2}$	0
91.2	−1	+1	79.5	0	$\sqrt{2}$
77.4	+1	+1	81.2	0	$-\sqrt{2}$

TABLE 10.2
ANALYSIS OF VARIANCE

Source	SS	d.f.	MS
Raw total	84,653.24	12	
Mean	84,436.96	1	
Corrected total	216.28	11	
First-order terms	16.37	2	8.18
Second-order terms	171.96	3	57.32
Lack of fit	16.18	3	5.39
Error	11.77	3	3.92
Combined residual	27.95	6	4.66

10.5 Central Composite Designs

These second-order designs for k factors are composed of three sets of points.

(i) A 2^k factorial design, or a suitable fraction, with $x_i = \pm 1$. These are called the cube points, and there are n_c of them. This portion of the design is sometimes called a *measure polytope*;
(ii) A set of star or axial points (*cross polytope*). There are $n_a = 2k$ such points, two on each axis at a distance α from the origin. Their coordinates are $(\pm\alpha, 0, 0, \ldots), (0, \pm\alpha, 0, \ldots)\ldots$;
(iii) n_0 center points.

For such a design we have

$$a = N = n_c + n_a + n_0, \qquad b = n_c + 2\alpha^2, \qquad c = n_c, \qquad d = 2\alpha^4.$$

The rotatability condition becomes

$$n_c + 2\alpha^4 = 3n_c, \qquad \alpha^4 = n_c.$$

For orthogonality $N \cdot n_c = (n_c + 2\alpha^2)^2$, or,

$$n_0 + n_a = \frac{4\alpha^2(n_c + \alpha^2)}{n_c}.$$

For both orthogonality and rotatability, $\alpha^2 = \sqrt{(n_c)}$, and so $n_0 = 4(\sqrt{(n_c)} + 1) - 2k$. If $\sqrt{(n_c)}$ is an integer, we can obtain both orthogonality and rotatability by proper choice of n_0.

If $k = 2$ we have $n_c = 4$, $\alpha = \sqrt{2}$. The cube points and the star points form a regular octagon with vertices on a circle of radius $\sqrt{2}$. From the results of the previous section we recall that eight center points are needed for orthogonality. If $k = 4$, we have $\alpha = 2$, $n_0 = 12$.

If $k = 5$, we may take $n_c = 16$ (the half replicate). Then $n_0 = 10$, $\alpha = 2$.

If $\sqrt{(n_c)}$ is irrational, we must sacrifice either rotatability or orthogonality. One possibility is to choose the points so as to attain orthogonality while coming as close as we can to rotatability. For $k = 3$, rotatability requires $\alpha^4 = 8$. Then, for orthogonality as well, we should need $n_0 + 6 = 15.5$. We should, therefore, take $n_0 = 9$ and choose α to give orthogonality, i.e., so that $(8 + 6 + 9) \cdot 8 = (8 + 2\alpha^2)^2$ or $\alpha = 1.662$.

It will already be apparent to the reader that in these designs, just as in the polygon of the previous section, the price of orthogonality with rotatability is taking several center points. Some experimenters will not be happy about devoting such a large portion of their experimental effort to duplicating a single point.

It is not necessary that the cube portion of the design consist of a complete 2^k

design; the use of the half replicate has already been mentioned. In seeking to reduce the number of points in a composite design, we may note that the response at a center point estimates β_0; the difference between the average of the responses at the two star points on the ith axis and the average at the center points gives an estimate of β_{ii}, and the difference between the responses at the two axial points gives an estimate of $2\alpha\beta_i$. Thus the center points and the star points give estimates of all but the interaction terms β_{hi}, and the latter have to be estimated from the cube points. We may therefore use a fraction, if there is one that will allow all two-factor interactions to be estimated, assuming the main effects to be known. If we are prepared to suppress all interactions with three or more factors, we may thus use any 2^{k-p} fraction that contains no more than one 2 f.i. in any alias set. (If there is a single 2 f.i. among the defining contrasts, that will contradict our simplifying assumption that $\sum x_{jh}x_{ji} = 0$ for all pairs and complicate the calculations, but the parameters will still be estimable.) Hartley (1959) suggests using, when $k = 6$, the quarter replicate defined by $I = ABC = DEF = ABCDEF$. Consider, for example, the AB contrast. If we suppress higher order interactions, its expectation is $16(\beta_{12} + \beta_3)$, and we already have an estimate of β_3 from the star points.

Aia, et al. (1961) used a central composite design with $k = 3$, $\alpha = 1.6818$, and six center points in an experiment on the precipitation of stoichiometric dihydrate of calcium hydrogen orthophosphate ($CaHPO_4.2H_2O$). Among the responses that they considered was the percentage yield. The three factors used were the mole ratio of NH_3 to $CaCl_2$ in the calcium chloride solution, the addition time in minutes of the NH_3-$CaCl_2$ mix, and the starting pH of the $NH_4H_2PO_4$ solution used.

The scaled factors were

$$x_1 = [(NH_3)/(CaCl_2) - 0.85]/0.09, \qquad x_2 = (t - 50)/24,$$

$$x_3 = (pH - 3.5)/0.9.$$

The fitted second-order equation was

$$\hat{y} = 76.03 + 5.49x_1 - 0.71x_2 + 10.18x_3 - 0.88x_1x_2 - 1.46x_1x_3 - 0.29x_2x_3$$

$$+0.69x_1^2 + 0.43x_2^2 - 7.18x_3^2.$$

They later decided to drop the x_2 terms from the model. The new fitted equation was then

$$\hat{y} = 76.39 + 5.49x_1 + 10.18x_3 - 1.46x_1x_3 + 0.64x_1^2 - 7.22x_3^2.$$

The changes in the quadratic coefficients and the constant term when x_2 is omitted are small. There is no change in the coefficients of x_1 and x_3 because these terms are orthogonal to the x_2 term and the other terms.

The data and the analysis of variance table follow.

x_1	x_2	x_3	y	x_1	x_2	x_3	y
−1	−1	−1	52.8	0	−1.6818	0	80.6
1	−1	−1	67.9	0	+1.6818	0	77.5
−1	1	−1	55.4	0	0	−1.6818	36.8
1	1	−1	64.2	0	0	+1.6818	78.0
−1	−1	1	75.1	0	0	0	74.6
1	−1	1	81.6	0	0	0	75.9
−1	1	1	73.8	0	0	0	76.9
1	1	1	79.5	0	0	0	72.3
−1.6818	0	0	68.1	0	0	0	75.9
+1.6818	0	0	91.2	0	0	0	79.8

TABLE 10.3
ANALYSIS OF VARIANCE

Source	SS	d.f.
Total	770.96	19
First-order terms	610.93	3
Second-order terms	135.15	6
Lack of fit	18.71	5
Error	6.17	5

10.6 Orthogonal Blocking

We now discuss breaking up the response surface design into m blocks; the wth block contains n'_w points. The block effects are additive constants, and there is no interaction between blocks and factors. We may modify our model accordingly and write

$$E(y_j) = \beta_0 + \sum_{w=1}^{m} \beta_{0w}(z_{jw} - \bar{z}_w) + \sum_h \beta_h x_{jh} + \sum_{\substack{h \\ h<i}} \sum_i \beta_{hi} x_{jh} x_{ji} + \sum_h \beta_{hh} x_{jh}^2,$$

where $\beta_0 + \beta_{0w}$ is the expected response at the origin in the wth block; β_{0w} is the wth block effect, and we impose the side condition $\sum_w \beta_{0w} n'_w = 0$; z_{jw} is a dummy variable which takes the value $+1$ if the jth point is in the wth block, and is zero otherwise. We denote by $\bar{z}_w$ the average n'_w/N.

It is desirable that the block effects be orthogonal to the parameters of the response surface. This is achieved if, and only if, the corresponding design vectors are perpendicular, i.e., if for each w

$$\sum_j (z_{jw} - \bar{z}_w) x_{jh} = 0 \tag{10.6}$$

for each h,

$$\sum_j (z_{jw} - \bar{z}_w)x_{jh}x_{ji} = 0 \tag{10.7}$$

for each pair $h, i, h \neq i$,

$$\sum_j (z_{jw} - \bar{z}_w)x_{jh}^2 = 0 \tag{10.8}$$

for each h. Since we are confining our attention to designs for which $\sum_j x_h = \sum_j x_{jh}x_{ji} = 0$ and $\sum_j x_{jh}^2 = b$, the first two conditions become

$$\sum_j z_{jw}x_{jh} = 0, \tag{10.6$'$}$$

$$\sum z_{jw}x_{jh}x_{ji} = 0. \tag{10.7$'$}$$

The third condition is, for all the points in the wth block,

$$\sum_j x_{jh}^2 = n'_w b/N.$$

The set of points in each block must thus comprise an orthogonal first-order design with $\sum x_{jh}^2$ proportional to the block size.

An example for $k = 2$ is provided by the regular hexagon with an even number of center points. This design may be split into two blocks of size $N/2$ by dividing the hexagon into two equilateral triangles, and letting each block consist of one of the triangles together with half the center points.

Thus, with two center points we might have

	x_1	x_2		x_1	x_2
Block 1	1.000	0.000	Block 2	−1.000	0.000
	−0.500	0.866		0.500	0.866
	−0.500	−0.866		0.500	−0.866
	0.000	0.000		0.000	0.000

Indeed, if we take six center points and put three of them in each block, we can simultaneously obtain orthogonal blocking, orthogonality (of the estimates of quadratic coefficients), and rotatability.

A central composite design can be divided into two rotatable first-order designs, namely the cube points and the axial points. Suppose that this is done and that n_{0c} of the center points are included in the block with the cube points and $n_{0a} = n_0 - n_{0c}$ are included with the axial points.

For the blocking to be orthogonal we must have

$$n_c = (n_c + n_{0c})b/N, \qquad 2\alpha^2 = (2k + n_{0a})b/N.$$

Eliminating b/N,

$$(2k + n_{0a})n_c = 2\alpha^2(n_c + n_{0c}).$$

A convenient example is the case $k = 4$ and $n_c = 16$. We have already seen that for orthogonality and rotatability $n_0 = 12$ and $\alpha^2 = 4$. Orthogonal blocking is then obtained by putting $n_{0c} = 8$ and $n_{0a} = 4$.

The cube points may be divided into two half replicates confounding the four-factor interaction. Each half replicate is an orthogonal first-order design, and so we can add four center points to each 2^{4-1} and thereby have a design for $n = 16 + 8 + 12 = 36$ divided into three orthogonal blocks of twelve points each.

The octagonal design with four center points given as an example in the section on polygonal designs can be run in two orthogonal blocks. The blocks correspond to the columns in the listing of the data. If we consider the octagonal design as a central composite design for $k = 2$ with $\alpha = 4$ and $n_c = 4$, the first block consists of the "cube" points together with two center points. The second block contains the star points plus the two remaining center points.

10.7 Steepest Ascent and Canonical Equations

So far we have talked about designs for fitting first- and second-order models without concerning ourselves about the problem of where or when to fit them. It will be convenient for us to confine ourselves for the moment to two factors, x_1 and x_2; the generalization to k factors will be clear. How do we decide about the levels of the factors? There are two points to be remembered.

1. We are not really concerned about investigating the response over the entire sample space of points (x_1, x_2) at which the plant will operate. We are interested only in operating in the subset of the sample space in which the response is optimized. The term *optimizing the response* is not as simple as it sounds. In a practical situation there may be confusion over what response is to be optimized. The engineer wishes to optimize the yield of some product. The comptroller's department employees in the head office have in mind optimizing profit, and they may calculate it by a formula unknown to the researchers. There are, for example, some situations in which it is better to maximize the total yield than the percentage yield. It might, perhaps, be more desirable to pass 200 gallons of feed stock per hour through a plant converting only 60 per cent, than to pass 100 gallons per hour with 80 per cent conversion. For convenience we shall think in terms of maximizing conversion or yield, recognizing, however, that in some cases other responses may be of greater interest.

If we are interested in running the plant in the area of maximum yield, then we are interested only in investigating closely the area of the sample space in which the yields are high. We do not care to devote any more time than is necessary to exploring regions of no economic interest. We envision a two-dimensional factor space, with the yield as the third dimension rising out of the x_1, x_2 plane. Hopefully, this three-dimensional figure $y = f(x_1, x_2)$ has a single peak, and we would like to find it. Having found it we shall operate the plant at those conditions, and the stockholders will be content. Perhaps the peak is unattainable—it might require so high a temperature that we should burst a boiler—but we shall do the best we can.

2. Even if we can find a set of optimum conditions for operating the plant, we may not be able to maintain them. Suppose, for example, that we have a polymerization process using an acid catalyst. A simplified description of the process might be that the feed stock is passed through a reactor containing acid of strength x_1 at a temperature x_2, and a fraction y of it is converted.

The first-order model gives an equation

$$y = \beta_0 + \beta_1 x_1 + \beta_2 x_2.$$

If β_1 and β_2 are both positive, the message seems to be clear. We should increase the acid strength and the temperature as much as we can without wrecking the equipment. As the plant runs, water and contaminants leak into the reactor and the overall acid strength drops. All we can do is to continue to run the plant at as high a temperature as possible until it is decided that the acid strength has fallen far enough. Then the old acid is removed, the reactor is charged with new acid, and we start again.

Consider, however, a process such as the polymerization of propylene. The monomer is fed into the reactor. The dimer is unstable, and the product, or effluent, consists of unconverted monomer together with a mixture of trimer, tetramer, and pentamer (i.e., polymers in which the molecules consist of three, four, or five molecules of propylene, respectively). Suppose that we want to optimize the yield of tetramer. For any given acid strength we can vary the operating temperature. If the temperature is too low we get mostly trimer and little of the heavier polymers. On the other hand, if the reactor is too hot the polymerization is so active that we have a lot of pentamer. The ideal, therefore, for maximizing the yield of tetramer is to find an operating temperature that is neither too hot nor too cold. For any given acid strength, an approximation of the response curve $y = f(x_2)$ by a parabola with a maximum at some intermediate temperature seems appropriate. Similarly, for a fixed operating temperature we find that when the acid is too dilute there is too much trimer, and when the acid is too strong there is too much pentamer. Again a quadratic approximation for $y = f(x_1)$ is appropriate. The second-order model $y = f(x_1, x_2)$ is justified, at least in the immediate area of the optimal value of y, or the peak of the mountain. It is not enough, however, for us to know where the peak is. As the acid grows weaker, we shall have to adjust the temperature upward to compensate for the loss in strength. We should like to avoid overcompensation or undercompensation.

In short, our objective should be to find the peak and map the area around it so that we can find a path that will minimize loss, i.e., a path to follow by adjusting x_2 so that as x_1 decreases we shall lose height as slowly as possible.

10.8 The Method of Steepest Ascent

In seeking the neighborhood of the peak we begin by fitting a first-order design in the region where the plant is currently operating. On the basis of this experiment we estimate the direction of steepest ascent. We march in this direction until we find that we are no longer climbing. At that stage we perform another first-order experiment and perhaps arrive at a new direction in which to resume our climb toward the peak. On the other hand, we may find from this latest first-order experiment that we are already either near the peak or near a saddle point, in which case we augment the first-order design by further points to make it into a second-order design.

We illustrate this by referring back to the experiment that led to Hunter's octagonal design, which we mentioned earlier. There were two factors, reaction time and reaction temperature.

Hunter began his experiment at a reaction time of sixty minutes and a reaction temperature of 160°. He elected to make changes of ± 10 minutes in the time and $\pm 10°$ in the temperature, and ran a duplicated 2^2 factorial with a duplicated center point,

$$x_1 = \frac{\text{time} - 60}{10}, \qquad x_2 = \frac{\text{temp} - 160}{10}.$$

The data were

x_1	x_2	y
+1	+1	68.0, 66.8
+1	−1	60.3, 61.7
−1	+1	62.2, 64.6
—1	−1	54.3, 56.5
0	0	60.3, 62.3

The fitted first-order model was

$$\hat{y} = 61.7 + 2.4x_1 + 3.6x_2.$$

The contours of $\hat{y}$ are a series of parallel lines. In general we have a series of parallel hyperplanes $y = \beta_0 + \sum_i \beta_i x_i$. The direction of steepest ascent, i.e., the direction in which we cut across the contours most rapidly, is parallel to the normal to the hyperplane. This is along the line through the center of the first-order model with direction cosines proportional to β_i. In the example this procedure calls for moving 2.4 units in the x_1 direction for every 3.6 units in the x_2 direction. The length of the steps has to be decided by the experimenter on the basis of his experience.

A second 2^2 factorial with two center points was carried out, centered at a reaction time of ninety minutes and a temperature of 205°. The coded levels were

$$x_1 = \frac{\text{time} - 90}{10}, \qquad x_2 = \frac{\text{temp} - 205}{10}.$$

This design is the first column of the octagonal design given earlier. The fitted model including the x_1x_2 term is

$$y = 78.0 + 2.0x_1 + 1.4x_2 - 5.1x_1x_2,$$

and we see that the interaction coefficient is considerably larger than either of the main effect terms.

The design was therefore augmented to the octagon with four center points by adding the second column of points. In this instance the experimenter did not divide the design into two blocks. The analysis of variance table and the fitted equation have already been given.

10.9 The Canonical Form of the Equations

We are going to confine our discussion to second-order equations. Third-order models may be appropriate in some special cases, but the additional complications that cubic curves bring make a discussion of them infeasible here.

Corresponding to each value of $\hat{y}$ is the second-degree function $f(x_1, x_2) =$ const in k variables. When $k = 2$ we have the conic sections; for $k = 3$ we have the quadric surfaces.

Confining ourselves to the case $k = 2$, we have for the contour $\hat{y} = \hat{y}_0$ the general conic

$$F(x_1, x_2) = \beta_{11}x_1^2 + \beta_{12}x_1x_2 + \beta_{22}x_2^2 + \beta_1x_1 + \beta_2x_2 + \gamma = 0,$$

where $\gamma = \beta_0 - y_0$.

For various values of γ this equation gives a family of concentric conics. Their center is given by the solution to the equations

$$\partial F/\partial x_1 = 0, \qquad \partial F/\partial x_2 = 0$$

or $2\beta_{11}x_1 + \beta_{12}x_2 + \beta_1 = 0$, $\beta_{12}x_1 + 2\beta_{22}x_2 + \beta_2 = 0$.

In the simplest case the contours form a family of concentric ellipses with the center being the peak of the mountain. A conic is an ellipse if the discriminant $4\beta_{11}\beta_{22} - \beta_{12}^2$ is positive, and a hyperbola if the discriminant is negative.

Box and Youle (1955) give an example in which two factors, both concentrations of reactants, are considered. The fitted second-order equation is

$$\hat{y} = 78.56 + 0.50x_1 - 0.21x_2 - 2.31x_1^2 - 2.15x_2^2 + 4.08x_1x_2.$$

The concentrations of the reactants were measured in grams per liter and

$$x_1 = \frac{c_1 - 55}{5}, \qquad x_2 = \frac{c_2 - 35}{5},$$

where c_1 is the concentration of reactant A, and c_2 the concentration of reactant B. Experiments were conducted in the region $-1 \leq x_1 \leq +1$, $-1 \leq x_2 \leq +1$.

The discriminant, $4(2.31 \times 2.15) - 4.08^2$, is positive, and the contours of $\hat{y}$ are a family of concentric ellipses with the center given by

$$0.50 = 4.62x_1^* - 4.08x_2^*, \qquad -0.21 = -4.08x_1^* + 4.30x_2^*,$$

whence

$$x_1^* = 0.41, \qquad x_2^* = 0.34.$$

The value of the predicted (maximum) response at the center is $\hat{y} = 78.63$. To derive the canonical form of the equation, we find the latent roots of the matrix of second-order coefficients.

This is the matrix $\mathbf{B} = (b_{ij})$ where $b_{ii} = \beta_{ii}$ and $b_{ij} = \beta_{ij}/2$ $(i \neq j)$. In this case

$$\mathbf{B} = \begin{bmatrix} -2.31 & 2.04 \\ 2.04 & -2.15 \end{bmatrix},$$

and the roots are $\lambda_1 = -4.27$, $\lambda_2 = -0.19$.

The corresponding latent vectors of unit length are

$$\mathbf{z}_1 = \begin{bmatrix} 0.72 \\ -0.69 \end{bmatrix}, \qquad \mathbf{z}_2 = \begin{bmatrix} 0.69 \\ 0.72 \end{bmatrix}$$

and the canonical form of the equation is

$$\hat{y} = 78.63 - 4.27z_1^2 - 0.19z_2^2,$$

where $z_1 = 0.72(x_1 - x_1^*) - 0.69(x_2 - x_2^*) = 0.72x_1 - 0.69x_2 - 0.06$ and $z_2 = 0.69x_1 + 0.72x_2 - 0.53$.

In the new coordinate system the major axis of the ellipses is the line $z_1 = 0$. We have already seen that the optimum operating conditions are at the center $x_1 = x_1^* = 0.41$ and $x_2 = 0.34$. If, however, we are obliged to move away from the center because of a change in concentration of one of the two reactants, we shall lose yield most slowly if we can adjust x_1 and x_2 so that z_1 is kept at zero. This means that if, for example, a change δx_1 occurs in x_1, we should adjust for this by making a change δx_2 in x_2, where

$$0.72\delta x_1 - 0.69\delta x_2 = 0.$$

In the particular example this is almost equivalent to $\delta x_2 = \delta x_1$, which suggests that here the important factor might be the difference in concentrations rather than the actual concentrations themselves.

10.10 Other Surfaces

In the example given by Hunter the second-order equation was

$$\hat{y} = 87.3750 - 1.3838x_1 + 0.3625x_2 - 2.1438x_1^2 - 3.0938x_2^2 - 4.8750x_1x_2.$$

This is also a family of ellipses, but in this case the center is at $x_1 = -3.7369$, $x_2 = +3.0028$, which is well outside the area of experimentation. It may even be outside the range in which the plant can be operated. In the area where the experiments took place the response surface is a rising ridge, and further experimentation might take place in the direction of a line toward the center.

The canonical form is

$$\hat{y} = 90.50 - 5.10z_1^2 - 0.14z_2^2,$$

where $z_1 = 0.64x_1 + 0.77x_2 + 0.06$ and $z_2 = -0.77x_1 + 0.64x_2 - 4.79$. Increases in the value of z_1^2 bring about sharp decreases in the yield.

When the contours are hyperbolas, the response surface is a saddle surface with the center of the family at the center of the saddle. Moving away from the center along one of the axes will produce an increase in the yield. Moving away along the other axis will result in a drop in the yield. The experiment by Aia et al. (1961) gave a saddle surface.

As the discriminant approaches zero, the family of curves approaches a family of degenerate parabolas, and the contours become a series of pairs of parallel lines. This is a ridge surface, and the optimum procedure would be to find the highest of the parallel ridges and to remain on it.

With three factors the geometrical representation becomes more difficult. There are several worked examples in the literature, particularly in the paper by Box and Youle (1955) and the book by Davies (1956). Further examples appear from time to time in chemical and agronomic journals. Examples of third- and fourth-order models are not found as often because the designs and the interpretation of the results are, on the whole, too complicated for general use by engineers. Third- and fourth-order rotatable designs are discussed by Gardner, et al. (1959), and in a series of papers by Draper (1960*a*, *b*; 1961; 1962).

10.11 Evolutionary Operation

The discussion so far has been concerned with the design of specific experiments, particularly with experiments that might be carried out in a pilot plant by a research and development group. Once an actual plant is in operation, little experimentation usually takes place. If the plant can be persuaded to operate under conditions that produce profitable yields, the attention of the operators is directed toward keeping the plant working at those conditions and maintaining the status quo. Perhaps it would be possible to improve the yield somewhat, but, on the other hand, tinkering around with the operating variables might cause a drop in production. This defensive approach is bolstered by the conviction that the research personnel will have to make large changes in the operating variables in order to obtain meaningful results.

A new plant may produce at a lower yield than had been expected because of problems of scale up. If a plant to produce a thousand barrels a day is to be designed on the basis of a pilot plant that produces two barrels a day, some dimensions will have to be increased five hundred fold, some by $\sqrt{500}$, and some by the cube root. When the actual plant is built, it may well happen that the actual optimum operating conditions are not those that were obtained on the pilot plant, and the yield will fall short of the projections. Little will be gained by asking the research staff to generate more data on the pilot plant. If they did a reasonable job the first time, further data will only serve to confirm their earlier results. What is needed at this stage is not more information about how to run the pilot plant but information about optimizing production on the full-size plant that has already been built.

Evolutionary operation was proposed by Box (1957) as an operating procedure for nudging a plant toward optimum conditions without causing dramatic disturbances and catastrophic cutbacks in production. It is designed as a method of operation that can be carried out by the plant personnel themselves with a minimum of supervision, or interference, from the research people. Indeed, it is recommended as a standard operating procedure, whether or not there is concern about the performance of the plant.

An evolutionary operation program for a plant contains two basic steps:

(i) A routine of systematic small changes is introduced in the levels of the operating variables being considered;
(ii) The information generated is fed back to the plant foreman or other supervisor.

Box and Hunter (1959) show how the latter objective is facilitated by using simplified calculations and emphasizing graphical presentation. They give an example of an application to a batch process in which the two operating variables being monitored are temperature and reaction time. We shall use part of their example. There is nothing in the theory to prevent us from monitoring several variables but, for practical purposes, two or three variables are enough to handle under normal circumstances. Box and Hunter also discuss the case in which three variables are monitored.

The operating cycle is a cycle of five points that form a 2^2 factorial with a center point. If the plant is currently operating at a temperature of 120° with a reaction time of sixty minutes, and if fluctuations of $\pm 10°$ in the temperature and ± 10

minutes in the time are contemplated, the cycle would consist of the five points

120°, 60 min;
110°, 50 min;
130°, 70 min;
110°, 70 min;
130°, 50 min.

The cycle is repeated several times. Each time the responses y_1, y_2, y_3, y_4, y_5 are recorded and their averages are posted. The main effects and the interaction of the two factors are calculated, and when one or more of these three effects becomes significant, decisions about changes in operating conditions are made by the plant foreman.

The most important feature in step (ii) is the information board. Table 10.4

TABLE 10.4
YIELDS DURING THE FIRST FOUR CYCLES

	Conditions				
Cycle	1	2	3	4	5
1	63.7	62.8	63.2	67.2	60.5
2	62.1	65.8	65.5	67.6	61.3
3	59.6	62.1	62.0	65.3	64.1
4	63.5	62.8	67.9	62.6	61.7

TABLE 10.5
INFORMATION BOARD—CYCLE 4

Requirement	Per cent yield Maximize			Other Responses
|	61.9		64.6	
T E M P		62.2		
|	63.4		65.7	
	—TIME—			

Error Limits for Averages		±2.3
Effects with	time	+2.5 ± 2.3
95% error	temperature	−1.3 ± 2.3
Limits	$t \times T$	0.2 ± 2.3
	Change in mean	1.4 ± 2.0
Standard deviation		2.3
Prior estimate of σ		1.8

contains the data given by Box and Hunter for the yields during the first four cycles of their process. Table 10.5 is the format of an information board at the end of the fourth cycle. The information board is easily kept up to date, and the data are clearly and simply displayed on it.

The calculation procedures recommended by Box and Hunter make use of the range to provide an estimate of the standard deviation, σ. We discuss this in the next section.

10.12 The Method of Calculation

Table 10.6 shows a calculation sheet at the end of the fourth cycle. Similar sheets for the first three cycles are included in the paper by Box and Hunter. The calculation of the averages and of the estimates of the effects is straightforward. The calculation of σ is based upon the use of the range. This is easier to compute than the usual quadratic estimate, and for the range of five random variables the efficiency relative to the quadratic estimate is 0.955.

Let $y_{i,j}$ denote the value of y_i at the jth cycle, and let $\bar{y}_{i,j}$ denote the average of the values of y_i for the first j cycles. The entries in row (iv) of the table for the nth cycle are the observed values of $y_{i,n} - \bar{y}_{i,n-1}$ with variance $\sigma^2\{1 + (n - 1)^{-1}\} = n\sigma^2/(n - 1)$. Their range, R, is independent of any blocking that may have occurred inadvertently between cycles. The estimate of the standard deviation is obtained by multiplying the range by a factor, c_k, which depends upon the number of observations included. In the example the range is $4.1 - (-4.3) = 8.4$, and it is the range of five observations. The corresponding factor c_k may be found in the standard tables such as the Biometrika tables of Pearson and Hartley (1954). For $k = 5$ we have $c_5 = 0.430$; for a 2^3 design with one center point we need $c_9 = 0.337$, and with two center points, $c_{10} = 0.325$; $c_4 = 0.486$. Then $c_k R$ estimates $\sigma\{n/(n - 1)\}^{1/2}$, and so the desired unbiased estimate of σ is

$$\sigma s = Rc_k\sqrt{\left(\frac{n-1}{n}\right)} = R \times f_{k,n}.$$

Values of $f_{k,n}$ for $2 \le n \le 10$ and $k = 4, 5, 9, 10$ follow.

$n =$	2	3	4	5	6	7	8	9	10
$k = 4$	0.34	0.40	0.42	0.43	0.44	0.45	0.45	0.46	0.46
5	0.30	0.35	0.37	0.38	0.39	0.40	0.40	0.40	0.41
9	0.24	0.27	0.29	0.30	0.31	0.31	0.31	0.32	0.32
10	0.23	0.26	0.28	0.29	0.30	0.30	0.30	0.31	0.31

TABLE 10.6

CALCULATION SHEET—END OF FOUR CYCLES

Calculation of Averages

Operating Conditions	1	2	3	4	5
(i) Previous cycle sum	185.4	190.7	190.7	200.1	185.9
(ii) Previous cycle average	61.8	63.6	63.6	66.7	62.0
(iii) New observations	63.5	62.8	67.9	62.6	61.7
(iv) Differences (ii) less (iii)	−1.7	0.8	−4.3	4.1	0.3
(v) New sums	248.9	253.5	258.6	262.7	247.6
(vi) New averages $\bar{y}_i$	62.2	63.4	64.6	65.7	61.9

Calculation of σ

Previous sum s	$= 3.66$
Previous average s	$= 1.83$
New s = range × $f_{k,n}$	$= 3.10$
Range	$= 8.4$
New sum s	$= 6.76$
New average s = (sum)/$(n-1)$	$= 2.25$

Calculation of Effects

A (time) effect	$= \frac{1}{2}(\bar{y}_3 + \bar{y}_4 - \bar{y}_2 - \bar{y}_5)$	$= +2.5$
B (temperature) effect	$= \frac{1}{2}(\bar{y}_3 + \bar{y}_5 - \bar{y}_2 - \bar{y}_4)$	$= -1.3$
$A \times B$	$= \frac{1}{2}(\bar{y}_2 + \bar{y}_3 - \bar{y}_4 - \bar{y}_5)$	$= +0.2$
Change in mean	$= \frac{1}{5}(\bar{y}_2 + \bar{y}_3 + \bar{y}_4 + \bar{y}_5 - 4\bar{y}_1)$	$= +1.4$

Calculation of Limits

For new average	$2\sigma/\sqrt{n}$	$= \pm 2.25$
For new effects	$2\sigma/\sqrt{n}$	$= \pm 2.25$
For change in mean	$1.78\sigma/\sqrt{n}$	$= \pm 2.00$

Exercises

1. Show that the polygonal design retains its properties even if rotated through an angle α, i.e., it is still a rotatable design and n center points are still needed for orthogonality.
2. Prove that for any rotatable design, satisfying the conditions of the early part of the chapter, section 10.2,

$$\sigma^{-2}V(\hat{y}_0) = p + (b^{-1} + 2q)\rho^2 + (s + t)\rho^4.$$

3. Prove that, in the notation of Box and Hunter, a design is both orthogonal and rotatable if, and only if,

$$p = (k + 2)/(2N), \qquad q = -1/(2N\lambda_2), \qquad t = 1/(2N\lambda_2).$$

4. Prove that for prediction based on a 3^3 factorial design with the vertices of the cube at $(\pm 1, \pm 1, \pm 1)$ the variance of a predicted response is a minimum at the points where $3x_1^2 = 3x_2^2 = 3x_3^2 = 1$.
5. Find the variance of an estimated response at the point (x_1, x_2, x_3) for Debaun's designs (2), (3), (5), and (6).
6. Show that for fitting a first-order model in three dimensions, with the points to be taken on the surface of a sphere with unit radius, the vertices of a regular tetrahedron with any orientation is an optimal design for four points.
7. Consider a 3^2 design with $k = 2$ that has been rotated through 45° so that the design points are $(0, 0)$, $(\pm\sqrt{2}, 0)$, $(0, \pm\sqrt{2})$, and so on. Analyze the design and compare the variance contours with those of a 3^2 design in the usual orientation.
8. A regular icosahedron inscribed in a sphere of radius 3 has twelve vertices given by $(0, \pm a, +b)$, $(\pm b, 0, \pm a)$, $(\pm a, \pm b, 0)$ where $a = 1.473$ and $b = 0.911$. Consider this set of points together with some center points as a second-order design.
9. (Continuation). Another regular figure is the dodecahedron with twenty vertices given by $(0, \pm c^{-1}, \pm c)$, $(\pm c, 0, \pm c^{-1})$, $(\pm c^{-1}, \pm c, 0)$, $(\pm 1, \pm 1, \pm 1)$ where $c = 1.618$. Consider this set of points plus some center points as a second-order design.
10. (Continuation). Show that the dodecahedron with $5n$ center points can be divided into five orthogonal blocks of $4 + n$ points each. Each block is a regular tetrahedron plus n center points.

CHAPTER 11

Incomplete Block Designs

In Chapter 3 we considered the randomized complete block design. In that design v varieties or t treatments were tested in b blocks, and each block contained exactly one observation on each treatment, or one plot with each variety. The advantage of this design is that because the variance between plots in the same block is smaller than the variance between plots in different blocks, we obtain an increase in the precision of comparisons between varieties by being able to eliminate the block differences. An obvious problem arises when the blocks available are not large enough to accommodate one plot with each variety. We are then led to the consideration of incomplete block designs: incomplete inasmuch as the blocks do not each contain a complete set of varieties.

Suppose that in the gasoline experiment there had been seven gasolines to be tested in cars but because of a shortage of time each car could be used only for three gasolines. Clearly, each gasoline cannot be tested in each car. Our interest is in comparing gasolines, and we might try to allot the gasolines to the cars in such a way as to have each pair of gasolines appear together in the same number of cars. One way of doing this is to take $35 = \binom{7}{3}$ cars and to give to each car a different combination of three gasolines. Each pair of gasolines will appear together in five of the cars. A smaller design, in which each pair appears together in one car, is given by

$$ABE, \quad CDE, \quad ACF, \quad BDF, \quad ADG, \quad BCG, \quad EFG,$$

where the letters denote gasolines and each set of three gasolines is assigned to a different car. Such designs are called balanced incomplete block designs.

This chapter will begin with a discussion of balanced incomplete block designs. This will be followed by sections on the general incomplete block design, in which the blocks may be of different sizes, and where there may be no semblance of balance. We shall also discuss such topics as the recovery of interblock information. We have a choice of notations. We may talk either of t treatments or of v varieties. In this chapter we shall follow the more modern usage of talking about t treatments, although we shall still call the experimental units to which they are applied, plots. We shall thus have t treatments applied to several plots in b blocks. In the succeeding chapters we shall discuss partially balanced association schemes and the construction of incomplete block designs. In those chapters we shall employ the older usage of v varieties because v is the usual notation in that area.

11.1 Balanced Incomplete Block Designs

These designs were introduced by Yates (1936). They are designs for t treatments in b blocks. Each block contains k plots, $k < t$. Each plot contains a single treatment and each treatment appears in r plots. We say that each treatment has r replications, or is replicated r times. No treatment appears more than once in any block. Each pair of treatments appears in the same block λ times. We shall sometimes use the abbreviations BIB design or BIBD.

The five parameters b, t, r, k, and λ are not independent. They are integers subject to the following restrictions:

(i) $N = rt = bk$, where N is the total number of plots;
(ii) $\lambda(t - 1) = r(k - 1)$;
(iii) $b \geq t$.

A design with $b = t$ and $r = k$ is said to be symmetric.

To establish restriction (ii) we consider any treatment, such as A. The treatment A appears in r blocks, in each of which are also $(k - 1)$ other treatments. There are, therefore, $r(k - 1)$ plots, which are in the same block as a plot containing A. These plots have to contain the remaining $(t - 1)$ treatments exactly λ times each. Hence, $\lambda(t - 1) = r(k - 1)$. The four parameter values $b = 12, k = 4, t = 8$, and $r = 6$ satisfy restrictions (i) and (iii). However, restriction (ii) gives $\lambda = 18/7$, which is not an integer. Hence, there is no balanced incomplete block design with those values of b, t, r, and k.

The incidence matrix of a design is the matrix $\mathbf{N} = (n_{ij})$ of t rows and b columns; n_{ij} is the number of times that the ith variety occurs in the jth block. For a balanced incomplete block design, n_{ij} is either one or zero, and $n_{ij}^2 = n_{ij}$. The matrix $\mathbf{NN}'$ has t rows and t columns. It is of considerable interest in the development of the theory of incomplete block designs. If we denote the elements of $\mathbf{NN}'$ by q_{ih} we see that $q_{ii} = \sum_j n_{ij}^2$, and $q_{ih} = \sum_j n_{ij}n_{hj}$, $(i \neq h)$. For the balanced incomplete block design we have $q_{ii} = r$ and $q_{ih} = \lambda$, $(i \neq h)$; we may therefore write

$$\mathbf{NN}' = (r - \lambda)\mathbf{I}_t + \lambda\mathbf{J}_t,$$

where $\mathbf{I}_t$ is the identity matrix of order t and $\mathbf{J}_t$ is a square matrix of order t with every element unity.

To evaluate the determinant of $\mathbf{NN}'$ we subtract the first column from each of the other columns, and then add all the other rows to the first row. It is now readily seen that

$$|\mathbf{NN}'| = (r - \lambda)^{t-1}(r + (t - 1)\lambda) = (r - \lambda)^{t-1}rk > 0.$$

Thus $\mathbf{NN}'$ has rank t, and so $\mathbf{N}$ has rank t. However, $\mathbf{N}$ is a matrix of t rows and b columns so that $r(\mathbf{N}) \leq \min(b, t)$. Hence, $t \leq \min(b, t)$, and the inequality (iii) is proved. This inequality is due to Fisher (1940). The proof that has been presented follows that of Bose (1949).

If the design is symmetric, $b = t$ and $\mathbf{N}$ is square. In that case $|\mathbf{NN}'| = |\mathbf{N}|^2$, and so $(r - \lambda)^{t-1}r^2$ is a perfect square. This implies that when t is even, $(r - \lambda)$ must be a square. Consider the set of parameters $t = b = 22, r = k = 7, \lambda = 2$; t is even but $r - \lambda = 5$, which is not a square, and so no design with these parameters exists. This result [Shrikhande (1950)] shows that the conditions given are necessary, but not sufficient, for the existence of a balanced incomplete block design.

11.2 Some Balanced Incomplete Block Designs

For any pair of integers $t, k(k < t)$, a balanced incomplete block design can be obtained by taking all possible combinations of k out of t treatments. For these designs

$$b = \binom{t}{k}, \qquad r = \binom{t-1}{k-1}, \qquad \lambda = \binom{t-2}{k-2}.$$

There are, however, smaller designs for some values of t and k. The design given at the beginning of the chapter with $t = b = 7$, $r = k = 3$, and $\lambda = 1$ is an example.

A design with $t = 6, b = 10, r = 5, k = 3$, and $\lambda = 2$ is

$$ABE, \quad ABF, \quad ACD, \quad ACF, \quad ADE, \quad BCD, \quad BCE, \quad BDF, \quad CEF, \quad DEF.$$

A design for $t = 9, b = 12, r = 4, k = 3$, and $\lambda = 1$ can be obtained by using orthogonal Latin squares. (The design is employed in the example in the next section.) A design with $t = 16$, $b = 20$, $r = 5$, $k = 4$, and $\lambda = 1$ may be constructed in a similar fashion. Both designs are from the orthogonal series of Yates. Their derivation will be discussed in Chapter 13. The second design follows. We denote the treatments by 1, 2, . . ., 16. The treatments are written in five 4 × 4 arrays and each array forms a complete replicate. The rows are the blocks.

I	II	III	IV	V
1 2 3 4	1 5 9 13	1 6 11 16	1 7 12 14	1 8 10 15
5 6 7 8	2 6 10 14	2 5 12 15	2 8 11 13	2 7 9 16
9 10 11 12	3 7 11 15	3 8 9 14	3 5 10 16	3 6 12 13
13 14 15 16	4 8 12 16	4 7 10 13	4 6 9 15	4 5 11 14

This design has at least two applications outside the field of experimental design. In each of the five groups of four blocks every treatment appears exactly once. The design may, therefore, be used in the following situation to allocate bridge players to tables. Suppose that sixteen bridge players are gathered together for an evening and that there is enough time for five games to be played at each table. How do we arrange the players so that each player plays five games; four tables are in continuous use and no player has to sit out for a time; and every player has the opportunity to play one game at the same table as every other player. The solution is to let replications correspond to games and treatments to players. Thus, during the fourth game of the evening the players at the first table are 1, 7, 12, and 14; at the second table they are 2, 8, 11, and 13, and so on.

The second application was found in Great Britain before the Second World War in professional motorcycle racing on cinder tracks, a sport which is still popular there. The tracks are about 350 yards in circumference. Four riders take part in each four-lap race, and an evening's program calls for about twenty races. Usually the evening is devoted to team races between various teams in a league, but occasionally the riders race as individuals. The following situation then occurs. There are sixteen riders; they are to be arranged, four at a time, in such a way that each rider races five times, and, during the evening, he races once against each of the other fifteen men. This is achieved by using the balanced incomplete block design and letting treatments represent riders and blocks represent races. Unfortunately, the program cannot be arranged so that no rider is obliged to ride in two consecutive races.

A design with $t = 8$, $b = 14$, $r = 7$, $k = 4$, and $\lambda = 3$ is obtained when a 2^3 design is repeated seven times in two blocks of four runs each time, confounding in turn the three main effects and the four interactions. Writing, 1, 2, 3, 4, 5, 6, 7, 8 for (1), *a*, *b*, *ab*, *c*, *ac*, *bc*, *abc*, respectively, the design is

1 3 5 7	1 2 5 6	1 2 3 4	1 4 5 8
2 4 6 8	3 4 7 8	5 6 7 8	2 3 6 7
1 3 6 8	1 2 7 8	1 4 6 7	
2 4 5 7	3 4 5 6	2 3 5 8	

The design that we have just given for $t = 16$ could be derived by adding $d, ad, bd, \ldots = 9, 10, 11, \ldots$. Then the first replicate is obtained by dividing the 2^4 factorial into four blocks, confounding C, D, CD. The effects confounded in the other replicates are A, B, AB; AC, BD, $ABCD$; ABC, AD, BCD; BC, ABD, ACD.

The problems of the existence and construction of balanced incomplete block designs have been of considerable interest to both statisticians and algebraists. (They will be the topic of Chapter 13.) The major paper on the construction of designs is by Bose (1939). The Fisher and Yates tables contain a listing of almost all the known designs for $r \leq 10$ and $k \leq 10$. The restrictions on the parameters have led to the investigation of other types of incomplete block design, notably the partially balanced incomplete block designs of Bose and Nair (1939); these designs will be discussed in the next chapter.

11.3 The Analysis of Balanced Incomplete Block Designs

We consider now the usual method of analyzing balanced incomplete block designs. This is called the *intrablock analysis* because the block differences are eliminated and the estimates of all contrasts in the treatment effects can be expressed in terms of comparisons between plots in the same block.

We have the following model:

$$y_{ij} = \mu + \tau_i + \beta_j + e_{ij},$$

where y_{ij} is the observation, if there is one, on the ith treatment in the jth block, τ_i is the effect of the ith treatment, β_j is the jth block effect, and e_{ij} is the random error with the usual assumptions. The mean μ and the treatment effects τ_i are unknown constants. The block effects may be fixed or random; we treat them as if they are fixed. It is important to note that the model assumes that there is no interaction between treatments and blocks.

The normal equations are

$$G = N\hat{\mu} + r\sum_i \hat{\tau}_i + k\sum_j \hat{\beta}_j, \tag{11.1}$$

$$T_i = r\hat{\mu} + r\hat{\tau}_i + \sum_j n_{ij}\hat{\beta}_j, \tag{11.2}$$

$$B_j = k\hat{\mu} + \sum_i n_{ij}\hat{\tau}_i + k\hat{\beta}_j, \tag{11.3}$$

where T_i is the sum of all the observations on the ith treatment and B_j is the sum of all the observations in the jth block.

We now eliminate the block effects from Equation 11.2 and obtain

$$kT_i - \sum_j n_{ij}B_j = rk\hat{\tau}_i - r\hat{\tau}_i - \sum_j \sum_h{}' n_{ij}n_{hj}\hat{\tau}_h, \tag{11.4}$$

where the prime denotes summation over all values of h except $h = i$.

We rewrite this as

$$kQ_i = r(k-1)\hat{\tau}_i - \lambda \sum{}' \hat{\tau}_h = (r(k-1) + \lambda)\hat{\tau}_i - \lambda \sum \hat{\tau}_h.$$

We call Q_i the adjusted treatment total for the ith treatment; it is easily seen that $\sum_i Q_i = 0$. Imposing the side condition $\sum_h \hat{\tau}_h = 0$, and recalling that $r(k-1) = \lambda(t-1)$, we obtain

$$kQ_i = \lambda t \hat{\tau}_i.$$

If $\psi = \sum c_i\tau_i$ is any contrast in the treatments, we have

$$\hat{\psi} = k \sum c_i Q_i/(\lambda t).$$

The adjusted treatment totals are not independent. To derive the variance $V(Q_i)$ we note that

$$kQ_i = (k-1)T_i - \left(\sum_j n_{ij}B_j - T_i\right).$$

The expression $\sum n_{ij}B_j - T_i$ is now the sum of $r(k-1)$ observations on treatments other than the ith treatment. Then

$$k^2V(Q_i) = r(k-1)^2\sigma^2 + r(k-1)\sigma^2,$$

and

$$V(Q_i) = r(k-1)\sigma^2/k.$$

It is clear from symmetry that $\operatorname{cov}(Q_h, Q_i)$ is the same for all pairs h, i $(h \neq i)$. Then

$$0 = V\left(\sum_i Q_i\right) = tV(Q_i) + t(t-1)\operatorname{cov}(Q_h, Q_i),$$

and so

$$\operatorname{cov}(Q_h, Q_i) = -\frac{r(k-1)\sigma^2}{k(t-1)} = -\lambda\sigma^2/k.$$

It follows that

$$V(\hat{\tau}_h - \hat{\tau}_i) = 2k\sigma^2/(\lambda t).$$

Indeed, we may write the covariance matrix of the Q_i as

$$\operatorname{cov} \mathbf{Q} = (\lambda t \mathbf{I} - \lambda \mathbf{J})\sigma^2/k.$$

Then, writing $\psi = \mathbf{c}'\boldsymbol{\tau}$ where $\boldsymbol{\tau}$ is the vector of treatment effects and $\mathbf{c}'\mathbf{1} = \mathbf{1}'\mathbf{c} = 0$, we have

$$V(\hat{\psi}) = \frac{k^2}{\lambda^2 t^2} V(\mathbf{c}'\mathbf{Q}) = \frac{k^2\sigma^2}{\lambda^2 t^2} \cdot \frac{\lambda t}{k} \mathbf{c}'\mathbf{c} = \left(\sum_i c_i^2\right) k\sigma^2/(\lambda t).$$

The analysis of variance table is slightly complicated by the fact that treatments are not orthogonal to blocks. We take the sum of squares for regression $\hat{\mu}G + \sum_i T_i\hat{\tau}_i + \sum_j B_j\hat{\beta}_j$ and then subtract the sums of squares for the mean G^2/N (which we write as C), and for blocks (ignoring treatments), $\sum B_j^2/k - C$. The remainder is called the sum of squares for treatments (adjusted for blocks). We write it as $S_{t\,(\mathrm{adj})}$. Then

$$S_{t\,(\mathrm{adj})} = \sum T_i\hat{\tau}_i + \sum B_j\hat{\beta}_j - \sum B_j^2/k + C.$$

But $kB_j\hat{\beta}_j = B_j^2 - kB_j\hat{\mu} - B_j \sum n_{ij}\hat{\tau}_i$, and so

$$S_{t(\mathrm{adj})} = \sum_i \left(T_i - \sum n_{ij}B_j/k\right) \hat{\tau}_i = \sum Q_i\hat{\tau}_i = k \sum Q_i^2/(\lambda t).$$

TABLE 11.1
ANALYSIS OF VARIANCE

Source	SS	d.f.
Total	$\sum\sum\sum y_{ijk}^2 - C$	$N - 1$
Blocks (ignoring treatments)	$\sum B_j^2/k - C$	$b - 1$
Treatments (adjusted for blocks)	$\sum \hat{\tau}_i Q_i = k \sum Q_i^2/(\lambda t)$	$t - 1$
Residual	By subtraction	$N - t - b + 1$

The residual sum of squares is the sum of squares for intrablock error. In the F test for the hypothesis that all the treatment effects are zero, the mean square for treatments (adj) is tested against the error mean square.

The efficiency of an incomplete block design in which each treatment appears r times is measured relative to the complete block design with the same number of replicates. For a complete block design with r observations on each treatment, $V(\hat{\tau}_h - \hat{\tau}_i) = 2\sigma^2/r$. If, for an incomplete block design, the average of the variances of all contrasts $(\hat{\tau}_h - \hat{\tau}_i)$ is $2\sigma^2/a$, the efficiency of the design is defined to be $E = a/r$.

For the balanced incomplete block designs we have $a = \lambda t/k$, so that $E = \lambda t/(rk)$. It is clear that for any design $E < 1$. Some authors write $\hat{\tau}_i = Q_i/rE$ and

$S_{t(\text{adj})} = \sum Q_i^2/rE$. That sum of squares is also called by some authors the sum of squares for treatments eliminating blocks.

Under the normality assumption the hypothesis $\tau_h - \tau_i = 0$ may be tested by the t test or by Tukey's method; Scheffé's S method may also be used for testing the more complex contrasts.

EXAMPLE. The following data are taken from a dishwashing experiment [John (1961*b*)]. The treatments were detergents and the response observed was the number of plates washed under standard test conditions before the foam disappeared. In the experiment, detergent solutions are made up and plates soiled with a standard soil are washed one at a time until they are clean. The testing procedure calls for three basins to be used (i.e., for three treatments to be tested simultaneously), and the three operators wash at a common speed during a test. A block is thus a set of three operators testing three detergents simultaneously. We have $t = 9$, $b = 12$, $r = 4$, $k = 3$, and $\lambda = 1$. The data and the analysis of variance table follow. The letters denote treatments and the responses are given in parentheses.

1.	A (19)	B (17)	C (11)	7.	A (20)	E (26)	J (31)
2.	D (6)	E (26)	F (23)	8.	B (16)	F (23)	G (21)
3.	G (21)	H (19)	J (28)	9.	C (13)	D (7)	H (20)
4.	A (20)	D (7)	G (20)	10.	A (20)	F (24)	H (19)
5.	B (17)	E (26)	H (19)	11.	B (17)	D (6)	J (29)
6.	C (15)	F (23)	J (31)	12.	C (14)	E (24)	G (21)

TABLE 11.2
ANALYSIS OF VARIANCE

	d.f.	SS	MS
Total	35	1512.75	
Blocks (unadjusted)	11	412.75	37.52
Treatments (adjusted)	8	1086.81	135.85
Residual	16	13.19	0.82

The adjusted treatment totals ($Q_i = 3\hat{\tau}_i$) are given by $3Q_A = 3$, $3Q_B = -20$, $3Q_C = -56$, $3Q_D = -116$, $3Q_E = 53$, $3Q_F = 32$, $3Q_G = 15$, $3Q_H = -2$, and $3Q_J = 91$.

11.4 Resolvable Designs

An incomplete block design in which each treatment appears r times is said to be resolvable if the blocks can be divided into r groups in such a way that each group consists of a complete replication of the treatments, i.e., each group of blocks contains each treatment exactly once. The designs for $t = 8$ and $t = 16$, given earlier, and the design for $t = 9$, used in the example, are all resolvable designs. Bose (1942c) showed that the following condition is necessary for a balanced incomplete block design to be resolvable:

$$b \geq t + r - 1.$$

If $b = t + r - 1$, the design is said to be affine resolvable. Bose's condition can be derived from the weaker assumption that t is a multiple of k. It was subsequently derived from this assumption by Roy (1952), Mikhail (1960), and Murty (1961). We now give the proof by Murty.

Let $t = nk$ where n is an integer, and $b = nr$. We have

$$\frac{r}{t - 1} = \frac{\lambda}{k - 1} = \frac{r - \lambda}{t - k} = \frac{r - \lambda}{k(n - 1)}.$$

Then

$$\frac{r - \lambda}{k} = \frac{r(n - 1)}{t - 1} = \frac{rn - r}{t - 1} = \frac{b - r}{t - 1}.$$

But $r - \lambda = rk - t\lambda = k(r - n\lambda)$, and so $(r - \lambda)/k = r - n\lambda = (b - r)/(t - 1)$. All the parameters are integers, and so $r - n\lambda$ is an integer. Furthermore, $r > \lambda$, and so $(r - \lambda)/k > 0$. It follows that $(b - r)/(t - 1) \geq 1$, and so $b \geq r + t - 1$.

In the analysis of variance table for a resolvable design, the sum of squares for blocks (ignoring treatments) may be divided into a sum for replicates and a sum for blocks in replicates.

11.5 The General Incomplete Block Design

The balanced incomplete block design is a special case of the general incomplete design; in the latter the blocks need not be of the same size, nor need each treatment appear the same number of times. We shall still consider t treatments in b blocks. Suppose that the ith treatment is tested in r_i plots and that the jth block contains k_j plots, $(r_i, k_j > 0)$; we do not need to assume that $k_j < t$ for any or all j. As before, T_i denotes the total of all observations on the ith treatment; B_j is the sum of all the observations in the jth block; $\mathbf{T}$, $\mathbf{B}$ are the corresponding vectors of treatment and block totals.

We define two matrices, $\mathbf{R}$ and $\mathbf{K}$; $\mathbf{R}$ is a $t \times t$ matrix with diagonal elements $r_1, r_2, \ldots, r_t$; the off-diagonal elements are zero. We write this as $\mathbf{R} = \text{diag}(r_1, r_2, \ldots, r_t)$. The matrix $\mathbf{K}$ is defined as $\mathbf{K} = \text{diag}(k_1, k_2, \ldots, k_b)$. The incidence matrix of a design has already been defined as $\mathbf{N} = (n_{ij})$ where n_{ij} is the number of times that the ith treatment appears in the jth block. If $n_{ij} = 0$ or 1 for all pairs, the design is said to be binary. If $k_j = k$ for all j, the design is called proper. If $r_i = r$ for all i, the design is called equireplicate. The most commonly used designs are proper, binary, and equireplicate.

For all designs we have

$$\mathbf{R1}_t = \mathbf{N1}_b, \qquad \mathbf{K1}_b = \mathbf{N}'\mathbf{1}_t,$$

where $\mathbf{1}_n$ denotes a vector of n unit elements. For a proper design, $\mathbf{K} = k\mathbf{I}$; for an equireplicate design, $\mathbf{R} = r\mathbf{I}$. Some authors let $\mathbf{r}$ and $\mathbf{k}$ denote the vectors of r_i and k_j; they then write $\mathbf{r}^\delta$ and $\mathbf{k}^\delta$ for the matrices that we have called $\mathbf{R}$ and $\mathbf{K}$.

Let y_{ijm} denote the mth observation on the ith treatment in the jth block; for a binary design we omit the subscript m. The model is

$$y_{ijm} = \mu + \tau_i + \beta_j + e_{ijm},$$

where μ, τ_i, and β_j are defined in the same way as they were for the balanced incomplete block design. We denote by $\boldsymbol{\tau}$, $\boldsymbol{\beta}$, $\hat{\boldsymbol{\tau}}$, and $\hat{\boldsymbol{\beta}}$ the corresponding vectors of effects and their estimates. For the present, τ_i and β_j are assumed to be unknown constants.

The normal equations are

$$G = N\hat{\mu} + \sum_i r_i\hat{\tau}_i + \sum_j k_j\hat{\beta}_j, \tag{11.5}$$

$$T_i = r_i\hat{\mu} + r_i\hat{\tau}_i + \sum_j n_{ij}\hat{\beta}_j, \qquad i = 1, \ldots, t, \tag{11.6}$$

$$B_j = k_j\hat{\mu} + \sum_i n_{ij}\hat{\tau}_i + k_j\hat{\beta}_j, \qquad j = 1, \ldots, b. \tag{11.7}$$

Multiplying each equation of the set of equations (11.7) by the corresponding fraction n_{ij}/k_j for a given value of i and subtracting them all from the ith equation of the set (11.6) eliminates the block effects and μ, giving

$$Q_i = T_i - \sum_j n_{ij}B_j/k_j = r_i\hat{\tau}_i - \sum_h \left(\sum_j n_{ij}n_{hj}/k_j\right)\hat{\tau}_h. \tag{11.8}$$

In matrix notation the normal equations are

$$\begin{bmatrix} G \\ \mathbf{T} \\ \mathbf{B} \end{bmatrix} = \begin{bmatrix} N & \mathbf{1}_t'\mathbf{R} & \mathbf{1}_b'\mathbf{K} \\ \mathbf{R1}_t & \mathbf{R} & \mathbf{N} \\ \mathbf{K1}_b & \mathbf{N}' & \mathbf{K} \end{bmatrix} \begin{bmatrix} \hat{\mu} \\ \hat{\boldsymbol{\tau}} \\ \hat{\boldsymbol{\beta}} \end{bmatrix}. \tag{11.9}$$

Multiplying both sides of the equation on the left by

$$\mathbf{F} = \begin{bmatrix} N^{-1} & \mathbf{0} & \mathbf{0} \\ \mathbf{0} & \mathbf{I}_t & -\mathbf{NK}^{-1} \\ \mathbf{0} & -\mathbf{N}'\mathbf{R}^{-1} & \mathbf{I}_b \end{bmatrix}$$

and recalling that $\mathbf{R1}_t = \mathbf{N1}_b$, and $\mathbf{K1}_b = \mathbf{N}'\mathbf{1}_t$, we obtain the reduced equations

$$G/N = \hat{\mu} + \mathbf{1}'\mathbf{R}\hat{\boldsymbol{\tau}}/N + \mathbf{1}'\mathbf{K}\hat{\boldsymbol{\beta}}/N,$$

$$\mathbf{Q} = \mathbf{T} - \mathbf{NK}^{-1}\mathbf{B} = (\mathbf{R} - \mathbf{NK}^{-1}\mathbf{N}')\hat{\boldsymbol{\tau}},$$

$$\mathbf{B} - \mathbf{N}'\mathbf{R}^{-1}\mathbf{T} = (\mathbf{K} - \mathbf{N}'\mathbf{R}^{-1}\mathbf{N})\hat{\boldsymbol{\beta}}.$$

If we now impose the side conditions

$$\sum r_i\tau_i = \mathbf{1}'\mathbf{R}\boldsymbol{\tau} = 0 \qquad \text{and} \qquad \sum k_j\beta_j = \mathbf{1}'\mathbf{K}\boldsymbol{\beta} = 0,$$

the first of these equations becomes $\hat{\mu} = G/N = \bar{y}$.

The second set of equations may be written $\mathbf{A}\hat{\boldsymbol{\tau}} = \mathbf{Q}$. They are the adjusted intrablock equations. Some authors write $\mathbf{C}\hat{\boldsymbol{\tau}} = \mathbf{Q}$. The third set of equations are the counterparts for estimating $\boldsymbol{\beta}$. We note that $\mathbf{A1} = \mathbf{0}$ and $(\mathbf{K} - \mathbf{N}'\mathbf{R}^{-1}\mathbf{N})\mathbf{1} = \mathbf{0}$, and we now confine our investigations to the case in which $\mathbf{A}$ has rank $t - 1$ and $(\mathbf{K} - \mathbf{N}'\mathbf{R}^{-1}\mathbf{N})$ has rank $b - 1$. Such a design is said to be a connected design and under these conditions all the contrasts $\tau_i - \tau_{i'}$ and $\beta_j - \beta_{j'}$ are estimable.

To calculate the sum of squares for regression, we obtain $\hat{\boldsymbol{\beta}}$ from the equations

$$\mathbf{K1}\hat{\mu} + \mathbf{N}'\hat{\boldsymbol{\tau}} + \mathbf{K}\hat{\boldsymbol{\beta}} = \mathbf{B}$$

as

$$\hat{\boldsymbol{\beta}} = \mathbf{K}^{-1}\mathbf{B} - \mathbf{K}^{-1}\mathbf{N}'\hat{\boldsymbol{\tau}} - \mathbf{1}_b\hat{\mu}.$$

Then the sum of squares for regression is

$$\begin{aligned} G\hat{\mu} + \mathbf{B}'\hat{\boldsymbol{\beta}} + \mathbf{T}'\hat{\boldsymbol{\tau}} &= G\hat{\mu} + \mathbf{B}'\mathbf{K}^{-1}\mathbf{B} + (\mathbf{T}' - \mathbf{B}'\mathbf{K}^{-1}\mathbf{N}')\hat{\boldsymbol{\tau}} - G\hat{\mu} \\ &= \mathbf{B}'\mathbf{K}^{-1}\mathbf{B} + \mathbf{Q}'\hat{\boldsymbol{\tau}}. \end{aligned}$$

The analysis of variance table for the intrablock analysis follows:

TABLE 11.3
ANALYSIS OF VARIANCE

Source	SS	d.f.
Total	$\sum\sum\sum y_{ijm}^2 - G^2/N$	$N - 1$
Blocks (ignoring treatments)	$\mathbf{B}'\mathbf{K}^{-1}\mathbf{B} - G^2/N$	$b - 1$
Treatments (adjusted)	$\mathbf{Q}'\hat{\boldsymbol{\tau}}$	$t - 1$
Residual	By subtraction	$N - b - t + 1$

If there are some cells with more than one observation, i.e., $n_{ij} > 1$, we may take out of the residual a sum of squares within cells $\sum\sum\sum_{ijm}(y_{ijm} - y_{ij\cdot})^2$ for the error term and attribute the remainder to interaction between treatments and blocks. For binary designs we assume that there is no interaction.

11.6 The Adjusted Treatment and Block Totals

The adjusted treatment totals Q_i are linearly dependent since $\mathbf{Q}'\mathbf{1}_t = \sum Q_i = 0$. (Were this not so the set of equations $\mathbf{A}\hat{\boldsymbol{\tau}} = \mathbf{Q}$ would be inconsistent.) We have

$$\mathbf{Q}'\mathbf{1}_t = \mathbf{T}'\mathbf{1}_t - \mathbf{B}'\mathbf{K}^{-1}\mathbf{N}'\mathbf{1}_t = G - \mathbf{B}'\mathbf{K}^{-1}\mathbf{K}\mathbf{1}_b = G - G = 0.$$

Similarly, $(\mathbf{B} - \mathbf{N}'\mathbf{R}^{-1}\mathbf{T})'\mathbf{1}_b = 0$.

We now calculate the covariances of the adjusted totals. The basic form of the normal equations in the format of the general linear hypothesis (Chapter 2) is $\mathbf{X}'\mathbf{Y} = \mathbf{X}'\mathbf{X}\hat{\boldsymbol{\theta}}$ with $\text{cov}\,\mathbf{Y} = \mathbf{I}\sigma^2$ and $\text{cov}\,(\mathbf{X}'\mathbf{Y}) = \mathbf{X}'(\mathbf{I}\sigma^2)\mathbf{X} = \mathbf{X}'\mathbf{X}\sigma^2$. Thus, in the present case the square matrix on the right in the normal equations 11.9 is the covariance matrix of the totals G, $\mathbf{T}$, $\mathbf{B}$; call it $\mathbf{X}'\mathbf{X}$. The covariances of G/N, $\mathbf{Q}$, $(\mathbf{B} - \mathbf{N}'\mathbf{R}^{-1}\mathbf{T})$ are given by $\text{cov}\,(\mathbf{F}(G, \mathbf{T}', \mathbf{B}')') = \mathbf{F}\mathbf{X}'\mathbf{X}\mathbf{F}\sigma^2 =$

$$\begin{bmatrix} N^{-1} & \mathbf{0} & \mathbf{0} \\ \mathbf{0} & \mathbf{R} - \mathbf{N}\mathbf{K}^{-1}\mathbf{N}' & -(\mathbf{R} - \mathbf{N}\mathbf{K}^{-1}\mathbf{N}')\mathbf{R}^{-1}\mathbf{N} \\ \mathbf{0} & -(\mathbf{K} - \mathbf{N}'\mathbf{R}^{-1}\mathbf{N})\mathbf{K}^{-1}\mathbf{N}' & \mathbf{K} - \mathbf{N}'\mathbf{R}^{-1}\mathbf{N} \end{bmatrix} \sigma^2.$$

In particular we have $\text{cov}\,\mathbf{Q} = (\mathbf{R} - \mathbf{N}\mathbf{K}^{-1}\mathbf{N}')\sigma^2 = \mathbf{A}\sigma^2$, and $\text{cov}\,(\mathbf{B} - \mathbf{N}'\mathbf{R}^{-1}\mathbf{T}) = (\mathbf{K} - \mathbf{N}'\mathbf{R}^{-1}\mathbf{N})\sigma^2$.

11.7 Solving the Intrablock Equations

The reduced intrablock equations, $\mathbf{Q} = \mathbf{A}\hat{\boldsymbol{\tau}}$, may be solved by using either of the two methods for obtaining solution matrices that were discussed in Chapter 2. The side condition $\mathbf{H}\boldsymbol{\tau} = 0$ is usually taken with $\mathbf{H} = \mathbf{1}'$ or with $\mathbf{H} = \mathbf{1}'\mathbf{R}$; in any case, $\mathbf{H}\boldsymbol{\tau}$ is not a contrast, and $\mathbf{1}'\mathbf{H}' = \sum h_i \neq 0$. For an equireplicate design these side conditions are, of course, equivalent; for other designs the side condition $\mathbf{H}\boldsymbol{\tau} = \mathbf{1}'\mathbf{R}\boldsymbol{\tau} = 0$ (coupled with $\mathbf{1}'\mathbf{K}\boldsymbol{\beta} = 0$) gives $\hat{\mu} = \bar{y}$. A discussion of the two methods of solving the equation follows.

(a) Graybill, Kempthorne.

Let

$$\mathbf{A}^* = \begin{bmatrix} \mathbf{A} & \mathbf{H}' \\ \mathbf{H} & 0 \end{bmatrix} = \begin{bmatrix} \mathbf{B}_{11} & \mathbf{B}_{12} \\ \mathbf{B}_{21} & B_{22} \end{bmatrix}^{-1}.$$

Then $\mathbf{A}^*\mathbf{A}^{*-1} = \mathbf{I}$ gives four equations

$$\mathbf{AB}_{11} + \mathbf{H}'\mathbf{B}_{21} = \mathbf{I}, \tag{11.10}$$

$$\mathbf{AB}_{12} + \mathbf{H}'B_{22} = \mathbf{0}, \tag{11.11}$$

$$\mathbf{HB}_{11} = \mathbf{0}, \tag{11.12}$$

$$\mathbf{HB}_{12} = 1. \tag{11.13}$$

Multiplying Equations 11.10 and 11.11 on the left by $\mathbf{1}'$ and recalling that $\mathbf{1}'\mathbf{A} = \mathbf{0}$, we have

$$\mathbf{1}'\mathbf{H}'\mathbf{B}_{21} = \mathbf{1}' \qquad \text{and} \qquad \mathbf{1}'\mathbf{H}'B_{22} = 0,$$

so that $\mathbf{B}_{21} = \mathbf{1}'(\mathbf{1}'\mathbf{H}')^{-1} = \mathbf{1}'/(\sum h_i)$ and $B_{22} = 0$. If $\mathbf{H} = \mathbf{1}'\mathbf{R}$, $\mathbf{B}_{21} = \mathbf{1}'/N$; if $\mathbf{H} = \mathbf{1}'$, $\mathbf{B}_{21} = \mathbf{1}'/t$.

Multiplying Equation 11.10 on the left by $\mathbf{B}_{11}$ gives $\mathbf{B}_{11}\mathbf{AB}_{11} = \mathbf{B}_{11}$; multiplying on the right by $\mathbf{A}$ gives $\mathbf{AB}_{11}\mathbf{A} = \mathbf{A}$. It follows that $\mathbf{AB}_{11}$ and $\mathbf{B}_{11}\mathbf{A}$ are idempotent. Taking the transpose of Equation 11.10 and noting that $\mathbf{B}_{11}$ and $\mathbf{A}$ are symmetric, we have, after multiplying on the right by $\hat{\boldsymbol{\tau}}$,

$$\mathbf{B}_{11}\mathbf{A}\hat{\boldsymbol{\tau}} + \mathbf{B}_{21}'\mathbf{H}\hat{\boldsymbol{\tau}} = \hat{\boldsymbol{\tau}};$$

but $\mathbf{A}\hat{\boldsymbol{\tau}} = \mathbf{Q}$ and $\mathbf{H}\hat{\boldsymbol{\tau}} = 0$, and so we have obtained a solution vector $\mathbf{B}_{11}\mathbf{Q} = \hat{\boldsymbol{\tau}}$, and $\mathbf{B}_{11}$ is a solution matrix. Furthermore,

$$\operatorname{cov} \hat{\boldsymbol{\tau}} = \operatorname{cov} \mathbf{B}_{11}\mathbf{Q} = \mathbf{B}_{11}\mathbf{AB}_{11}\sigma^2 = \mathbf{B}_{11}\sigma^2.$$

(b) Scheffé, Plackett.

Let $\mathbf{P}_1 = (\mathbf{A} + \mathbf{H}'\mathbf{H})^{-1}$. Then $\hat{\boldsymbol{\tau}} = \mathbf{P}_1\mathbf{Q}$ is a solution vector, for we have

$$\mathbf{Q} = \mathbf{A}\hat{\boldsymbol{\tau}} = (\mathbf{A} + \mathbf{H}'\mathbf{H})\hat{\boldsymbol{\tau}}.$$

It was shown in Chapter 2 that $\mathbf{P}_1 = \mathbf{B}_{11} + \mathbf{J}/(\sum h_i)^2$.

Let $\mathbf{P}$ be any solution matrix, i.e., let $\hat{\boldsymbol{\tau}} = \mathbf{PQ}$ be any solution to the equations $\mathbf{Q} = \mathbf{A}\hat{\boldsymbol{\tau}}$. Then, if $\psi = \mathbf{c}'\boldsymbol{\tau}$ is a contrast, $\hat{\psi} = \mathbf{c}'\mathbf{PQ}$ and $V(\hat{\psi}) = \mathbf{c}'\mathbf{PAP}'\mathbf{c}\sigma^2$. But $\hat{\psi} = \mathbf{c}'\mathbf{B}_{11}\mathbf{Q}$ so that $\mathbf{c}'\mathbf{PQ} - \mathbf{c}'\mathbf{B}_{11}\mathbf{Q} = 0$ and

$$\mathbf{c}'(\mathbf{P} - \mathbf{B}_{11})\mathbf{Q} = 0$$

is an identity for all $\mathbf{c}$, $\mathbf{Q}$ subject to $\mathbf{c}'\mathbf{1} = 0$, $\mathbf{1}'\mathbf{Q} = 0$. We note therefore, that the general form of $\mathbf{P}$ is $\mathbf{P} = \mathbf{B}_{11} + \mathbf{u}_1\mathbf{1}' + \mathbf{1}\mathbf{u}_2'$ where $\mathbf{u}_1$ and $\mathbf{u}_2$ are arbitrary vectors. It follows that

$$\begin{aligned} V(\hat{\psi}) &= \mathbf{c}'(\mathbf{B}_{11} + \mathbf{u}_1\mathbf{1}' + \mathbf{1}\mathbf{u}_2')\mathbf{A}(\mathbf{B}_{11} + \mathbf{1}\mathbf{u}_1' + \mathbf{u}_2\mathbf{1}')\mathbf{c}\sigma^2 \\ &= \mathbf{c}'\mathbf{B}_{11}\mathbf{AB}_{11}\mathbf{c}\sigma^2 = \mathbf{c}'\mathbf{B}_{11}\mathbf{c}\sigma^2 = \mathbf{c}'\mathbf{P}\mathbf{c}\sigma^2. \end{aligned}$$

This result implies that if $\mathbf{P}$ is any solution matrix we may, in computing the variance of the estimate $\hat{\psi}$ of a contrast ψ, act as if $\mathbf{P}$ were the covariance matrix of the estimates $\hat{\boldsymbol{\tau}}$.

For the balanced incomplete block design we have

$$k\mathbf{A} = (r(k-1)+\lambda)\mathbf{I}_t - \lambda\mathbf{J}_t = \lambda t\mathbf{I}_t - \lambda\mathbf{J}_t.$$

Using the first method,

$$\mathbf{A}^* = \frac{1}{k}\begin{bmatrix} \lambda t\mathbf{I} - \lambda\mathbf{J} & k\mathbf{1} \\ k\mathbf{1}' & 0 \end{bmatrix},$$

$$\mathbf{A}^{*-1} = \begin{bmatrix} \dfrac{k\mathbf{I}}{\lambda t} - \dfrac{k\mathbf{J}}{\lambda t^2} & \mathbf{1}/t \\ \mathbf{1}'/t & 0 \end{bmatrix},$$

so that

$$\mathbf{B}_{11} = \frac{k}{\lambda t}\left(\mathbf{I}_t - \frac{1}{t}\mathbf{J}_t\right),$$

and

$$\hat{\boldsymbol{\tau}} = \mathbf{B}_{11}\mathbf{Q} = k\mathbf{Q}/(\lambda t).$$

In using the second method it is easier to take $\mathbf{H} = \sqrt{(\lambda/k)}\mathbf{1}'$ so that $\mathbf{H}'\mathbf{H} = \lambda\mathbf{J}/k$ and $\mathbf{A} + \mathbf{H}'\mathbf{H} = \lambda t\mathbf{I}/k$. Then $(\mathbf{A} + \mathbf{H}'\mathbf{H})^{-1} = k\mathbf{I}/\lambda t$ and $\hat{\boldsymbol{\tau}} = k\mathbf{Q}/(\lambda t)$. Taking the side condition as $\mathbf{H}'\boldsymbol{\tau} = \mathbf{1}'\boldsymbol{\tau}$, we have $\mathbf{A} + \mathbf{H}'\mathbf{H} = (\lambda t\mathbf{I} + (k-\lambda)\mathbf{J})/k$, and $(\mathbf{A} + \mathbf{H}'\mathbf{H})^{-1} = \{k\mathbf{I} - (k-\lambda)\mathbf{J}/t\}/(\lambda t)$.

Side conditions other than $\mathbf{H}' = \mathbf{1}$ or $\mathbf{H}' = \mathbf{R1}$ are sometimes useful. Consider a design in which a single plot with a control treatment is added to each block of a balanced incomplete block design. The first design in this chapter for seven treatments in blocks of three would become a design for eight treatments in blocks of four. Denoting the control treatment by O, the new design would be

$$OABE,\quad OCDE,\quad OACF,\quad OBDF,\quad OADG,\quad OBCG,\quad OEFG.$$

Let t, b, r, k', and λ be the parameters of the balanced design. The whole design has b blocks of $k = k' + 1$ plots each, $r_O = b$, $r_i = r\ (i > 0)$. Then, using the first row and column for the control treatment,

$$\mathbf{A} = \frac{1}{k}\begin{bmatrix} bk' & -r\mathbf{1}_t' \\ -r\mathbf{1}_t & (r+\lambda t)\mathbf{I}_t - \lambda\mathbf{J}_t \end{bmatrix}.$$

We choose $\mathbf{H} = (h_0, h\mathbf{1}_t')$, in order to make $\mathbf{A} + \mathbf{H}'\mathbf{H}$ a diagonal matrix; but

$$\mathbf{H}'\mathbf{H} = \begin{bmatrix} h_0^2 & hh_0\mathbf{1}_t' \\ hh_0\mathbf{1}_t & h^2\mathbf{J}_t \end{bmatrix},$$

and so we wish to have $hh_0 = r/k$ and $h^2 = \lambda/k$, or,

$$h = \sqrt{(\lambda/k)}, \qquad h_0 = r/\sqrt{(k\lambda)}.$$

With this choice of $\mathbf{H}$ we have

$$\mathbf{A} + \mathbf{H}'\mathbf{H} = \frac{1}{k}\begin{bmatrix} bk' + r^2/\lambda & \mathbf{0} \\ \mathbf{0} & (r + \lambda t)\mathbf{I}_t \end{bmatrix},$$

whence

$$\hat{\tau}_0 = \frac{k\lambda Q_o}{r(r + \lambda t)}, \quad \hat{\tau}_i = \frac{kQ_i}{r + \lambda t}, \qquad (i > 0).$$

11.8 Connected Designs

The idea of connected designs is due to Bose. A block and a treatment are said to be associated if the treatment appears in the block. Two treatments, A and B, are said to be connected if we can form a chain of treatments and blocks: treatment—block—treatment—block—· · ·—treatment, beginning with A and ending with B, such that every block in the chain is associated with both the treatments adjacent to it. The relationship *A connected to B* is an equivalence relation which forms disjoint equivalence classes for the treatments. A design is said to be connected if there is only one equivalence class, i.e., if every pair of treatments is connected.

The following design for four blocks I, II, III, IV and seven treatments is connected:

I *ABCD*, II *BCE*, III *DE*, IV *EFG*.

A and G are connected by the chain A—I—B—II—E—IV—G. Denoting the effects of $A, B, \ldots,$ by $\tau_1, \tau_2, \ldots,$ we see that $\tau_1 - \tau_7$ is estimable. We follow the chain and consider the contrast

$$\begin{aligned} z &= y_{11} - y_{21} + y_{22} - y_{52} + y_{54} - y_{74}; \\ E(z) &= \tau_1 + \beta_1 - \tau_2 - \beta_1 + \tau_2 + \beta_2 - \tau_5 - \beta_2 + \tau_5 + \beta_4 - \tau_7 - \beta_4 \\ &= \tau_1 - \tau_7. \end{aligned}$$

The following design with $t = b = 6$, which is the union of two disjoint designs having no treatment in common, is disconnected:

AB, *AC*, *BC*, *DE*, *DF*, *EF*.

11.9 The Recovery of Interblock Information

Yates (1940) pointed out that under certain circumstances a second set of estimates of the treatment effects, called the *interblock estimates*, can be obtained.

In discussing these estimates we shall confine our consideration to proper designs and we shall write the variance of the intrablock errors e_{ij} as σ_e^2.

The additional assumption to be made is that the block effects β_j are random variables distributed with zero means and variance σ_b^2; they are uncorrelated, and are also not correlated with the intrablock errors e_{ij}. Under these circumstances the block totals can be regarded as a set of observations:

$$B_j = k\mu + \sum_i n_{ij}\tau_i + \left(k\beta_j + \sum_i e_{ij}\right).$$

We may write the combined error term as f_j. Then $E(f_j) = 0$ and $\sigma_f^2 = V(f_j) = k^2\sigma_b^2 + k\sigma_e^2$. The interblock estimates are obtained by minimizing

$$\sum_j \left(B_j - k\tilde{\mu} - \sum_i n_{ij}\tilde{\tau}_i\right)^2,$$

where $\tilde{\mu}$ and $\tilde{\tau}_i$ denote the interblock estimates; $\hat{\mu}$ and $\hat{\tau}_i$ will be used to denote the intrablock estimates.

Consider first the balanced incomplete block design. The normal equations for interblock estimation are

$$G = N\tilde{\mu} + \sum_i r\tilde{\tau}_i,$$

$$\sum_j n_{ij}B_j = rk\tilde{\mu} + r\tilde{\tau}_i + \lambda \sum_h{}' \tilde{\tau}_h, \qquad (h \neq i).$$

We shall denote $\sum n_{ij}B_j$ by T_i'. With the side condition $\mathbf{1}'\boldsymbol{\tau} = 0$, the first equation gives $\tilde{\mu} = G/N$, and the second equation becomes

$$T_i' = (r - \lambda)\tilde{\tau}_i + rk\tilde{\mu},$$

whence

$$\tilde{\tau}_i = \frac{T_i' - rk\tilde{\mu}}{r - \lambda}.$$

In the example of the balanced incomplete block design given earlier, $r = 4$, $k = 3$, $r - \lambda = 3$. Then $T_i' - 12\tilde{\mu} = T_i' - G/3 = T_i' - 233 = 3\tilde{\tau}_i$.

The interblock and intrablock estimates of the treatment effects follow.

T_i'	234	221	215	194	253	247	234	233	266
$\tilde{\tau}_i$	0.33	−4.00	−6.00	−13.00	6.67	4.67	0.33	0.00	11.00
$\hat{\tau}_i$	0.33	−2.22	−6.22	−12.89	5.89	3.55	1.67	−0.22	10.11

It should be noted that $\sum T_i' = kG = trk\tilde{\mu}$ so that $\sum \tilde{\tau}_i \equiv 0$.

In the general case, we minimize

$$(\mathbf{B} - k\tilde{\mu}\mathbf{1}_b - \mathbf{N}'\tilde{\boldsymbol{\tau}})'(\mathbf{B} - k\tilde{\mu}\mathbf{1}_b - \mathbf{N}'\tilde{\boldsymbol{\tau}})$$

and obtain the normal equations

$$\begin{bmatrix} G \\ \mathbf{NB} \end{bmatrix} = \begin{bmatrix} bk & \mathbf{1}'\mathbf{R} \\ \mathbf{R1}k & \mathbf{NN}' \end{bmatrix} \begin{bmatrix} \tilde{\mu} \\ \tilde{\boldsymbol{\tau}} \end{bmatrix}.$$

Multiplying on the left by

$$\begin{bmatrix} 1 & \mathbf{0} \\ \dfrac{-\mathbf{R1}}{b} & \mathbf{I} \end{bmatrix}$$

gives the equations

$$\begin{bmatrix} G \\ \mathbf{NB} - \dfrac{G\mathbf{R1}}{b} \end{bmatrix} = \begin{bmatrix} bk & \mathbf{1}'\mathbf{R} \\ \mathbf{0} & \mathbf{NN}' - \dfrac{\mathbf{RJR}}{b} \end{bmatrix} \begin{bmatrix} \tilde{\mu} \\ \tilde{\boldsymbol{\tau}} \end{bmatrix}$$

or

$$\mathbf{N}(\mathbf{B} - G\mathbf{1}/b) = (\mathbf{NN}' - \mathbf{RJR}/b)\tilde{\boldsymbol{\tau}}.$$

If we apply the side condition $\mathbf{1}'\mathbf{R}\boldsymbol{\tau} = 0$ the equations become

$$G/N = \tilde{\mu} \qquad \text{and} \qquad \mathbf{N}(\mathbf{B} - G\mathbf{1}/b) = \mathbf{NN}'\tilde{\boldsymbol{\tau}},$$

so that

$$\tilde{\boldsymbol{\tau}} = (\mathbf{NN}')^{-1}\mathbf{N}(\mathbf{B} - G\mathbf{1}/b) = (\mathbf{NN}')^{-1}\mathbf{NB} - \mathbf{1}(G/bk);$$

the existence of the interblock estimates requires that $\mathbf{NN}'$ exist, and hence that the incidence matrix $\mathbf{N}$ has rank t.

We now have two sets of estimates: $\hat{\boldsymbol{\tau}} = \mathbf{B}_{11}\mathbf{Q}$ (or some other function of $\mathbf{Q}$), and $\tilde{\boldsymbol{\tau}}$, which is a function of $\mathbf{B}$ and G. Because, for any i and any j, $\text{cov}\,(Q_i, B_j) = 0$, it follows that the two sets of estimates are orthogonal, and, under normality, independent.

11.10 Combining Intrablock and Interblock Estimates

Consider a contrast $\psi = \mathbf{c}'\boldsymbol{\tau}$ in the treatments. If $\mathbf{P}$ is a solution matrix for the intrablock equations, we have

$$\hat{\psi} = \mathbf{c}'\mathbf{PQ} \qquad \text{and} \qquad V(\hat{\psi}) = \mathbf{c}'\mathbf{Pc}\sigma_e^2.$$

The interblock estimate of ψ is

$$\tilde{\psi} = \mathbf{c}'\tilde{\boldsymbol{\tau}} = \mathbf{c}'(\mathbf{NN}')^{-1}\mathbf{NB};$$

the term $G\mathbf{1}/b$ drops out because ψ is a contrast and $V(\tilde{\psi}) = \mathbf{c}'(\mathbf{NN}')^{-1}\mathbf{c}\sigma_f^2$.

We now wish to combine these two uncorrelated estimates of ψ to obtain an unbiased joint estimate with minimum variance. We shall use the following well-known property of combined estimates.

Let θ_1 and θ_2 be unbiased estimates of a parameter θ with variances v_1 and v_2 and zero covariance. Consider the class of combined estimates of the form $u_1\theta_1 + u_2\theta_2$. In this class the minimum variance unbiased estimate of θ is $(w_1\theta_1 + w_2\theta_2)/(w_1 + w_2)$ where the weights are $w_1 = v_1^{-1}$ and $w_2 = v_2^{-1}$; i.e., the individual estimates θ_1 and θ_2 are given weights inversely proportional to their variances.

In the present case we should take for the combined estimate of ψ,

$$\psi^* = (w_1\hat{\psi} + w_2\tilde{\psi})/(w_1 + w_2),$$

where $w_1 = 1/\mathbf{c}'\mathbf{P}\mathbf{c}\sigma_e^2$ and $w_2 = 1/\mathbf{c}'(\mathbf{NN}')^{-1}\mathbf{c}\sigma_f^2$. This simplifies to

$$\psi^* = \frac{\hat{\psi}V(\tilde{\psi}) + \tilde{\psi}V(\hat{\psi})}{V(\hat{\psi}) + V(\tilde{\psi})}.$$

If we knew σ_e^2 and σ_f^2, we could now obtain ψ^*. The original method of Yates for the balanced incomplete block design is to use this weighted estimate, but with the weights computed by substituting for σ_e^2 and σ_f^2 estimates obtained from the data. Rao (1947*a*) derived combined estimates under the assumption that the e_{ij} and the β are normally distributed. For the balanced incomplete block design the methods of Yates and of Rao give the same estimates. Rao's derivation will be given later in this chapter.

We shall now confine our discussion to the case of balanced incomplete block designs. In that case, $\mathbf{P} = k\mathbf{I}/(\lambda t)$ and $V(\hat{\psi}) = \mathbf{c}'\mathbf{c}k\sigma_e^2/(\lambda t)$. For the interblock estimates, $\mathbf{NN}' = (r - \lambda)\mathbf{I}_t + \lambda\mathbf{J}_t$, and

$$(\mathbf{NN}')^{-1} = \frac{\mathbf{I}}{r - \lambda} - \frac{\lambda\mathbf{J}}{(r - \lambda)(r + (t - 1)\lambda)} = \frac{\mathbf{I}}{r - \lambda} - \frac{\lambda\mathbf{J}}{(r - \lambda)rk},$$

but $\mathbf{c}'\mathbf{J} = \mathbf{0}$, and so we have

$$\tilde{\psi} = \frac{\mathbf{c}'\mathbf{NB}}{r - \lambda} = \frac{\sum c_i T_i'}{r - \lambda}, \qquad V(\tilde{\psi}) = \frac{\mathbf{c}'\mathbf{c}\sigma_f^2}{r - \lambda}.$$

The minimum variance joint estimate is thus

$$\psi^* = (w_1\hat{\psi} + w_2\tilde{\psi})/(w_1 + w_2),$$

where $w_1 = \lambda t/k\sigma_e^2$ and $w_2 = (r - \lambda)/\sigma_f^2$, and

$$V(\psi^*) = \frac{\mathbf{c}'\mathbf{c}}{w_1 + w_2}.$$

As the first step toward an estimate of σ_f^2 we obtain an estimate of σ_b^2. In the intrablock analysis, the sum of squares for regression was separated into three components; they were the sums of squares for the mean (G^2/N), blocks (ignoring treatments), and treatments (adjusted for blocks).

An alternative is to subdivide into sums for the mean, treatments (ignoring blocks), and blocks (adjusted for treatments).

TABLE 11.4

Analysis of Variance

Source	SS	d.f.
Total	$\sum\sum y_{ij}^2 - G^2/N$	$N - 1$
Treatments (ignoring blocks)	$r^{-1}\sum T_i^2 - G^2/N$	$t - 1$
Blocks (adjusted for treatments)	By subtraction	$b - 1$
Intrablock error	From intrablock analysis	$N - b - t + 1$

The sum of squares for blocks (adjusted for treatments) is

$$S_{\text{bl adj}} = \frac{\sum_j B_j^2}{k} + \frac{k\sum_i Q_i^2}{\lambda t} - \frac{\sum_i T_i^2}{r}.$$

Scheffé's method of obtaining the expectation of this sum of squares is to consider for convenience the special case in which μ and τ_i are all zero because they do not appear in the result. Then B_j, Q_i, and T_i all have zero means, so that the expected values of their squares are equal to their variances, giving

$$E(S_{\text{bl adj}}) = \frac{bV(B_j)}{k} + \frac{kV(Q_i)}{\lambda} - \frac{tV(T_i)}{r}.$$

But,

$$V(B_j) = \sigma_f^2 = k^2\sigma_b^2 + k\sigma_e^2, \qquad V(Q_i) = r(k-1)\sigma_e^2/k,$$
$$V(T_i) = r(\sigma_b^2 + \sigma_e^2),$$

and, substituting above,

$$E(S_{\text{bl adj}}) = \sigma_b^2(bk - t) + \sigma_e^2(b - 1).$$

Writing E_e and E_b for the mean squares for intrablock error and for blocks (adjusted for treatments) respectively, we obtain

$$\hat{\sigma}_e^2 = E_e, \qquad \hat{\sigma}_b^2 = \frac{(E_b - E_e)(b - 1)}{(N - t)}.$$

We may now write

$$\psi^* = \hat{\psi} + \frac{w_2(\tilde{\psi} - \hat{\psi})}{w_1 + w_2} = \hat{\psi} + \frac{(\tilde{\psi} - \hat{\psi})k(r - \lambda)\sigma_e^2}{\lambda t\sigma_f^2 + \sigma_e^2 k(r - \lambda)} = \hat{\psi} + \frac{(\tilde{\psi} - \hat{\psi})(r - \lambda)\sigma_e^2}{kt\sigma_b^2 + rk\sigma_e^2},$$

and replace σ_e^2 and σ_b^2 by $\hat{\sigma}_e^2$ and $\hat{\sigma}_b^2$. This is essentially what Yates (1940) did. Graybill and Seshadri (1960) have shown that this estimate is unbiased; Graybill and Weeks (1959) have shown that it is based on a set of minimal sufficient statistics.

11.11 Some Computational Formulae for Combined Estimates

Yates (1940) also gave some formulae for calculating the adjusted yields, Y_i; $Y_i = r(\tau_i^* + \mu^*)$. We shall make use of the following identity:

$$rk - \lambda t = r(t - k)/(t - 1) = r - \lambda.$$

The intrablock estimate of $\tau_i + \mu$ is

$$\hat{\tau}_i + \hat{\mu} = \frac{kQ_i}{\lambda t} + \frac{G}{N} = \frac{T_i}{r} + T_i\left(\frac{k}{\lambda t} - \frac{1}{r}\right) - \frac{T_i'}{\lambda t} + \frac{G}{N}$$

$$= \frac{T_i}{r} + \frac{T_i(rk - \lambda t) - rT_i' + \lambda G}{\lambda rt} = \frac{T_i}{r} + \frac{W_i}{\lambda t(t - 1)},$$

where $W_i = (t - k)T_i - (t - 1)T_i' + (k - 1)G$. Similarly,

$$\tilde{\tau}_i + \tilde{\mu} = \frac{T_i' - rk\tilde{\mu}}{r - \lambda} + \tilde{\mu} = \frac{T_i}{r} - \frac{T_i(r - \lambda) - rT_i' + \lambda G}{r(r - \lambda)} = \frac{T_i}{r} - \frac{W_i}{r(t - k)}.$$

Thus, we can think of $W_i/\lambda t(t - 1)$ and $W_i/r(t - k)$ as the intrablock and inter-block adjustments to the raw treatment means T_i/r. We now combine the two estimates, using the weights $w_1 = \lambda t/(k\sigma_e^2)$ and $w_2 = (r - \lambda)/\sigma_f^2$ to obtain the estimate $(\tau_i^* + \mu^*) = (T_i + \theta W_i)/r = Y_i/r$, where

$$\theta = r\left(\frac{w_1}{\lambda t(t - 1)} - \frac{w_2}{r(t - k)}\right)(w_1 + w_2)^{-1}$$

$$= \frac{\sigma_f^2 - k\sigma_e^2}{t(k - 1)\sigma_f^2 + k(t - k)\sigma_e^2} = \frac{\sigma_b^2}{t(k - 1)\sigma_b^2 + (t - 1)\sigma_e^2}. \qquad (11.14)$$

We may rearrange Equation 11.14 and obtain

$$\sigma_f^2 = k(1 + (t - k)\theta)\sigma_e^2/(1 - t(k - 1)\theta),$$

and, after some further algebraic manipulation,

$$V(\psi^*) = \mathbf{c}'\mathbf{c}(w_1 + w_2)^{-1} = \mathbf{c}'\mathbf{c}(1 + (t - k)\theta)\sigma_e^2/r.$$

When E_b is large by comparison to E_e we have, taking the limit as $E_b \to \infty$, $\theta = 1/\{t(k - 1)\}$; then $(1 + (t - k)\theta)\sigma_e^2/r$ becomes $k\sigma_e^2/(\lambda t)$, which gives the variance of the intrablock estimator.

Substituting $\hat{\sigma}_e^2$, $\hat{\sigma}_b^2$ for σ_e^2, σ_b^2 as before, we have

$$\begin{aligned}\theta &= \frac{(E_b - E_e)(b - 1)}{t(k - 1)(b - 1)(E_b - E_e) + (N - t)(t - 1)E_e} \\ &= \frac{(E_b - E_e)(b - 1)}{t(k - 1)(b - 1)E_b + (t - k)(b - t)E_e}.\end{aligned}$$

For the symmetric design with $b = t$, θ takes the simpler form

$$\theta = E_b - E_e/t(k - 1)E_b.$$

If $E_b < E_e$, it is usual to take $\theta = 0$. This is equivalent to putting $E_b = E_e$ in the formula and to acting as if there were no block effects; thus the design is analyzed as a completely randomized design.

Yates also considered the situation in which the blocks can be broken up into c groups of b/c blocks each, each group consisting of r/c complete replications of the treatments; r/c is an integer. Although Equation 11.14 still holds, the estimate of σ_b^2 is changed. We return to Scheffé's method of calculation and add a term for the sum of squares between groups in the analysis of variance table. Let G_g denote the group total for the gth group. Then

$$S_{\text{bl adj}} = \sum_k \frac{B_j^2}{k} - \sum_g \frac{G_g^2}{N/c} + \frac{k \sum Q_i}{\lambda t} - \frac{\sum T_i^2}{r} + \frac{G^2}{N}.$$

Assuming that τ_i, μ are all zero, and also that there is no group effect,

$$V(G_g) = bV(B_j)/c \qquad \text{and} \qquad V(G) = bV(B_j),$$

where $V(B_j) = \sigma_f^2$. Then

$$\begin{aligned}E(S_{\text{bl adj}}) &= \frac{k}{\lambda} V(Q_i) - \frac{t}{r} V(T_i) + \frac{b(b - c + 1)}{N} V(B_j) \\ &= (k(b - c + 1) - t)\sigma_b^2 + (b - c)\sigma_e^2,\end{aligned}$$

so that

$$\sigma_b^2 = (E_b - E_e)(b - c)/\{N - t - k(c - 1)\}.$$

If the design is resolvable, the blocks can be arranged to form single replicates. In this case, $c = r$, and the previous result simplifies to

$$\sigma_b^2 = (E_b - E_e)r/\{k(r - 1)\}.$$

Substituting in Equation 11.14, we obtain for the general case

$$\theta = \frac{(E_b - E_e)(b - c)}{t(k - 1)(b - c)E_b + (t - k)(b - c - t + 1)E_e},$$

and, for the resolvable case,

$$\theta = \frac{(E_b - E_e)r}{rt(k - 1)E_b + k(b - r - t + 1)E_e}.$$

A special case is the balanced lattice. This is the balanced design with $t = k^2$, $r = k + 1$, and $b = k(k + 1)$. We have already seen examples for $t = 9$ and $t = 16$. In this case, the formula is further simplified to

$$\theta = (E_b - E_e)/k^2(k - 1)E_b.$$

EXAMPLE. James and Bancroft (1951) carried out a 2^3 experiment in a balanced incomplete block design. They took clover plants, split them into two halves, and applied a different treatment combination to each half. A block is a pair of half plants. The response observed was the percentage of hard seed; they considered using the arcsine transformation on the data, but decided against it. The treatments in the design were the 2^3 treatment combinations in a factorial experiment, with factors calcium (C), phosphorus (P), and potassium (K). The treatments represented by $A, B, \ldots, H$ are, respectively, (1), k, p, pk, c, ck, cp, cpk. The design is resolvable and the data are arranged in seven replications; the parameters of the design are $t = 8$, $b = 28$, $r = 7$, $k = 2$, and $\lambda = 1$.

I		II		III		IV	
A 38	*B* 29	*A* 37	*C* 27	*A* 15	*D* 23	*A* 3	*E* 13
C 49	*D* 28	*B* 37	*H* 50	*B* 47	*G* 64	*B* 45	*C* 36
E 32	*F* 29	*D* 90	*E* 89	*C* 35	*F* 39	*D* 11	*G* 24
G 64	*H* 32	*F* 28	*G* 71	*E* 22	*H* 18	*F* 39	*H* 37

V		VI		VII	
A 23	*F* 39	*A* 66	*G* 68	*A* 28	*H* 30
B 21	*D* 14	*B* 23	*F* 46	*B* 10	*E* 40
C 18	*H* 10	*C* 22	*E* 28	*C* 32	*G* 33
E 23	*G* 53	*D* 23	*H* 39	*D* 18	*F* 23

The various totals for each treatment and the estimates of the treatment effects are

	A	B	C	D	E	F	G	H
T_i	210	212	219	207	247	243	377	216
T_i'	439	500	439	467	472	482	616	447
Q_i	−9.5	−38.0	−0.5	−26.5	11.0	2.0	69.0	−7.5
Y_i	220	186	234	198	257	244	366	225
W_i	118	−297	172	−96	109	15	−119	98
$\hat{\tau}_i$	−2.4	−9.5	−0.1	−6.6	2.8	0.5	17.2	−1.9
$\tilde{\tau}_i$	−7.3	2.9	−7.3	−2.6	−1.8	−0.1	22.2	−6.0
τ_i^*	−3.0	−7.9	−1.0	−6.2	2.2	0.4	17.8	−2.3

The intrablock and interblock analysis of variance table follows:

TABLE 11.5
ANALYSIS OF VARIANCE

Source	SS (Intrablock)	SS (Interblock)	d.f.	MS
Total	19,123.99	19,123.99	55	
Replicates	4,977.11	4,977.11	6	
Blocks	10,428.37	8,994.25	21	428.30 E_b
Treatments	1,794.72	3,228.84	7	
Intrablock error	1,923.79	1,923.79	21	91.61 E_e

The weighting factor θ is given by the formula for the resolvable case:

$$\theta = \frac{7(428.30 - 91.61)}{56 \times 428.30 + 28 \times 91.61} = 0.08876.$$

Then $W_i = 6T_i - 7T_i' + G$ and $Y_i = T_i + \theta W_i$. The estimates have just been given; $\tau_i^* = Y_i/7 - \hat{\mu}$ where $\hat{\mu} = G/56 = 34.48$.

In the factorial experiment, the treatment contrasts are of primary interest. When the joint estimates are used, the C contrast is $Y_5 + Y_6 + Y_7 + Y_8 - Y_1 - Y_2 - Y_3 - Y_4 = 254$. Dividing by $8r$ gives the estimate, 4.54, for β_c^*, the main effect of C. (James and Bancroft called $2\beta_1$ the main effect and estimated it by 9.1). The hypothesis $\beta_c = 0$ may be tested by the F test under normality. The numerator of the test statistic is $254^2/56 = 1152.1$; the denominator is $(1 + (t - k)\theta)E_e = 140.4$. In this particular experiment, the only two significant effects were the main effects of C and K; for the latter, $\hat{\beta}_k = -4.0$, with mean square 896.0.

If the intrablock analysis is used, the estimate of the main effect of C is $\hat{\beta}_c = k \sum \pm Q_i/(8\lambda t) = 2(149)/64 = 4.66$. The corresponding sum of squares is $(149) \times k/(8\lambda t) = 693.8$. This is tested against E_e, and, again, C and K are the only two significant effects. The sums of squares for the several contrasts add up to the sum of squares for treatments, adjusted for blocks. They are

C	P	K	CP	CK	PK	CPK
1152.1	240.3	896.0	73.1	126.0	301.8	283.5

11.12 Other Combined Estimates

In previous sections the estimate of σ_b^2 was obtained from E_b. This is not the only method. Yates (1940) subdivided the sum of squares for blocks ignoring treatments into a treatments component and a remainder. The interblock estimates, $\tilde{\tau}$, were obtained by least squares analysis from the set of b observations B_j. The total sum of squares $\sum_j B_j^2 - G/b$ was then divided into two components: *regression*, $\sum_i T_i'\tilde{\tau}_i$, and *remainder*, with $t - 1$ and $b - t$ d.f., respectively. The remainder has expectation $(b - t)\sigma_f^2$. The total is $S_{\text{bl ig tr}}$ multiplied by k. Hence, dividing each of the component sums of squares by k, we have a subdivision of $S_{\text{bl ig tr}}$ into a treatment component $\sum_i T_i'\tau_i/k$ and a remainder. The mean square of the latter component is an unbiased estimate of σ_f^2/k with $(b - t)$ degrees of freedom.

The ten different sums of squares that have been considered during this discussion of balanced incomplete block designs are collected in table 11.6, together with their expectations, under the assumptions that treatments are fixed with the side condition $\sum_i \tau_i = 0$ and that blocks are random. Under normality they have, with the proper divisors, chi-square distributions, and the F test can be used to test hypotheses, $\boldsymbol{\tau} = \mathbf{0}$, and $\sigma_b^2 = 0$.

S_1, S_2, S_3, S_6, and S_9 appear in the intrablock analysis and are distributed independently under normality. S_4 and S_5 are independent of S_1, S_2, S_6, and S_9 under normality. S_1, S_2, S_7, S_8, and S_9 appear in the interblock analysis and are independent under normality.

Graybill and Deal (1959) use as their combined estimator $(T_i + \theta W_i)/r$, with θ given by (11.14) and σ_e^2 and σ_b^2 estimated by E_e and $S_5/(b - t)$. By an argument that depends heavily upon the assumption of normality, they show that when the β_j are normally distributed their estimator is unbiased and is uniformly better than either $\hat{\psi}$ or $\tilde{\psi}$ in the sense that $V(\psi^*) \leq \min(V(\hat{\psi}), V(\tilde{\psi}))$ for all values of σ_e^2 and σ_b^2 when $b \geq t + 10$ or when $b = t + 9$ and $N - b - t + 1 \geq 18$. Seshadri (1963*a*,*b*) also discusses alternate combined estimates of this type.

11.13 Maximum Likelihood Estimation

We now consider the maximum likelihood method under normality, given by Rao (1947*a*). We shall assume in the derivation that we have a general incomplete block design, the only restriction being that it is proper; it does not have to be

TABLE 11.6
ANALYSIS OF VARIANCE

	Source	SS	d.f.	ESS
S_0	Raw total	$\mathbf{Y}'\mathbf{Y}$	N	$N\mu^2 + N\sigma_e^2 + r \sum \tau_i^2 + N\sigma_b^2$
S_1	Mean	G^2/N	1	$N\mu^2 + \sigma_e^2 + k\sigma_b^2$
S_2	Total	$S_0 - S_1$	$N - 1$	$(N - 1)\sigma_e^2 + r \sum \tau_i^2 + (N - k)\sigma_b^2$
S_3	Blocks (unadjusted)			
	divided into	$k^{-1} \sum B_j^2 - S_1$	$b - 1$	$(b - 1)\sigma_e^2 + (r - \lambda) \sum \tau_i^2/k + (N - k)\sigma_b^2$
S_4	Treatment component	$[\sum T_i'^2/k(r - \lambda)] - [S_1 rk/(r - \lambda)]$	$t - 1$	$(t - 1)\sigma_e^2 + (r - \lambda) \sum \tau_i^2/k + (tk - k)\sigma_b^2$
S_5	Remainder	$S_3 - S_4$	$b - t$	$(b - t)\sigma_e^2 + (N - tk)\sigma_b^2 = (b - t)\sigma_f^2/k$
S_6	Treatments (adjusted)	$k\mathbf{Q}'\mathbf{Q}/\lambda t$	$t - 1$	$(t - 1)\sigma_e^2 + \lambda t \sum \tau_i^2/k$
S_7	Treatments (unadjusted)	$r^{-1} \sum T_i^2 - S_1$	$t - 1$	$(t - 1)\sigma_e^2 + r \sum \tau_i^2 + (t - k)\sigma_b^2$
S_8	Blocks (adjusted)	$S_3 + S_6 - S_7$	$b - 1$	$(b - 1)\sigma_e^2 + (N - t)\sigma_b^2$
S_9	Intrablock error	$S_2 - S_3 - S_6 = S_2 - S_7 - S_8$	$N - b - t + 1$	$(N - b - t + 1)\sigma_e^2$

equireplicate. Although we are, by our notation y_{ij}, tacitly assuming a binary design, that assumption is not necessary. We suppose that our model is

$$y_{ij} = \mu + \tau_i + \beta_j + e_{ij},$$

where β_j, e_{ij} are random normal variables and μ and τ_i are unknown constants.

We have $V(y_{ij}) = \sigma_b^2 + \sigma_e^2$, $\text{cov}\,(y_{ij}, y_{i'j'}) = \sigma_b^2$ if $j = j'$ and zero if $j \neq j'$.

If $\mathbf{V}$ is the covariance matrix of the observations $\mathbf{Y}$, the maximum likelihood estimates are obtained by minimizing the quadratic form $[\mathbf{Y}' - E(\mathbf{Y}')]\mathbf{V}^{-1}[\mathbf{Y} - E(\mathbf{Y})]$ of the multivariate normal distribution of the y_{ij}.

If the observations in $\mathbf{Y}$ are arranged by blocks, $\mathbf{V}$ consists of a series of square matrices $\sigma_e^2\mathbf{I}_k + \sigma_b^2\mathbf{J}_k$ along the main diagonal and zeros elsewhere. Then $\mathbf{V}^{-1}$ consists of matrices $\mathbf{I}_k/\sigma_e^2 - [\sigma_b^2\mathbf{J}_k/\sigma_e^2(\sigma_e^2 + k\sigma_b^2)]$ along the main diagonals and zeros elsewhere. The quadratic form to be minimized in order to obtain the maximum likelihood estimates may now be written

$$L = \frac{1}{\sigma_e^2}\sum_i\sum_j (y_{ij} - \mu^* - \tau_i^*)^2 - \frac{\sigma_b^2}{\sigma_e^2(\sigma_e^2 + k\sigma_b^2)}\sum_j\left(B_j - k\mu^* - \sum_i n_{ij}\tau_i^*\right)^2.$$

The asterisks denote the estimates of the parameters; in particular, $\mu^* = G/N$.

In matrix notation, the second member is

$$\left(\frac{1}{k\sigma_e^2} - \frac{1}{\sigma_f^2}\right)(\mathbf{B} - k\mu^*\mathbf{1} - \mathbf{N}'\boldsymbol{\tau}^*)'(\mathbf{B} - k\mu^*\mathbf{1} - \mathbf{N}'\boldsymbol{\tau}^*).$$

Differentiating, and recalling that $\mathbf{R1} = \mathbf{N1}$, we have

$$\frac{\partial L}{\partial \tau^*} = \frac{1}{\sigma_e^2}[\mathbf{T} - \mathbf{R1}\mu^* - \mathbf{R}\boldsymbol{\tau}^*] - \left(\frac{1}{k\sigma_e^2} - \frac{1}{\sigma_f^2}\right)[\mathbf{NB} - \mathbf{R1}k\mu^* - \mathbf{NN}'\boldsymbol{\tau}^*] = \mathbf{0},$$

whence

$$\frac{1}{\sigma_e^2}[\mathbf{Q} - \mathbf{A}\boldsymbol{\tau}^*] + \frac{1}{\sigma_f^2}\left[\mathbf{NB} - \frac{\mathbf{R}G\mathbf{1}}{b} - \mathbf{NN}'\boldsymbol{\tau}^*\right] = \mathbf{0}.$$

The first term alone gives the intrablock estimates; the second term gives the interblock estimates.

Writing $w_1 = 1/\sigma_e^2$ and $w_2 = k/\sigma_f^2$ gives

$$\left(w_1\mathbf{A} + \frac{w_2}{k}\mathbf{NN}'\right)\boldsymbol{\tau}^* = \left(w_1\mathbf{R} - \frac{(w_1 - w_2)}{k}\mathbf{NN}'\right)\boldsymbol{\tau}^*$$

$$= \left[w_1\mathbf{Q} + \frac{w_2}{k}\mathbf{N}\left(\mathbf{B} - \frac{G\mathbf{1}}{b}\right)\right].$$

The individual equations may be written

$$P_i = R_i\left(\frac{k-1}{k}\right)\tau_i^* - \sum_{h \neq i} \frac{\Lambda_{hi}\tau_h^*}{k},$$

where, for binary designs,

$$P_i = w_1 Q_i + \frac{w_2}{k}\left(\sum_j n_{ij}B_j - w(r_i G/b)\right),$$

$$R_i = r_i(w_1 + w_2/(k-1)), \qquad \Lambda_{hi} = \lambda_{hi}(w_1 - w_2),$$

and $\lambda_{hi} = \sum n_{hj}n_{ij}$. For a binary design, λ_{hi} is the number of times that the hth and ith treatments appear in the same block. The combined estimates can thus be obtained by substituting P_i, R_i, and Λ_{hi} for Q_i, r_i, and λ_{hi} in the solutions to the intrablock equations. As before, we substitute for σ_e^2 and σ_b^2 estimates from the intrablock and interblock analyses of variance.

For the balanced incomplete block design, the methods of Yates and of Rao give the same combined estimates. This is not, however, true in the general case [see Sprott (1956*a*)]. It should be noted that the use of Rao's method does not require that the matrix $\mathbf{NN}'$ be of full rank. We shall use this method in the next chapter for the simple lattice design in which n^2 treatments are tested in a design with $k = n$, $b = 2n$, and $r = 2$, in this case, $b < t$.

11.14 The General Balanced Design

A general incomplete block design is said to be balanced if, with $\sum \tau = 0$, in the intrablock analysis, $V(\hat{\tau}_h - \hat{\tau}_i)$ is the same for all pairs of treatments h and i, $h \neq i$. Atiqullah (1961) showed that this implied that if the covariance matrix of the estimates is $\mathbf{V}\sigma^2$, the diagonal elements of $\mathbf{V}$ are all equal and the off-diagonal elements are all equal.

Solving the normal equations by the first method with $\mathbf{H} = \mathbf{1}'$, we have $\operatorname{cov} \hat{\boldsymbol{\tau}} = \mathbf{B}_{11}\sigma^2$ so that $\mathbf{V} = \mathbf{B}_{11}$ and $\mathbf{V1} = \mathbf{B}_{11}\mathbf{1} = \mathbf{0}$. Denote the elements of $\mathbf{V}$ by v_{hi}. Then, if the design is balanced,

$$V(\hat{\tau}_h - \hat{\tau}_i) = (v_{hh} + v_{ii} - 2v_{hi})\sigma^2 = d\sigma^2$$

for all pairs h, i, where d is some constant. Summing over all i, $1 \leq i \leq t$, we have

$$(t-1)d = (t-1)v_{hh} + \sum{}' v_{ii} - 2\sum{}' v_{hi},$$

where the prime denotes summation over all i, except $i = h$. However, since $\mathbf{V1} = \mathbf{0}$, $\sum_i v_{hi} = 0$, and the equation becomes

$$(t-1)d = tv_{hh} + \operatorname{tr}(\mathbf{V}),$$

where tr (**V**) is the trace of the matrix **V**. It follows that

$$v_{hh} = \{(t - 1)d - \text{tr}(\mathbf{V})\}/t$$

for each h, and the diagonal elements of **V** are all equal. Furthermore, all the off-diagonal elements of **V** are also equal, and, since $v_{hh} + (t - 1)v_{hi} = 0$, **V** must be of the form

$$\mathbf{V} = c(\mathbf{I}_t - \mathbf{J}_t/t),$$

where c is some constant.

If the side condition $\mathbf{H}\boldsymbol{\tau} = \mathbf{1}'\boldsymbol{\tau} = 0$ is imposed in solving the equations, we have $\mathbf{AB}_{11} = \mathbf{I} - \mathbf{H}'\mathbf{B}_{21} = \mathbf{I} - \mathbf{J}/t$. Suppose $\mathbf{B}_{11} = \mathbf{V} = c(\mathbf{I} - \mathbf{J}/t)$. Then

$$\mathbf{AB}_{11} = c\mathbf{A} - c\mathbf{AJ}/t = c\mathbf{A},$$

and

$$\mathbf{A} = c^{-1}\mathbf{AB}_{11} = c^{-1}(\mathbf{I} - \mathbf{J}/t).$$

Conversely, if $\mathbf{A} = c(\mathbf{I} - \mathbf{J}/t)$, we have $\mathbf{B}_{11}\mathbf{A} = c\mathbf{B}_{11}$ and $\mathbf{B}_{11} = c^{-1}(\mathbf{I} - \mathbf{J}/t)$.

We have thus shown that a necessary and sufficient condition for a design to be balanced is that **A** shall have its diagonal elements equal and its off-diagonal elements equal.

An equivalent condition is that the nonnegative latent roots of **A** should be equal [Rao (1958)]. It follows that if a proper binary design is balanced, it must be equireplicate and be a balanced incomplete block design as defined in the previous sections. If, however, the requirement that the design be binary, or the requirement that the design be proper, is relaxed, balanced designs can be obtained that are not equireplicate [John (1964*c*)].

We now generalize the definition of the efficiency of a design and compare the average variance of a comparison $(\hat{\tau}_h - \hat{\tau}_i)$ to the variance in a completely randomized design with no block effects and $r = \bar{r} = N/t$. We then have

$$E = 2\sigma^2/\bar{r}\bar{v},$$

where

$$\bar{v} = \{t(t - 1)\}^{-1} \sum\sum (v_{hh} + v_{ii} - 2v_{hi})\sigma^2,$$

and so $E = N^{-1}t\delta$, where $\delta = (t - 1)/\text{tr}(\mathbf{V})$.

It is now appropriate to consider how E might be maximized, and how E might depend upon the choice of the design. The problem becomes one of minimizing tr (**V**), which is the sum of the latent roots of $\mathbf{B}_{11}$. The rank of $\mathbf{B}_{11}$ is the same as the rank of **A**, which, so long as we confine our attention to connected designs, is $t - 1$; both $\mathbf{B}_{11}$ and **A** have a single zero latent root. It can be shown that the nonzero latent roots of $\mathbf{B}_{11}$ are the reciprocals of the nonzero roots of $\mathbf{A}$ with the same multiplicities. Thus, if the nonzero roots of $\mathbf{A}$ are denoted by θ_i and each root is counted as many times as its multiplicity,

$$\delta = (t - 1)\left\{\sum \theta_i^{-1}\right\}^{-1},$$

which is the harmonic mean of the nonzero roots. We can proceed to find an upper bound for δ.

The nonzero latent roots of $\mathbf{A}$ are all positive, and so their harmonic mean is not greater than their arithmetic mean. Hence,

$$(t - 1)\delta \leq \operatorname{tr}(\mathbf{A}) = \sum_i \left(r_i - \sum_j n_{ij}^2/k_j\right) = N - \sum_j m_j/k_j,$$

where $m_j = \sum_i n_{ij}^2$. The elements n_{ij} are either zero or positive integers with $\sum_i n_{ij} = k_j$ so that $m_j \geq k_j$ and $m_j \geq k_j^2/t$. Hence, $m_j \geq \max(k_j, k_j^2/t)$.

There are two cases to be considered. If $t \geq k_j$, then $k_j \geq k_j^2/t$ and $\min(m_j)$ occurs when k_j of the treatments appear once each in the block. If $k_j = pt + q$ where p is a positive integer, and $0 \leq q < t$, $\min(m_j)$ occurs when q treatments appear $p + 1$ times each, and the other treatments appear p times; in this case

$$m_j \geq q(p + 1)^2 + (t - q)p^2 = p^2t + 2pq + q \geq t^{-1}(pt + q)^2 = k_j^2/t.$$

We have, therefore, $m_j/k_j \geq 1$ if $k_j < t$, and $m_j/k_j \geq k_j/t$ if $k_j \geq t$.

Suppose that b' of the blocks have $k_j < t$, and that there are N' plots in all the blocks that contain t or more plots. Then

$$\operatorname{tr}(\mathbf{A}) \leq N - b' - t^{-1}N' \leq N - b$$

and $E \leq E_0 = \{t(N - b') - N'\}/\{N(t - 1)\}$.

If we now restrict ourselves to designs with $k_j < t$ for all j, we have $b' = b$ and $N' = 0$ so that

$$E_0 = \{t(N - b)\}/\{N(t - 1)\}.$$

For a proper design, with all $k_j = k$, this simplifies to

$$E_0 = t(k - 1)/\{k(t - 1)\},$$

which is the efficiency factor for a balanced incomplete block design if one exists. This derivation of E_0 is due to J. Roy (1958), and proves a conjecture made by Kempthorne (1956*a*) that in the class of proper designs of block size k with b blocks and t treatments the most efficient design is the balanced incomplete block design, if one exists. Nothing in the derivation restricts the designs under consideration to being binary.

11.15 Extended Incomplete Block Designs

In some experimental situations we have $t < k < 2t$. A reasonable procedure in this case is to let each block consist of a single replication of the t treatments together with a block from an incomplete block design with block size $k - t$. We consider the case in which the incomplete block design is balanced [John (1963)]. For example, with $t = 4$, $b = 3$, and $k = 7$, we could take the design *ABCDABC, ABCDABD, ABCDACD, ABCDBCD.*

Suppose that there are t treatments in b blocks of size k, and that each treatment appears r times. The parameters of the balanced incomplete block design are $t, b, r' = r - b, k' = k - t$, and λ. The diagonal elements of $\mathbf{NN}'$ are

$$\sum_j n_{ij}^2 = 4(r - b) + (2b - r) = 3r - 2b;$$

the off-diagonal elements are

$$\begin{aligned}\sum_j n_{hj}n_{ij} &= 4\lambda + 2(r - b - \lambda) + 2(r - b - \lambda) + (b - 2(r - b - \lambda) - \lambda)\\ &= 2r - b + \lambda.\end{aligned}$$

Writing $\lambda' = b + \lambda$ and $\lambda^* = 2r - b + \lambda$, we have

$$\mathbf{NN}' = (r - \lambda')\mathbf{I}_t + \lambda^*\mathbf{J}_t,$$

and

$$k\mathbf{A} = (rk - r + \lambda')\mathbf{I}_t - \lambda^*\mathbf{J}_t = (t\mathbf{I} - \mathbf{J})\lambda^*,$$

which differs from the normal equations for the balanced incomplete block design only in the substitution of λ^* for λ.

Imposing the side condition $\mathbf{H}'\boldsymbol{\tau} = (\lambda^*/k)^{1/2}\mathbf{1}'\boldsymbol{\tau} = 0$, we have $\mathbf{A} + \mathbf{H}'\mathbf{H} = t\lambda^*\mathbf{I}_t/k$, and $\hat{\tau}_i = kQ_i/(t\lambda^*)$. Then $V(\hat{\tau}_h - \hat{\tau}_i) = \{(t - 1)k/t^2\lambda^*\}\sigma_e^2$, and the efficiency of the design is $E = t\lambda^*/(rk)$.

The residual sum of squares can be divided into two components: the sum of squares between duplicates and a remainder attributable to interaction between blocks and treatments. The analysis of variance table follows. Interblock estimates and combined estimates may be derived in the same manner as they were for the balanced incomplete block design.

TABLE 11.7
ANALYSIS OF VARIANCE

Source	d.f.	SS
Total	$N - 1$	$\sum\sum y^2 - G^2/N$
Blocks (unadjusted)	$b - 1$	$\sum B_j^2/k - G^2/N$
Treatments (adjusted)	$t - 1$	$k\sum Q_i^2/(t\lambda^*)$
Intrablock error	$b(k - t)$	(Between duplicates) $= \sum (\text{diff})^2/2$
Interaction	$(b - 1)(t - 1)$	By subtraction

Exercises

1. Is the equation

$$E(S_{\text{bl adj}}) = \sigma_b^2(bk - t) + \sigma_e^2(b - 1)$$

true for any binary, proper, equireplicate design? Is it true if any of the three restrictions, binary, proper, and equireplicate, are removed?

2. A design for two treatments has p pairs of blocks. Each block has three plots. In each pair one block has two plots with A and one with B, and the other has two plots with B and a single plot with A. Analyze the design. How would your analysis be modified if $p = mq$, and the p pairs are actually m pairs at each level of a factor with q levels?
3. Apply the methods of this chapter to the case of a randomized complete block design in which the first treatment is missing from the first block.
4. In a resolvable balanced incomplete block design consider the first block in the first replication. Let m_{ij} denote the number of treatments that this block has in common with the jth block of the ith replication ($1 < i \leq \text{r}, 1 \leq j \leq b/r$). Show that the average of the m_{ij} is $m = k^2/v$, and that their variance is given by

$$\sum_i \sum_j (m_{ij} - m)^2 = k(t - k)(b - t - r + 1)/\{n^2(r - 1)(t - 1)\},$$

where $n = t/k$. For an affine resolvable design $m_{ij} = m$ for all i, j [Bose (1942c)].

5. Find the combined estimates for the extended block design, Sec. 11.15.

CHAPTER 12

Partially Balanced Incomplete Block Designs

Much of our emphasis in the last chapter was upon balanced incomplete block designs. Unfortunately, such designs do not exist for all combinations of parameters that we might wish to use. In particular, the requirement that λ be an integer imposes a severe restriction upon the number of treatments for which balanced designs of a reasonable size can be found. It will be recalled that if there are t treatments each repeated r times in blocks of size k, then $\lambda = r(k - 1)/(t - 1)$. If $(t - 1)$ is a prime, r has to be a multiple of $(t - 1)$, and then there is the further restriction that $b = rt/k$ has to be an integer. For example, if $t = 32$, r must be a multiple of 31; if, in addition, k is odd, and thus prime to 32, b has to be a multiple of 31×32.

The characteristic property of a balanced incomplete block design is that $V(\hat{\tau}_h - \hat{\tau}_i)$ has the same value for all pairs, h, i, of treatments. An alternative to requiring balance is to find a design in which the condition of equal variances is modified somewhat. This can be achieved by the use of partially balanced designs. In a partially balanced design with two associate classes, $V(\hat{\tau}_h - \hat{\tau}_i)$ may take either of two values, depending upon whether the hth and ith treatments are first or second associates of each other. When there are m associate classes, $V(\hat{\tau}_h - \hat{\tau}_i)$ may take any of m values.

Partially balanced incomplete block designs (*PBIB* designs) were introduced by Bose and Nair (1939). Their idea has been developed considerably since then, largely by Bose and his students. In 1952 Bose and Shimamoto introduced the concept of partially balanced association schemes and the definition of a partially balanced design was rephrased in terms of this concept. We shall use the definition given by Bose (1963) in a review paper.

The symbol v will be used to denote the number of treatments. This will bring our notation into line with that used by Bose and his students. The treatments may be denoted either by numbers or by letters. A partially balanced association scheme with m classes is a relationship between the treatments, satisfying the following conditions:

(i) Any two treatments are either 1st, 2nd, ..., or mth associates, and the relation of association is symmetrical. Thus, if α and β are two treatments and α is an ith associate of β, then β is an ith associate of α;

(ii) Each treatment has exactly n_i ith associates, and the number n_i is independent of the treatment chosen;

(iii) If α and β are ith associates, the number of treatments that are both jth associates of α and kth associates of β is p_{jk}^i, and is independent of the pair of ith associates chosen. Hence, $p_{jk}^i = p_{kj}^i$.

The numbers v, n_i, p_{jk}^i are called the parameters of the first kind. The parameters p_{jk}^i may conveniently be written in the form of m matrices; $\mathbf{P}_i$ ($i = 1, 2, \ldots, m$) has p_{jk}^i as the element in the jth row and the kth column.

A partially balanced incomplete block design is a design based upon such a scheme. It is a proper, binary, equireplicate design, having b blocks of k plots each with every treatment repeated r times, and the additional requirement that if α and β are ith associates, then they appear together in exactly λ_i blocks where λ_i is independent of the particular pair of ith associates chosen. It was originally thought that the λ_i should all be different, but Nair and Rao (1942) pointed out that this was not necessary. Furthermore, some of the λ_i may be zero. We shall usually write $PBIB(m)$ to denote a design with m associate classes. The numbers b, r, k, and λ_i are called the parameters of the second kind.

There are several relationships between the parameters of a partially balanced incomplete block design and the corresponding scheme. Clearly,

$$rv = bk, \qquad \sum_i n_i\lambda_i = r(k-1), \qquad \sum_i n_i = v - 1.$$

It is also readily shown that

$$\sum_k p_{jk}^i = n_j, \qquad \text{if} \qquad i \neq j; \qquad \sum_k p_{jk}^i = n_j - 1, \qquad \text{if} \qquad i = j,$$

and also that

$$n_i p_{jk}^i = n_j p_{ik}^j = n_k p_{ij}^k.$$

To derive the last relationship, consider a treatment α. Let G_i denote the set of ith associates of α, and G_j the set of jth associates. Each treatment in G_i has exactly p_{jk}^i kth associates in G_j; similarly, each treatment in G_j has exactly p_{ik}^j kth associates in G_i. Thus, the number of pairs of kth associates that can be obtained by taking one treatment from G_i and another from G_j is, on the one hand, $n_i p_{jk}^i$ and, on the other, $n_j p_{ik}^j$.

Most of the research carried out so far has been concerned with designs having two associate classes. Chapter 14 will be devoted to the construction of designs from association schemes. Bose and Shimamoto (1952) classified all the known *PBIB*(2) designs into five types. Simple designs are those for which either λ_1 or $\lambda_2 = 0$. The other designs correspond to one of the following four association schemes: group divisible, triangular, Latin square, and cyclic. Extensive collections of designs are given by Bose and Shimamoto (1952), and by Bose, Clatworthy, and Shrikhande (1954). For $m = 2$, the relationships between the parameters just given take the simpler form:

$$n_1 + n_2 = v - 1, \qquad n_1\lambda_1 + n_2\lambda_2 = r(k - 1),$$

$$p_{11}^1 + p_{12}^1 = n_1 - 1, \qquad p_{21}^1 + p_{22}^1 = n_2,$$

$$p_{11}^2 + p_{12}^2 = n_1, \qquad p_{21}^2 + p_{22}^2 = n_2 - 1,$$

$$n_1 p_{12}^1 = n_2 p_{11}^2, \qquad n_1 p_{22}^1 = n_2 p_{12}^2.$$

12.1 The Group Divisible Association Scheme

This is a scheme for $v = mn$ treatments where m and n are integers. The treatments are divided into m groups of n treatments each. Treatments in the same group are first associates; treatments in different groups are second associates. This corresponds to writing the v symbols for the treatments in a rectangular array of m rows and n columns, and letting the rows be the groups. For this scheme $v = mn$, $n_1 = n - 1$, and $n_2 = n(m - 1)$,

$$\mathbf{P}_1 = \begin{bmatrix} n-2 & 0 \\ 0 & n(m-1) \end{bmatrix}, \qquad \mathbf{P}_2 = \begin{bmatrix} 0 & n-1 \\ n-1 & n(m-2) \end{bmatrix}.$$

It can be shown that a necessary and sufficient condition for a *PBIB*(2) association scheme to be group divisible is that p_{12}^1 (or p_{12}^2) $= 0$. It is customary to abbreviate group divisible to *GD*.

As an example of a *GD* design, consider the following design for $v = 6$:

$$1, 2, 3; \quad 3, 4, 5; \quad 2, 5, 6; \quad 1, 2, 4; \quad 3, 4, 6; \quad 1, 5, 6.$$

The three groups are 1 and 2, 3 and 4, 5 and 6. We have $\lambda_1 = 2$, $\lambda_2 = 1$.

We shall show in Chapter 14 that the following inequalities hold for all *GD* designs:

$$r \geq \lambda_1, \qquad rk - \lambda_2 v \geq 0.$$

Bose and Connor (1952) have divided the *GD* designs into three classes:

(i) Singular designs having $r = \lambda_1$;
(ii) Semiregular designs having $r > \lambda_1$ and $rk = \lambda_2 v$;
(iii) Regular designs having $r > \lambda_1$ and $rk > \lambda_2 v$.

A trivial example of a singular design is obtained by letting each row of the rectangular array constitute a block.

12.2 Intrablock Analysis of *GD* Designs

The intrablock equations for a *GD* design are easily solved. In Chapter 11 we derived the intrablock equations, $\mathbf{Q} = \mathbf{A}\hat{\boldsymbol{\tau}}$, for the general incomplete block design. The reduced equation for the hth treatment is

$$Q_h = \sum_j n_{hj}\hat{\tau}_h - \sum_i \sum_j n_{hj}n_{ij}\hat{\tau}_i/k.$$

Writing $S_1(\hat{\tau}_h) = \sum_s (\hat{\tau}_s)$ where the summation is made over all the first associates of the hth treatment and defining $S_2(\hat{\tau}_h)$ similarly for the second associates, we have

$$\begin{aligned} kQ_h &= r(k-1)\hat{\tau}_h - \lambda_1 S_1(\hat{\tau}_h) - \lambda_2 S_2(\hat{\tau}_h) \\ &= \{r(k-1) + \lambda_1\}\hat{\tau}_h - \lambda_1\{\hat{\tau}_h + S_1(\hat{\tau}_h)\} - \lambda_2 S_2(\hat{\tau}_h). \end{aligned}$$

Suppose now that the treatments are labeled in order so that treatments 1 through n form the first group, treatments $n + 1$ through $2n$ form the second group, and so on. The matrix $\mathbf{A}$ then consists of square submatrices

$$\{r(k-1) + \lambda_1\}\mathbf{I}_n - \lambda_1\mathbf{J}_n$$

along the main diagonal and submatrices $-\lambda_2\mathbf{J}_n$ off the diagonal. We take the side condition $\sum \tau_i = 0$ in the form $\mathbf{H}\boldsymbol{\tau} = 0$ where $\mathbf{H} = h\mathbf{1}$ and $h^2 = \lambda_2/k$. Then $k(\mathbf{A} + \mathbf{H}'\mathbf{H})$ consists of square matrices $\{r(k-1) + \lambda_1\}\mathbf{I}_n + (\lambda_2 - \lambda_1)\mathbf{J}_n$ along the diagonal and zero elsewhere. The solution matrix $(\mathbf{A} + \mathbf{H}'\mathbf{H})^{-1}$ thus consists of submatrices

$$\begin{aligned} &\frac{k}{r(k-1) + \lambda_1}\left[\mathbf{I} - \frac{(\lambda_2 - \lambda_1)\mathbf{J}}{\{r(k-1) + \lambda_1 + n\lambda_2 - n\lambda_1\}}\right] \\ &\qquad\qquad = \frac{k}{r(k-1) + \lambda_1}\left[\mathbf{I} - \frac{(\lambda_2 - \lambda_1)\mathbf{J}}{v\lambda_2}\right] \end{aligned}$$

along the main diagonal and zero elsewhere.

It follows that

$$\{r(k-1) + \lambda_1\}\hat{\tau}_h = kQ_h - k(\lambda_2 - \lambda_1)\{Q_h + S_1(Q_h)\}/(v\lambda_2)$$

where $S_1(Q_h)$ is the sum $\sum_s Q_s$ for all the first associates of the hth treatment.

The expression $Q_h + S_1(Q_h)$ is the sum of the Q_s for all the treatments in the group. We also have

$$V(\hat{\tau}_h - \hat{\tau}_i) = \frac{2k\sigma^2}{r(k-1)+\lambda_1},$$

if h and i are first associates, and

$$V(\hat{\tau}_h - \hat{\tau}_i) = \frac{2k\sigma^2}{r(k-1)+\lambda_1}\left[1 - \frac{(\lambda_2 - \lambda_1)}{mn\lambda_2}\right],$$

if they are second associates. In the example given earlier with $v = 6$, $m = 3$, $n = 2$, we have

$$\hat{\tau}_1 = (7Q_1 + Q_2)/16, \qquad \hat{\tau}_2 = (Q_1 + 7Q_2)/16,$$

$$V(\hat{\tau}_1 - \hat{\tau}_2) = 3\sigma^2/4, \qquad V(\hat{\tau}_1 - \hat{\tau}_3) = 7\sigma^2/8.$$

12.3 The Triangular Scheme

There are $v = n(n-1)/2$ treatments. The association scheme is obtained by writing the treatments as the elements above the diagonal of a square array of side n, leaving the diagonal blank, and repeating the treatments below the diagonal in order to obtain a symmetric array. Thus, if $n = 5$, we have

$$\begin{matrix} * & 1 & 2 & 3 & 4 \\ 1 & * & 5 & 6 & 7 \\ 2 & 5 & * & 8 & 9 \\ 3 & 6 & 8 & * & 10 \\ 4 & 7 & 9 & 10 & * \end{matrix}$$

Treatments appearing in the same row, or column, are first associates. Thus the first associates of 1 are 2, 3, 4, 5, 6, 7, and the second associates are 8, 9, 10. For this scheme

$$n_1 = 2(n-2), \qquad n_2 = (n-2)(n-3)/2,$$

$$\mathbf{P}_1 = \begin{bmatrix} n-2 & n-3 \\ n-3 & (n-3)(n-4)/2 \end{bmatrix}, \qquad \mathbf{P}_2 = \begin{bmatrix} 4 & 2(n-4) \\ 2(n-4) & (n-4)(n-5)/2 \end{bmatrix}$$

The following design has $b = 10$, $k = r = 4$, $\lambda_1 = 1$, $\lambda_2 = 2$.

2, 10, 6, 7; 10, 1, 2, 5; 7, 3, 8, 2; 6, 2, 9, 4; 1, 9, 10, 8;

5, 4, 3, 10; 8, 7, 4, 1; 3, 5, 7, 9; 9, 6, 1, 3; 4, 8, 5, 6.

Another way of deriving this association scheme is to let one treatment correspond to each of the ordered pairs of integers (x, y), $0 < x < y \leq n$. Then two treatments are first associates if their representations have an integer in common. The number pairs are equivalent to the coordinates in the upper triangle of the array mentioned before. Thus we can associate 1 with (1, 2), 2 with (1, 3), and so on. Then the first associates of (1, 2) would be (1, 3), (1, 4), (1, 5), (2, 3), (2, 4), and (2, 5).

12.4 The Latin Square Scheme (L_i)

There are $v = n^2$ treatments. For the L_2 scheme the treatments are arranged in a square array, and two treatments are first associates if they appear in the same row or column. For the L_i scheme, $i > 2$, we take a set of $i - 2$ mutually orthogonal Latin squares and superimpose them on the array. Then two treatments are first associates if they appear in the same row or column of the array, or if they correspond to the same letter in one of the squares.

For this scheme $n_1 = i(n - 1)$, $n_2 = (n - i + 1)(n - 1)$,

$$\mathbf{P}_1 = \begin{bmatrix} i^2 - 3i + n & (i - 1)(n - i + 1) \\ (i - 1)(n - i + 1) & (n - i)(n - i + 1) \end{bmatrix},$$

$$\mathbf{P}_2 = \begin{bmatrix} i(i - 1) & i(n - i) \\ i(n - i) & (n - i)^2 + (i - 2) \end{bmatrix}.$$

For $n = 3$ we may take the array

$$\begin{matrix} 1 & 2 & 3 \\ 4 & 5 & 6 \\ 7 & 8 & 9 \end{matrix}$$

and obtain the following L_2 design:

1 2 3, 4 5 6, 7 8 9, 1 4 7, 2 5 8, 3 6 9,

in which $\lambda_1 = 1$ and $\lambda_2 = 0$. This is a simple lattice design; it is further discussed later in this chapter.

12.5 The Cyclic Association Scheme

The treatments are denoted by the integers $0, 1, \ldots, v - 1$. The first associates of treatment i are $i + d_1, i + d_2, \ldots, i + d_{n_1}$, (mod v), where $d_1, \ldots, d_{n_1}$ are integers satisfying the following conditions:

(i) The d_j are all different, and $0 < d_j < v$ for each j;

(ii) Among the $n_1(n_1 - 1)$ differences $d_j - d_{j'}$ (mod v), each of the integers $d_1, d_2, \ldots, d_{n_1}$ occurs g times, and each of the other n_2 positive integers less than v occurs h times.

Then $n_1 g + n_2 h = n_1(n_1 - 1)$, $g = p_{11}^1$, $h = p_{11}^2$,

$$\mathbf{P}_1 = \begin{bmatrix} g & n_1 - g - 1 \\ n_1 - g - 1 & n_2 - n_1 + g + 1 \end{bmatrix},$$

$$\mathbf{P}_2 = \begin{bmatrix} h & n_1 - h \\ n_1 - h & n_2 - n_1 + h - 1 \end{bmatrix}.$$

As an example consider the following design for $v = 5$:

0 1 2, 1 2 3, 2 3 4, 3 4 0, 4 0 1.

Here $n_1 = n_2 = 2$, $\lambda_1 = 2$, $\lambda_2 = 1$, $d_1 = 1$, and $d_2 = 4$. The differences $d_j - d_{j'}$ are $4 - 1 \equiv 3$ and $1 - 4 \equiv 2 \pmod 5$, so that $g = 0$ and $h = 1$. The first associates of 2 are $2 + 1 \equiv 3$ and $2 + 4 \equiv 1$.

Much less attention has been paid to this scheme than to the three schemes mentioned earlier. A fifth association scheme, the negative Latin square scheme, has been discovered by Mesner (1967). (See section 15.1.)

12.6 The Intrablock Equations for *PBIB*(2) Designs

We have already seen, when we considered *GD* designs, that the hth intrablock equation may be written

$$kQ_h = r(k - 1)\hat{\tau}_h - \lambda_1 S_1(\hat{\tau}_h) - \lambda_2 S_2(\hat{\tau}_h). \tag{12.1}$$

Writing $S_1(Q_h)$ and $S_2(Q_h)$ for the sums of the Q_s for the first and second associates of the hth treatment, we have

$$kS_1(Q_h) = -\lambda_1 n_1 \hat{\tau}_h + S_1(\hat{\tau}_h)\{r(k - 1) - \lambda_1 p_{11}^1 - \lambda_2 p_{12}^1\} + S_2(\hat{\tau}_h)\{-\lambda_1 p_{11}^2 - \lambda_2 p_{12}^2\}, \tag{12.2}$$

$$kS_2(Q_h) = -\lambda_2 n_2 \hat{\tau}_h + S_1(\hat{\tau}_h)\{-\lambda_1 p_{12}^1 - \lambda_2 p_{22}^1\} + S_2(\hat{\tau}_h)\{r(k - 1) - \lambda_1 p_{12}^2 - \lambda_2 p_{22}^2\}. \tag{12.3}$$

There are two methods of continuing the analysis. They are given by Rao (1947*a*), and by Bose and Shimamoto (1952).

Rao's Solution: We eliminate $S_2(\hat{\tau}_h)$ by adding to Equation 12.1 the quantity

$$\lambda_2 \sum_i \hat{\tau}_i = \lambda_2\{\hat{\tau}_h + S_1(\hat{\tau}_h) + S_2(\hat{\tau}_h)\} \equiv 0,$$

and adding $(\lambda_1 p_{11}^2 + \lambda_2 p_{12}^2) \sum_i \hat{\tau}_i$ to Equation 12.2.

Then, after some simplification,

$$kQ_h = A_{12}\hat{\tau}_h + B_{12}S_1(\hat{\tau}_h), \qquad kS_1(Q_h) = A_{22}\hat{\tau}_h + B_{22}S_1(\hat{\tau}_h),$$

where

$$A_{12} = r(k-1) + \lambda_2, \qquad A_{22} = (\lambda_2 - \lambda_1)p_{12}^2,$$

$$B_{12} = \lambda_2 - \lambda_1, \qquad B_{22} = r(k-1) + \lambda_2 + (\lambda_2 - \lambda_1)(p_{11}^1 - p_{11}^2),$$

whence, writing $A_{12}B_{22} - A_{22}B_{12} = \Delta_r$,

$$\Delta_r \hat{\tau}_h = k\{B_{22}Q_h - B_{12}S_1(Q_h)\}. \tag{12.4}$$

(We use Δ_r for Rao's Δ; Bose and Shimamoto use Δ for a different quantity.)

This gives a solution matrix **P** with $p_{ii} = B_{22}k\Delta_r^{-1}$ and $p_{hi} = -kB_{12}\Delta_r^{-1}$ if h and i are first associates, $p_{hi} = 0$ if h and i are second associates.

Hence, $V(\hat{\tau}_h - \hat{\tau}_i) = 2k(B_{22} + B_{12})\Delta_r^{-1}\sigma^2$ if h and i are first associates and $V(\hat{\tau}_h - \hat{\tau}_i) = 2kB_{22}\Delta_r^{-1}\sigma^2$ if h and i are second associates.

EXAMPLE 1. The triangular design given in the previous section has $v = b = 10$, $r = k = 4$, $\lambda_1 = 1$, $\lambda_2 = 2$, $n_1 = 6$, and $n_2 = 3$,

$$\mathbf{P}_1 = \begin{bmatrix} 3 & 2 \\ 2 & 1 \end{bmatrix}, \qquad \mathbf{P}_2 = \begin{bmatrix} 4 & 2 \\ 2 & 0 \end{bmatrix},$$

$A_{12} = 14$, $A_{22} = 2$, $B_{12} = 1$, $B_{22} = 13$, and $\Delta_r = 180$. Then $45\hat{\tau}_1 = 13Q_1 - \sum_{i=2}^{7} Q_i$.

$$V(\hat{\tau}_1 - \hat{\tau}_2) = 28\sigma^2/45, \qquad V(\hat{\tau}_1 - \hat{\tau}_8) = 26\sigma^2/45.$$

EXAMPLE 2. For the *GD* design, $A_{12} = B_{22} = 7$, $B_{12} = A_{22} = -1$, and $\Delta_r = 48$. Then $\hat{\tau}_1 = 3(7Q_1 + Q_2)/48$, $V(\hat{\tau}_1 - \hat{\tau}_2) = 36\sigma^2/48$, and $V(\hat{\tau}_1 - \hat{\tau}_3) = 42\sigma^2/48$.

The Solution of Bose and Shimamoto: It is convenient to write Equations 12.2 and 12.3 as

$$\begin{aligned} kS_1(Q_h) &= -\lambda_1 n_1 \hat{\tau}_h + a_{11}S_1(\hat{\tau}_h) + a_{12}S_2(\hat{\tau}_h), \\ kS_2(Q_h) &= -\lambda_2 n_2 \hat{\tau}_h + a_{21}S_2(\hat{\tau}_h) + a_{22}S_2(\hat{\tau}_h). \end{aligned} \tag{12.5}$$

Consider the linear combination $L_h = k^2Q_h + c_1kS_1(Q_h) + c_2kS_2(Q_h)$. Substituting from Equations 12.5 and 12.1,

$$L_h = rk(k - 1)\hat{\tau}_h - (c_1\lambda_1n_1 + c_2\lambda_2n_2)\hat{\tau}_h + (a_{11}c_1 + a_{21}c_2 - \lambda_1k)S_1(\hat{\tau}_h)$$
$$+ (a_{12}c_1 + a_{22}c_2 - \lambda_2k)S_2(\hat{\tau}_h). \tag{12.6}$$

We choose c_1 and c_2 so that Equation 12.6 becomes

$$k^2Q_h + c_1kS_1(Q_h) + c_2kS_2(Q_h) = rk(k - 1)\hat{\tau}_h.$$

This requires that

$$-\lambda_1n_1c_1 - \lambda_2n_2c_2 = -\lambda_1k + a_{11}c_1 + a_{21}c_2 = -\lambda_2k + a_{12}c_1 + a_{22}c_2,$$

giving

$$\lambda_1k = (a_{11} + \lambda_1n_1)c_1 + (a_{21} + \lambda_2n_2)c_2,$$
$$\lambda_2k = (a_{12} + \lambda_1n_1)c_1 + (a_{22} + \lambda_2n_2)c_2.$$

Solving by determinants, $c_1 = D_1/D$, $c_2 = D_2/D$, where

$$D = (a_{11} + \lambda_1n_1)(a_{22} + \lambda_2n_2) - (a_{12} + \lambda_1n_1)(a_{21} + \lambda_2n_2)$$

and $D_1 = \lambda_1k(a_{22} + \lambda_2n_2) - \lambda_2k(a_{21} + \lambda_2n_2)$. After some simplification we obtain

$$D = (rk - r + \lambda_1)(rk - r + \lambda_2) + (\lambda_1 - \lambda_2)$$
$$\times \{r(k - 1)(p_{12}^1 - p_{12}^2) + \lambda_2p_{12}^1 - \lambda_1p_{12}^2\}.$$

Bose and Shimamoto write $k^2\Delta$ for D. We write their Δ as Δ_b. It can be shown that

$$\Delta_r = D = k^2\Delta_b.$$

Substituting for a_{21} and a_{22} in D_1, we obtain

$$k\Delta_bc_1 = \lambda_1(rk - r + \lambda_2) + (\lambda_1 - \lambda_2)(\lambda_2p_{12}^1 - \lambda_1p_{12}^2).$$

Similarly,

$$k\Delta_bc_2 = \lambda_2(rk - r + \lambda_1) + (\lambda_1 - \lambda_2)(\lambda_2p_{12}^1 - \lambda_1p_{12}^2).$$

The solution matrix **P** has

$$p_{ii} = \frac{k}{r(k - 1)}, \qquad p_{hi} = \frac{c_1}{r(k - 1)}, \qquad \text{or} \qquad \frac{c_2}{r(k - 1)},$$

according to whether h, i are first or second associates, so that

$$V(\hat{\tau}_i - \hat{\tau}_h) = \frac{2(k - c_1)\sigma^2}{r(k-1)} \quad \text{or} \quad \frac{2(k - c_2)\sigma^2}{r(k-1)}.$$

In our first example, $a_{11} = 5$, $a_{12} = -8$, $a_{21} = -4$, and $a_{22} = 10$. The equations to be solved are

$$4 = 11c_1 + 2c_2, \qquad 8 = -2c_1 + 16c_2,$$

whence

$$D = 16\Delta_b = 180, \qquad c_1 = 4/15, \qquad c_2 = 8/15,$$

$$V(\hat{\tau}_1 - \hat{\tau}_2) = \frac{28\sigma^2}{45}, \qquad V(\hat{\tau}_1 - \hat{\tau}_8) = \frac{26\sigma^2}{45}.$$

In practice it is convenient to write the solution as

$$\begin{aligned} rk(k-1)\hat{\tau}_h &= (k - c_1)kQ_h + (c_2 - c_1)kS_2(Q_h) \\ &= (k - c_2)kQ_h + (c_1 - c_2)kS_1(Q_h). \end{aligned}$$

Kapadia and Weeks (1964) give an alternate formula for computing the adjusted sum of squares for treatments in *GD* designs which provides a check on the computations.

12.7 Combined Intrablock and Interblock Estimates

In Chapter 11 we showed that the combined estimates of Yates and Rao were identical for *BIB* designs. This is not, in general, true for *PBIB* designs. Sprott (1956*a*) has shown that for the case of two associate classes the two methods give the same combined estimates only for *GD* designs.

In using the Rao maximum likelihood approach, Bose and Shimamoto define an additional constant H by

$$kH = (2rk - 2r + \lambda_1 + \lambda_2) + (p_{12}^1 - p_{12}^2)(\lambda_1 - \lambda_2).$$

Then, putting

$$w_1 = \frac{1}{\sigma_e^2}, \qquad w_2 = \frac{k}{\sigma_f^2} = \frac{1}{\sigma_e^2 + k\sigma_b^2},$$

$Q_i' = T_i - Q_i - G/v$ and $P_i = w_1Q_i + w_2Q_i'$, they obtain the combined estimators in the form

$$r\{w_2 + w_1(k-1)\}\tau_h^* = kP_h + d_1S_1(P_h) + d_2S_2(P_h),$$

where

$$d_j = \frac{c_j \Delta_b (w_1 - w_2)^2 + r\lambda_j w_2 (w_1 - w_2)}{\Delta_b (w_1 - w_2)^2 + rHw_2(w_1 - w_2) + r^2 w_2^2}.$$

The weights w_1 and w_2 are obtained by using the estimates of σ_e^2 and σ_f^2 from the analysis of variance tables.

12.8 The Simple Lattice Design

Lattice designs were first introduced by Yates (1937) and were called quasi-factorial designs at that time. They have been investigated extensively by Kempthorne, Federer, and their students. Detailed accounts of these designs and their analyses are given in the books by Kempthorne (1952) and Federer (1955). We shall consider here only the simple lattice, and that merely as a special case of a *PBIB*(2) design with the L_2 association scheme.

In the previous chapter we mentioned Yates' orthogonal series of balanced incomplete block designs with $v = n^2$, $k = n$, $r = n + 1$, $b = n(n + 1)$, and $\lambda = 1$. They are resolvable designs and their construction depends upon the existence of complete sets of orthogonal Latin squares of side n. Such a balanced incomplete block design is called a balanced lattice. Any two of the squares constitute a simple lattice; a set of three of the squares is a triple lattice. We should, however, phrase our definition differently in order to avoid excluding values such as $n = 6$, for which the balanced lattice does not exist, but simple and triple lattices do.

Let $v = n^2$. We arrange the treatments in a $n \times n$ array. In the first replicate, two treatments are in the jth block if, and only if, they are in the jth row of the array. In the second replicate, the blocks are given by the columns of the array. In the third replicate (for a triple lattice), a Latin square is superimposed on the array, and two treatments are in the same block if, and only if, they correspond to the same letter in the Latin square.

We now confine ourselves to the simple lattice consisting of the first two replicates. An example is the first two replicates of the dishwashing experiment in Chapter 11.

The simple lattice is a *PBIB*(2) design with the L_2 association scheme. Two treatments are first associates if they occur in the same row or column of the array. (The triple lattice is a *PBIB*(2) design with the L_3 scheme.) We have $v = n^2$, $b = 2n$, $r = 2$, $k = n$, $\lambda_1 = 1$, and $\lambda_2 = 0$.

Using Rao's method for the intrablock analysis,

$$A_{12} = 2(n - 1), \qquad A_{22} = -2(n - 2),$$

$$B_{12} = -1, \qquad B_{22} = n + 2, \qquad \Delta_r = 2n^2.$$

Then

$$2n\hat{\tau}_h = (n + 2)Q_h + S_1(Q_h).$$

Furthermore, $V(\hat{\tau}_h - \hat{\tau}_i) = n^{-1}(n + 1)\sigma^2$ for first associates, $n^{-1}(n + 2)\sigma^2$ for second associates, and the intrablock efficiency of the simple lattice is $(n + 1)/(n + 3)$. The estimate E_e of the intrablock error is obtained in the usual way from the intrablock analysis of variance table:

$$(n - 1)^2 E_e = \sum_h \sum_j y_{hj}^2 - n^{-1} \sum_j B_j^2 - \sum_h \hat{\tau}_h Q_h.$$

We consider now the combined analysis, incorporating the interblock information. We assume that the block effects β_j are a random sample from a normal population with mean zero and variance σ_b^2, and use Rao's maximum likelihood method, which was derived in the previous chapter.

To obtain the combined estimates τ^* we return to the intrablock equations and substitute P_h, R and Λ for Q_h, r, and λ_1 where

$$P_h = w_1 Q_h + n^{-1} w_2 \Big(\sum_j n_{hj} B_j - G/n\Big)$$

$$= w_1 T_h + (w_2 - w_1) \sum_j n_{hj} B_j / n - w_2 G/n^2,$$

$$R = 2w_1 + 2w_2/(n - 1), \qquad \Lambda = (w_1 - w_2).$$

Then, solving the equations, A_{12}, etc., become

$$A_{12}^* = R(n - 1) = 2(n - 1)w_1 + 2w_2, \qquad A_{22}^* = -2(n - 2)(w_1 - w_2),$$

$$B_{12}^* = -(w_1 - w_2), \qquad B_{22}^* = (n + 2)w_1 + (n - 2)w_2,$$

$$\Delta_r^* = 2n^2 w_1 (w_1 + w_2).$$

It follows that

$$\tau_h^* = [\{n(w_1 + w_2) + 2(w_1 - w_2)\}P_h + (w_1 - w_2)S_1(P_h)]/\Delta_r^*.$$

The weights w_1 and w_2 are obtained, as usual, by substituting for σ_e^2 and σ_b^2 their estimates from the analysis of variance tables. We have $\sigma_b^2 = 2(E_b - E_e)/n$ if the design is considered as being run in separate replications, and $\sigma_b^2 = (E_b - E_e) \times (2n - 1)n^2$, if no replications term is taken out.

Kempthorne and Federer (1948) have shown that the variance of a difference between the estimates of two treatments is

$$V(\tau_h^* - \tau_i^*) = \frac{(n + 1)w_1 + (n - 1)w_2}{n w_1 (w_1 + w_2)} \sigma^2,$$

for first associates, and

$$V(\tau_h^* - \tau_i^*) = \frac{(n + 2)w_1 + (n - 2)w_2}{n w_1 (w_1 + w_2)} \sigma^2,$$

for second associates. The average variance is

$$\bar{v} = \frac{(n+3)w_1 + (n-1)w_2}{(n+1)w_1(w_1+w_2)}\sigma^2.$$

The overall efficiency of the lattice relative to randomized complete blocks is thus $\bar{v}r/(2\sigma^2) = w_1\bar{v}$, which may be written

$$1 + \frac{2(w_1 - w_2)}{(n+1)(w_1+w_2)},$$

whence it is argued that the lattice design is more efficient than the randomized complete block design with the same experimental material.

12.9 Partial Diallel Cross Experiments

In the most common diallel crossing system with v lines, all $v(v-1)/2$ crosses between the lines are taken, but neither the parental lines nor the reciprocal crosses are included. This is the simple diallel cross, or method 4 of Griffing (1956), and it was discussed in Chapter 4. Because the number of crosses required increases rapidly with v, fractions, or partial diallel crosses, have been discussed by Kempthorne and Curnow (1961), Curnow (1963), and Fyfe and Gilbert (1963). In this section we consider the use of partially balanced incomplete block designs with two associate classes in the partial diallel cross. The analysis is similar to the interblock analysis.

We consider v inbred lines and associate them with the v treatments in the partially balanced association scheme. Each line is crossed with each of its second associates, but not with its first associates.

Let y_{ij} denote the response of the $i \times j$ cross. As before, our model is

$$y_{ij} = m + g_i + g_j + s_{ij} + e_{ij},$$

where g_i and g_j are the general combining abilities of the ith and jth lines, and s_{ij} is the specific combining ability; we assume that g_i, s_{ij}, and e_{ij} are independently (normally) distributed with zero means and variances σ_g^2, σ_s^2, and σ_e^2. The objectives of the experiment are to obtain estimates of the parameters g_i and the components of variance σ_g^2, σ_s^2, and σ_e^2.

We estimate the g_i by minimizing the sum of squares

$$S = \sum (y_{ij} - \hat{m} - \hat{g}_i - \hat{g}_j)^2,$$

with the side condition $\sum g_i = 0$. Then $\hat{m} = G/N = 2G/rv$ where G is the grand total of all the observations, $N = rv/2$ is the number of crosses and $r = n_2$, and

$$T_i = r\hat{m} + r\hat{g}_i + S_2(\hat{g}_i),$$

where T_i is the total of the responses on the crosses containing the ith line, and $S_2(\hat{g}_i)$ is the sum $\sum \hat{g}_j$ over all lines crossed with the ith line.

Writing $Q_i = T_i - 2G/v$, we immediately have

$$Q_i = r\hat{g}_i + S_2(\hat{g}_i).$$

Summing over all the first associates of i, and setting $n_2 = r$,

$$S_1(Q_i) = (r + p_{12}^1)S_1(\hat{g}_i) + p_{12}^2 S_2(\hat{g}_i).$$

Following the approach of Bose and Shimamoto in the intrablock analysis, we consider the quantity $pQ_i + n_2 S_1(Q_i)$, and choose p so that the coefficients of $S_1(\hat{g}_i)$ and $S_2(\hat{g}_i)$ are equal. We have, after some simplification,

$$p = r^2 + r(p_{12}^1 - p_{12}^2) = r(2r - v + 1) + (v - 1)p_{12}^1.$$

Then, since $\sum \hat{g}_i = 0, pQ_i + rS_1(Q_i) = s\hat{g}_i$ where

$$s = pn_2 - n_1 n_2 - p + n_2 p_{11}^2 = r^2(2r - v) + r(v - 2)p_{12}^1.$$

It follows that $V(\hat{g}_h - \hat{g}_i) = 2(p - r)\sigma^2/s$ if h and i are first associates, and $V(\hat{g}_h - \hat{g}_i) = 2p\sigma^2/s$ if h and i are second associates. It is interesting to note that comparisons between lines that are not crossed have smaller variances than comparisons between lines that are crossed.

If we assume, as we have tacitly done, only one observation on each cross, the variance σ^2 is the sum of σ_e^2 and σ_s^2 and $E(S_g) = \sigma^2 + c\sigma_g^2$. To obtain the coefficient, c, of σ_g^2 in this expression, we ignore the other terms and note that

$$Q_i = rg_i + S_2(g_i) - 2r\bar{g},$$

where $\bar{g}$ is the average of the general combining abilities of the lines used in the experiment. It follows that g_i is replaced by $g_i - \bar{g}$, and

$$E(T_i\hat{g}_i) = E[\{rg_i + S_2(g_i)\}\{g_i - \bar{g}\}] = \{r(v - 1) - r\}\sigma_g^2/v.$$

Then, restoring the other terms, we have

$$E(S_g) = E\left(\sum T_i g_i\right) = (v - 1)\sigma_e^2 + r(v - 2)\sigma_g^2.$$

If n observations are made on each cross, the following analysis of variance table is obtained. The sums of squares, which are readily found, are not included here.

TABLE 12.1
ANALYSIS OF VARIANCE

Source	d.f.	EMS
Total	$(nrv - 2)/2$	
General combining ability	$v - 1$	$\sigma_e^2 + n\sigma_s^2 + nr(v - 2)/(v - 1)\sigma_g^2$
Specific combining ability	$v(r - 2)/2$	$\sigma_e^2 + n\sigma_s^2$
Error	$(n - 1)rv/2$	σ_e^2

With a group divisible design, $p_{12}^1 = 0$ and the formulae take a more convenient form with

$$p = r(2r - v + 1), \qquad s = r^2(2r - v).$$

If, however, there are only two groups, $r = v/2$ so that $s = 0$, and the design is singular. On the other hand, if, instead of crossing each line with its second associates, we cross it with its first associates, we have

$$Q_i = (n - 1)\hat{g}_i + S_1(\hat{g}_i), \qquad S_1(Q_i) = (n - 1)\hat{g}_i + (2n - 3)S_1(\hat{g}_i),$$

so that

$$2(n - 1)(n - 2)\hat{g}_i = (2n - 3)Q_i - S_1(Q_i) = 2(n - 1)Q_i - [Q_i + S_1(Q_i)].$$

Thus the latter design is nonsingular except when $n = 2$. The design consists of breaking the set of lines into several subsets and making all possible crosses within each subset.

EXAMPLE. The following data were obtained in a drosophila selection experiment. There are ten lines and a triangular association scheme. There were fifteen crossings and the response observed was the number of fourth segment abdominal bristles. The data given are the sums of eight observations on each cross.

(1 × 8) 175, (1 × 9) 193, (1 × 10) 169, (2 × 6) 181, (2 × 7) 169,
(2 × 10) 165, (3 × 5) 190, (3 × 7) 180, (3 × 9) 185, (4 × 5) 179,
(4 × 6) 188, (4 × 8) 171, (5 × 10) 178, (6 × 9) 189, (7 × 8) 165.

For this design we have $p_{12}^1 = 2$, $r = 3$, and $v = 10$ so that $p = 9$ and $s = 12$. We work with the totals of eight observations, and modify the equations just derived to read

$$8s\hat{g}_i = (p - r)Q_i - rS_2(Q_i), \qquad \text{or} \qquad 32\hat{g}_i = 2Q_i - S_2(Q_i),$$

where $Q_i = T_i - G/5 = T_i - 535.4$. The estimates of the general combining abilities follow:

Line	1	2	3	4	5
T_i	537	515	555	538	547
Q_i	1.6	−20.4	19.6	2.6	11.6
$32\hat{g}_i$	19.4	−18.6	17.4	−4.6	24.4
$\hat{g}_i$	0.606	−0.581	0.544	−0.144	0.762

Line	6	7	8	9	10
T_i	558	514	511	567	512
Q_i	22.6	−21.4	−24.4	31.6	−23.4
$32\hat{g}_i$	31.4	−17.6	−31.6	19.4	−39.6
$\hat{g}_i$	0.981	−0.550	−0.987	0.606	−1.237

In the analysis of variance the correction for the mean was $G^2/120 = 59719.408$. The uncorrected sum of squares for crosses was 59870.375, and the uncorrected sum of squares for individuals $\sum y^2$ (not shown in the data summary) was 60,171. The point estimates of the components of variance were

$$\hat{\sigma}_e^2 = 2.863, \qquad \hat{\sigma}_s^2 = -0.036, \qquad \hat{\sigma}_g^2 = 0.597.$$

The analysis of variance table follows:

TABLE 12.2
ANALYSIS OF VARIANCE

Source	d.f.	SS	ms	EMS
Total	119	451.592		
(Between crosses)	14	150.967		
General combining ability	9	138.112	15.346	$\sigma_e^2 + 8\sigma_s^2 + 21.33\sigma_g^2$
Specific combining ability	5	12.855	2.571	$\sigma_e^2 + 8\sigma_s^2$
Individuals within crosses	105	300.625	2.863	σ_e^2

12.10 Choosing an Incomplete Block Design

The reader has seen in the last two chapters various association schemes for incomplete block designs. In the next three chapters he will see numerous designs and recipes for many more. An extensive listing of *PBIB*(2) designs is given in the tables by Bose, Clatworthy and Shrikhande (1954). A listing of balanced designs is found in the Fisher and Yates tables (1953). He might well ask how in a given experimental situation he should choose a design from this long list. It is an easier question to ask than to answer because there is no final answer.

Let us suppose that we have fixed upon the number of varieties v and decided that a proper, equireplicate design should be used, and that the intrablock analysis will be used. Kempthorne's conjecture tells us that for any given set of parameters v, k, N the most efficient incomplete block design is the balanced design, if there is one. For that design we have an efficiency factor

$$E_0 = \frac{\lambda v}{rk} = \frac{v(k-1)}{k(v-1)} = \frac{v}{v-1}\left(1 - \frac{1}{k}\right).$$

The efficiency depends only upon v and k. For a given v it is an increasing function of k.

We should, therefore, try to take k to be as large as we reasonably can. There are practical situations in which, although the block size could if necessary be increased so that $k = v$, the cost of doing so would be to introduce so much heterogeneity into the plots that the variance, σ^2, would be greatly increased.

The efficiency factor E is calculated by comparing the average of $V(\hat{\tau}_h - \hat{\tau}_i)$

to the average variance of $2\sigma^2/r$ that would be obtained with a complete block design. The variance σ^2 does not appear in E because it is assumed that σ^2 would be the same in both cases.

Once we have chosen k, E_0 becomes an upper bound upon the efficiency that we can obtain by picking and choosing designs. There are balanced designs available for any choice of v, k, for example, the unreduced design in which the blocks are the collection of all subsets of k out of the v varieties. Unfortunately, especially when $(v - 1)$ is prime, there may not be a balanced design with a small enough value of r to fall within the budget. (It will be recalled that if $v - 1$ is a prime and $k < v$, then, in order for λ to be an integer, $(r - 1)$ has to be a multiple of $v - 1$, and if, further, k and v have no common factor, r has also to be a multiple of k.) On the other hand, if $v - 1$ is prime, v itself is not, and so there is at least one *GD* scheme for the v varieties.

In choosing between partially balanced designs one might take into account the several efficiencies E_i where E_i denotes the efficiency of a comparison between two ith associates, as well as the overall efficiency E. In practice, except for *GD* designs, this is not often very useful. With a *GD* scheme the experimenter can to a certain extent control which varieties go into each group, and to that extent he can divide his varieties into groups of mutual first associates by some rational procedure. In the other schemes the associate classes are so interlocked that little such control exists. All that the experimenter can actually do is to find a design in which the E_i are more or less equal and which falls within his budgetary restrictions. Having the E_i nearly equal is roughly equivalent to having the λ_i nearly equal.

Within those guidelines the choice of a design is not really critical. J. A. John (1966) has published a listing of cyclic designs, i.e., designs obtained by cycling from initial blocks. He lists the designs for various values of k and v (he uses t for the number of treatments). It is interesting to note that for most of the designs for a given v and k, E has about the same value. Two examples serve to illustrate this.

In a cyclic design we start with an initial block and obtain a set of blocks from it by adding 1, 2, 3, . . . , in turn to each element of the initial block and reducing mod v. Some designs use several initial blocks. For $v = 6$ and $k = 4$, John lists four designs. Cycling on the initial block (0, 1, 2, 3) gives the design

$$0\,1\,2\,3,\quad 1\,2\,3\,4,\quad 2\,3\,4\,5,\quad 3\,4\,5\,0,\quad 4\,5\,0\,1,\quad 5\,0\,1\,2.$$

This is a *PBIB*(3) design. The first associates of i are $i + 1, i + 5$; the second associates are $i + 2$ and $i + 4$; the third associate is $i + 3$. We have $\lambda_1 = 3$, $\lambda_2 = 2$, $\lambda_3 = 2$, and $E = 0.8948$. Cycling on the initial block (0, 1, 2, 4) gives the *GD* design

$$0\,1\,2\,4,\quad 1\,2\,3\,5,\quad 2\,3\,4\,0,\quad 3\,4\,5\,1,\quad 4\,5\,0\,2,\quad 5\,0\,1\,3,$$

with $E = 0.8942$. The other two designs have $r = 6$. They are each formed by cycling on two initial blocks and discarding duplicate blocks. The first of these designs is formed from (0, 1, 2, 3) and (0, 1, 3, 4). It has three associate classes and $E = 0.8977$. The last design is formed from (0, 1, 2, 3) and (0, 1, 3, 4) and has $E = 0.8978$. The upper bound is $E_0 = 0.9000$.

John lists nineteen designs for $v = 12$ and $k = 4$. They are all partially balanced, although two of them have no fewer than six associate classes. Their efficiencies range from 0.8007 to 0.8166. The value of E_0 is 0.8182. Two of the designs are *GD* designs. The first has $r = 4$, $E = 0.8138$, and initial block (0, 1, 3, 7). The second has $r = 10$, $E = 0.8166$, and three initial blocks (0, 1, 2, 5), (0, 1, 3, 8), and (0, 2, 6, 8). There is no advantage in using the more complicated association schemes.

In summary, three guidelines are suggested. Take k as large as you reasonably can. Try to find a design with as much balance as possible. The third guideline is to choose, subject to the other considerations, the simplest design possible. This is sound advice in the design of any experiment. Keep it as simple as you can.

Exercises

1. Prove that a sufficient condition for a *PBIB*(2) scheme to be group divisible is that either $p_{12}^1 = 0$ or $p_{12}^2 = 0$.
2. Show that $\Delta_r = k^2\Delta_b$.
3. Derive the constant d_1 in the Bose–Shimamoto form of the combined intrablock and interblock estimates.
4. Prove that for a *GD* design the methods of Yates and Rao give the same combined estimates for comparing first associates.
5. Show that for the simple lattice the methods of Yates and Rao do not give the same combined estimates.
6. Derive the formula given for $V(\tau_h^* - \tau_i^*)$ in the simple lattice.

CHAPTER 13

The Existence and Construction of Balanced Incomplete Block Designs

Balanced incomplete block designs were introduced and discussed in Chapter 11 and partially balanced incomplete blocks in Chapter 12. This chapter and the next two are devoted to the construction of designs. The problems that we shall consider are essentially problems in combinatorial mathematics. Many of the contributions to the construction of balanced incomplete block designs have been made by algebraists who had no other interest in statistics. Some of their results antedate the discovery of balanced incomplete block designs by many years.

This chapter and the next two are quite different from the preceding chapters. They presuppose more knowledge of abstract algebra. The reader who is not interested in this topic can safely omit this and the next two chapters; if he wants a balanced incomplete design for a given practical situation he can get it easily enough, if it exists, from the Fisher and Yates tables. On the other hand, the research student in mathematical statistics will find in these chapters that there remain numerous questions to be answered. The algebraist who knows no statistics might find them interesting reading, and he will need only to refer back to Chapters 11 and 12; he does not need to know anything about the analysis of variance, or, indeed, why the statistician cares about these designs at all. In the interest of completeness, we shall begin by repeating some basic material from Chapter 11.

Balanced incomplete designs were introduced by Yates (1936) for use in agricultural experiments, and the agricultural terminology has remained in the jargon. A balanced incomplete block design, or *BIBD*, is a collection of b blocks, each consisting of k plots; v treatments, or varieties ($v > k$), are assigned to the plots (one treatment to each plot) in such a way that

(i) Each treatment appears exactly r times;
(ii) No treatment appears more than once in any block (a design with this property is called a binary design);
(iii) Each pair of treatments appears together in exactly λ blocks.

The five parameters b, v, r, k, λ are not independent. They satisfy the conditions:

(a) $bk = rv = N$, which is the total number of plots;
(b) $\lambda = r(k - 1)/(v - 1)$, which is an integer;
(c) $b \geq v$.

The inequality was proved by Fisher (1940), and is called Fisher's inequality. It was derived in Chapter 11, and the derivation is repeated here.

The 1938 edition of the Fisher and Yates tables contained a list of all sets of parameters satisfying conditions (a) and (b) with $r \leq 10$ for which designs were known to exist, in which case the design was given, and those for which it was not yet known whether or not designs existed. Bose (1939) solved all the unknown cases in the listing except for twelve. Later workers have reduced the number of unsolved cases in his list to two. The two open cases are $b = 69$, $v = 46$, $r = 9$, $k = 6$, $\lambda = 1$, and $b = 85$, $v = 51$, $r = 10$, $k = 6$, $\lambda = 1$; it is still not known whether designs exist in these cases. Rao (1961) published a list of all sets of parameters with $11 \leq r \leq 15$, satisfying conditions (a) and (b), together with known solutions, and Sprott (1962) did the same for $16 \leq r \leq 20$. Bose's 1939 paper is the major work in the field, and we shall draw heavily upon it in this chapter.

Associated with any incomplete block design is its incidence matrix $\mathbf{N} = (n_{ij})$. $\mathbf{N}$ is a matrix of v rows and b columns, and n_{ij} is the number of times that the ith variety appears in the jth block. For a *BIBD*, the elements of $\mathbf{N}$ are either zero or one, and $\mathbf{NN}' = (r - \lambda)\mathbf{I}_v + \lambda\mathbf{J}_v$ where $\mathbf{J}_v$ is the square matrix of order v, each element of which is unity.

The determinant $|\mathbf{NN}'| = (r - \lambda)^{v-1}(r + (v - 1)\lambda) = (r - \lambda)^{v-1}rk$, which is positive since $\lambda < r$.

Thus the rank of $\mathbf{NN}'$ is $r(\mathbf{NN}') = r(\mathbf{N}) = v$; but $r(\mathbf{N}) \leq \min(b, v)$, and so $b \geq v$.

If $b = v$ and $r = k$ the design is said to be symmetric. For such a design $\mathbf{N}$ is a square matrix and $|\mathbf{NN}'| = |\mathbf{N}|^2$. This implies that $(r - \lambda)^{v-1}$ is a square. Hence, if v is even, a necessary condition for the existence of a symmetric design is that $(r - \lambda)$ is a perfect square. This nonexistence theorem was derived independently by Schutzenberger (1949), Shrikhande (1950), and Chowla and Ryser (1950). It shows that conditions (a), (b), and (c) are not sufficient for the existence of a design. Consider, for example, the case $b = v = 22$ and $r = k = 7$; we have $\lambda = 2$, and so $r - \lambda = 5$, and no such design exists.

The simplest series of balanced incomplete block designs are the unreduced designs; these are the designs that consist of all possible combinations of k out of the v varieties. For these designs

$$b = \binom{v}{k}, \qquad r = \binom{v-1}{k-1}, \qquad \lambda = \binom{v-2}{k-2}.$$

It is not necessary to investigate cases for which $2k > v$. If a new design of b blocks and v varieties is formed by assigning to each block those varieties that do not appear in the corresponding block of the *BIBD*, the new design is also a *BIBD*; it is called the complementary design.

13.1 The Orthogonal Series

These two series of designs, introduced by Yates (1936), depend upon the existence of a series of $s + 1$ orthogonal Latin squares of side s; such sets of squares are known to exist whenever $s = p^m$ where p is a prime.

For the first series, $b = s(s + 1)$, $v = s^2$, $r = s + 1$, $k = s$, and $\lambda = 1$. The $s^2 + s$ blocks fall into $s + 1$ sets of s blocks each. In each set every treatment appears exactly once, and the set forms a complete replicate of all the varieties. A design that can be divided into complete replicates is called *resolvable*. Each set can be denoted by a square array of s^2 plots with the rows of the array corresponding to blocks. The varieties are denoted by the integers $1, 2, \ldots, s^2$.

In the first set the jth row, and hence, the jth block, consists of the varieties $(j - 1)s + 1, (j - 1)s + 2, \ldots, (j - 1)s + s = js$. The rows in the second set are the columns in the first set; the jth row consists of $j, j + s, j + 2s, \ldots, j + (s - 1)s$. The remaining sets make use of the orthogonal Latin squares. Consider one of the squares composed of letters $A, B, C, \ldots$. The corresponding set of blocks is obtained by superimposing the Latin square on the square array of the first set. Then the varieties in the first row of the new set are those that coincide with the letter A; the second row contains varieties coinciding with B, and so on.

We illustrate the procedure with an example for $s = 4$. A set of three orthogonal squares has already been obtained in the chapter on Latin squares, in which Bose's method of obtaining orthogonal sets was presented. The design, which was given in Chapter 11, is

1 2 3 4	1 5 9 13	1 6 11 16	1 7 12 14	1 8 10 15
5 6 7 8	2 6 10 14	2 5 12 15	2 8 11 13	2 7 9 16
9 10 11 12	3 7 11 15	3 8 9 14	3 5 10 16	3 6 12 13
13 14 15 16	4 8 12 16	4 7 10 13	4 6 9 15	4 5 11 14.

To show that the design is a *BIBD*, we note that any pair of varieties cannot appear together in more than one block. If they appear in the same block of the first set, they cannot appear in the same row of the second set, and vice versa; in neither case can they appear in the same block in one of the subsequent sets, for that would imply that the same letter appeared twice in the same row (or column) of a Latin square. Similarly, if a pair of varieties appears together in a block of the $(i + 2)$th set and also in a block of the $(h + 2)$th set, that contradicts the fact that the ith and hth squares are orthogonal. However, each variety appears $r = s + 1$ times, once in each set, and thus its plots appear in the same blocks as $r(k - 1) = s^2 - 1$ plots occupied by other varieties; if no variety is to appear more than once, these $s^2 - 1$ plots must be taken up by each of the other varieties appearing once each, and the design is a *BIBD* with $\lambda = 1$.

Conversely, the existence of a *BIBD* with parameters $b = s(s + 1)$, $v = s^2$, $r = s + 1$, $k = s$, and $\lambda = 1$ implies the existence of a set of $s - 1$ orthogonal squares of side s. This establishes the nonexistence of a design with $b = 42$, $v = 36$, $r = 7$, $k = 6$, and $\lambda = 1$ since there is not a complete set of Latin squares of side six.

To each member of the first series, *OS*1, there corresponds a member of the second series, *OS*2. These are designs for $v = s^2 + s + 1$ varieties. To each block of the ith set of the *OS*1 design an extra plot is added containing the $(s^2 + i)$th variety. One extra block is added, consisting of one plot with each of the new varieties. The resulting design is a *BIBD* with $b = v = s^2 + s + 1$, $r = k = s + 1$, and $\lambda = 1$. There is no such design for forty-three varieties. The designs of the first series are also called *balanced lattices*; any two of the replicate sets constitute a simple lattice.

13.2 Series of Designs Based on Finite Geometries

Let $s = p^n$ be a prime or a power of a prime. A point in the finite projective geometry $PG(N, s = p^n)$ is represented by a $(N + 1)$-tuple, each coordinate of which is an element of the Galois field $GF(s)$, and not all of which are zero. We write $x = (x_0, x_1, \ldots, x_N)$, $x_i \in GF(s)$. Two points, x and y, are identical if there exists an element θ of the field such that $x_i = \theta y_i$ for all i.

Altogether there are $s^{N+1} - 1$ such $(N + 1)$-tuples. They may be divided into groups of $(s - 1)$ members with each member of the same group representing the same point. Thus the geometry $PG(N, s)$ has $(s^{N+1} - 1)/(s - 1)$ distinct points.

Let x and y be two of these points. The line joining x and y is the set of all points $\lambda_1 x + \lambda_2 y$ where $\lambda_1, \lambda_2 \in GF(s)$. There are $(s + 1)$ distinct points on this line (or one-dimensional subspace), namely y, which corresponds to $\lambda_1 = 0$, $\lambda_2 = 1$, and the s points obtained by putting $\lambda_1 = 1$ and letting λ_2 take all s values in the field. The point x is joined by lines to each of the other points; each of these lines contains s points in addition to x, and none of the points lies on more than one of the lines. Hence, $[(s^{N+1} - 1)/(s - 1) - 1]/s = (s^N - 1)/(s - 1)$ lines pass through x. Altogether there are

$$\frac{s^{N+1} - 1}{s - 1} \cdot \frac{s^N - 1}{s - 1} \cdot \frac{1}{s + 1} = \frac{(s^{N+1} - 1)(s^N - 1)}{(s^2 - 1)(s - 1)}$$

lines, and each pair of points appears on one and only one line.

If we now let the points represent varieties and the lines represent blocks, the system of points and lines gives a *BIBD* with

$$v = \frac{s^{N+1} - 1}{s - 1}, \qquad b = \frac{(s^{N+1} - 1)(s^N - 1)}{(s^2 - 1)(s - 1)}, \qquad k = s + 1, \quad \lambda = 1.$$

If $N = 2$, we have the orthogonal series with $b = v = s^2 + s + 1$, $r = k = s + 1$.

The simplest example occurs when $s = N = 2$. Then s takes the values 0, 1, and the arithmetic is carried out mod 2. The seven points are A (100), B (010), C (001), D (011), E (101), F (110), G (111). The seven lines are $x_0 = 0$, BCD; $x_1 = 0$, ACE; $x_2 = 0$, ABF; $x_0 + x_1 \equiv 0$, CFG; $x_0 + x_2 \equiv 0$, BEG; $x_1 + x_2 \equiv 0$, ADG; $x_0 + x_1 + x_2 \equiv 0$, DEF. This design may also be obtained from the geometrical figure of the complete quadrangle, using as blocks the six sides and the line consisting of the three points in which the pairs of opposite sides meet.

We now extend these ideas from one-dimensional subspaces to m-dimensional subspaces. Let p_i, $i = 1, 2, \ldots, m + 1$ be a set of $(m + 1)$ points whose coordinate vectors are linearly independent; denote the vectors by p_i. The set of points whose coordinate vectors are linear combinations $\sum \lambda_i p_i$ forms an m-dimensional subspace of $PG(N, s)$. If $\lambda_i = \mu_i$ for all i, the two linear combinations $\sum \lambda_i y_i$ and $\sum \mu_i p_i$ denote the same point. Thus the sets of coefficients $\{\lambda_i\}$ themselves represent the points of $PG(m, s)$. Hence, the number of points in an m-dimensional subspace is the same as the number of points in $PG(m, s)$, namely $(s^{m+1} - 1)/(s - 1)$.

To obtain an ordered set of $(m - 1)$ linearly independent points from $PG(N, s)$, we may choose the first point in $(s^{N+1} - 1)/(s - 1)$ ways, and the second point in $\{(s^{N+1} - 1)/(s - 1)\} - 1$ ways; for the third point the choice is limited to those points that do not belong to the one-dimensional subspace generated by the first two points; it may be chosen in $\{(s^{N+1} - 1)/(s - 1)\} - \{(s^2 - 1)/(s - 1)\}$ ways. Similarly, the $(i + 1)$th point may be chosen in $\{(s^{N+1} - 1)/(s - 1)\} - \{(s^i - 1)/(s - 1)\} = \{s^i(s^{N-i+1} - 1)\}/(s - 1)$ ways. Hence there exist $\prod \{s^i(s^{N-i+1} - 1)/(s - 1)\}$ ordered sets of $(m - 1)$ independent points in $PG(N, s)$. By a similar argument there are $\prod \{s^i(s^{m-i+1} - 1)/(s - 1)\}$ ordered sets of $(m - 1)$ points in each $PG(m, s)$. Taking the quotient we have for the number of m-dimensional subspaces in $PG(N, s)$

$$\Phi(N, m, s) = \frac{(s^{N+1} - 1)(s^N - 1)\cdots(s^{N-m+1} - 1)}{(s^{m+1} - 1)(s^m - 1)\cdots(s - 1)}.$$

Each point belongs to

$$\frac{(s^{N+1} - 1)(s^N - 1)\cdots(s^{N-m+1} - 1)}{(s^{m+1} - 1)(s^m - 1)\cdots(s - 1)} \times \frac{(s^{m+1} - 1)}{(s^{N+1} - 1)} = \Phi(N - 1, m - 1, s)$$

subspaces and each pair of points belongs to $\Phi(N - 2, m - 2, s)$ subspaces.

We may now identify points with varieties and subspaces with blocks to obtain *BIB* designs with parameters

$$v = \binom{s^{N+1} - 1}{s - 1} = \Phi(N, 0, s); \qquad b = \Phi(N, m, s); \qquad r = \Phi(N - 1, m - 1, s);$$

$$k = \binom{s^{m+1} - 1}{s - 1} = \Phi(m, 0, 2); \qquad \lambda = \Phi(N - 2, m - 2, s),$$

where $\Phi(N, -1, s) = 1$. The subspaces, or m-flats, are obtained as the sets of points satisfying sets of $(N - m)$ linear restrictions of the form $\sum_i \lambda_i x_i = 0$.

EXAMPLES. $PG(3, 2)$. All arithmetic is mod 2; $N = 3, s = 2$. There are fifteen points.

1. 0001	2. 0010	3. 0011	4. 0100	5. 0101	6. 0110
7. 0111	8. 1000	9. 1001	10. 1010	11. 1011	12. 1100
13. 1101	14. 1110	15. 1111.			

For $m = 2$, each flat is defined by a single equation and the design has $b = v = 15$, $r = k = 7$, and $\lambda = 3$. The blocks are

$$\begin{array}{rl}
x_0 \equiv 0 & 1\ 2\ 3\ \ 4\ \ 5\ \ 6\ \ 7, \\
x_1 \equiv 0 & 1\ 2\ 3\ \ 8\ \ 9\ 10\ 11, \\
x_2 \equiv 0 & 1\ 4\ 5\ \ 8\ \ 9\ 12\ 13, \\
x_3 \equiv 0 & 2\ 4\ 6\ \ 8\ 10\ 12\ 14, \\
x_0 + x_1 \equiv 0 & 1\ 2\ 3\ 12\ 13\ 14\ 15, \\
x_0 + x_2 \equiv 0 & 1\ 4\ 5\ 10\ 11\ 14\ 15, \\
x_0 + x_3 \equiv 0 & 2\ 4\ 6\ \ 9\ 11\ 13\ 15, \\
x_1 + x_2 \equiv 0 & 1\ 6\ 7\ \ 8\ \ 9\ 14\ 15, \\
x_1 + x_3 \equiv 0 & 2\ 5\ 7\ \ 8\ 10\ 13\ 15, \\
x_2 + x_3 \equiv 0 & 3\ 4\ 7\ \ 8\ 11\ 12\ 15, \\
x_0 + x_1 + x_2 \equiv 0 & 1\ 6\ 7\ 10\ 11\ 12\ 13, \\
x_0 + x_1 + x_3 \equiv 0 & 2\ 5\ 7\ \ 9\ 11\ 12\ 14, \\
x_0 + x_2 + x_3 \equiv 0 & 3\ 4\ 7\ \ 9\ 10\ 13\ 14, \\
x_1 + x_2 + x_3 \equiv 0 & 3\ 5\ 6\ \ 8\ 11\ 13\ 14, \\
x_0 + x_1 + x_2 + x_3 \equiv 0 & 3\ 5\ 6\ \ 9\ 10\ 12\ 15.
\end{array}$$

If we take the intersections of pairs of two-flats, we obtain the design for $m = 1$ with $b = 35$ and $k = 3$. Equating any two linear combinations to zero also equates their sum to zero, and so the intersections are actually intersections of three two-flats. For example, $x_2 + x_3 \equiv 0$, $x_0 \equiv 0$, and $x_0 + x_2 + x_3 \equiv 0$ give the block 3, 4, 7; $x_0 \equiv 0$, $x_1 \equiv 0$, and $x_0 + x_1 \equiv 0$ give 1, 2, 3; $x_0 + x_3 \equiv 0$, $x_0 + x_1 + x_2 \equiv 0$, and $x_1 + x_2 + x_3 \equiv 0$ give the block 6, 11, and 13.

The $PG(2, s)$ designs with $m = 1$ are the $OS2$ series with $v = s^2 + s + 1$ and $\lambda = 1$. Other PG designs of interest are

$PG(4, 2)$	$v = 31$	$m = 1$	$k = 3$	$b = 155$	$r = 15$	$\lambda = 1$,
		$m = 2$	$k = 7$	$b = 155$	$r = 35$	$\lambda = 7$,
		$m = 3$	$k = 15$	$b = 31$	$r = 15$	$\lambda = 7$,
$PG(3, 3)$	$v = 40$	$m = 1$	$k = 4$	$b = 130$	$r = 13$	$\lambda = 1$,
		$m = 2$	$k = 13$	$b = 40$	$r = 13$	$\lambda = 4$,
$PG(3, 4)$	$v = 85$	$m = 1$	$k = 5$	$b = 357$	$r = 21$	$\lambda = 1$,
		$m = 2$	$k = 21$	$b = 85$	$r = 21$	$\lambda = 5$.

13.3 Designs Using Euclidean Geometries

If we remove from $PG(N, s)$ all the points in the $(N - 1)$ dimensional subspace $x_0 = 0$, we obtain the Euclidean geometry $EG(N, s)$. Dropping the now spurious coordinate $x_0 = 1$, the points are N-tuples, and two N-tuples x and y represent the same point if, and only if, $x_i = y_i$ for all i.

These Euclidean geometries provide BIB designs with parameters

$$v = \Phi(N, 0, s) - \Phi(N - 1, 0, s) = s^N,$$

$$k = \Phi(m, 0, s) - \Phi(m - 1, 0, \mathrm{s}) = s^m,$$

$$b = \Phi(N, m, s) - \Phi(N - 1, m, s) = s^{N-m}\Phi(N - 1, m - 1, s),$$

$$r = \Phi(N - 1, m - 1, s), \qquad \lambda = \Phi(N - 2, m - 2, s).$$

When $m = 1$ and $N = 2$, we have the orthogonal series $OS1$ with $v = s^2$, $k = s$, and $\lambda = 1$. Other EG designs of interest are

$EG(3, 2)$	$v = 8$	$m = 1$	$k = 2$	$b = 28$	$r = 7$	$\lambda = 1$,
		$m = 2$	$k = 4$	$b = 14$	$r = 7$	$\lambda = 3$,
$EG(4, 2)$	$v = 16$	$m = 1$	$k = 2$	$b = 120$	$r = 15$	$\lambda = 1$,
		$m = 2$	$k = 4$	$b = 140$	$r = 35$	$\lambda = 7$,
		$m = 3$	$k = 8$	$b = 30$	$r = 15$	$\lambda = 7$,
$EG(5, 2)$	$v = 32$	$m = 4$	$k = 16$	$b = 62$	$r = 31$	$\lambda = 15$,
$EG(3, 3)$	$v = 27$	$m = 1$	$k = 3$	$b = 117$	$r = 13$	$\lambda = 1$,
		$m = 2$	$k = 9$	$b = 39$	$r = 13$	$\lambda = 4$,
$EG(3, 4)$	$v = 64$	$m = 1$	$k = 4$	$b = 336$	$r = 21$	$\lambda = 1$,
		$m = 2$	$k = 16$	$b = 84$	$r = 21$	$\lambda = 5$.

13.4 Designs Derived from Difference Sets

Let G: $(x_0 = 0, x_1, \ldots, x_{v-1})$ be an Abelian group of v elements with addition as the group operation. We can associate each treatment with a different element of the group. If v is a prime, the set of residue classes mod v is a convenient group for our purposes; we associate the treatments with the integers $0, 1, \ldots, v - 1$, and reduce all sums mod v. Let $D = (d_1, d_2, \ldots, d_k)$ be a proper subset of G such that every element of G, except x_0, can be expressed in exactly λ ways as a difference $d_i - d_j$. Then D is called a *perfect difference set*, or, more conveniently, a *difference set*. If $D_i = (d_1 + x_i, d_2 + x_i, \ldots, d_k + x_i)$, $x_i \in G$, then D_i is also a difference set. It will be proved in a more general context that the collection of all v difference sets D_i forms a symmetric $BIBD$ with $b = v$ and $r = k$. These designs are called *cyclic designs*; their incidence matrices are *circulants*. The set D is called the *initial block*.

For $v = 7$ we may take the group of residue classes mod 7. Consider the set $D = (0, 1, 3)$; $d_1 - d_2 \equiv 6$, $d_2 - d_1 \equiv 1$, $d_1 - d_3 \equiv 4$, $d_3 - d_1 \equiv 3$, $d_2 - d_3 \equiv 5$, $d_3 - d_2 \equiv 2$, and each difference other than zero appears exactly once. We may, therefore, take D as an initial block. Adding 1 to each element of D gives $D_1 = (1, 2, 4)$; adding 2 gives (2, 3, 5), and so on. This gives a design for seven treatments in seven blocks of three. The blocks are

0, 1, 3;   1, 2, 4;   2, 3, 5;   3, 4, 6;   4, 5, 0;   5, 6, 1;   6, 0, 2.

The complementary design with $k = 4$ and $\lambda = 2$ is generated in a similar way from the initial block (0, 2, 3, 4).

A difference set for $v = 15$ and $k = 7$ can be obtained by representing each treatment by a pair of integers (x, y), $x = 0, 1, 2, 3, 4$, $y = 0, 1, 2$. Addition is defined by $(x_1, y_1) + (x_2, y_2) = (x_1 + x_2, y_1 + y_2)$ with $x_1 + x_2$ reduced mod 5, and $y_1 + y_2$ reduced mod 3. The design is

00, 10, 40, 01, 21, 31, 02;   01, 11, 41, 02, 22, 32, 00;   02, 12, 42, 00, 20, 30, 01;
10, 20, 00, 11, 31, 41, 12;   11, 21, 01, 12, 32, 42, 10;   12, 22, 02, 10, 30, 40, 11;
20, 30, 10, 21, 41, 01, 22;   21, 31, 11, 22, 42, 02, 20;   22, 32, 12, 20, 40, 00, 21;
30, 40, 20, 31, 01, 11, 32;   31, 41, 21, 32, 02, 12, 30;   32, 42, 22, 30, 00, 10, 31;
40, 00, 30, 41, 11, 21, 42;   41, 01, 31, 42, 12, 22, 40;   42, 02, 32, 40, 10, 20, 41.

If v is a power of a prime, we may use the Galois field $GF(v)$ in a manner similar to the double modulus system just mentioned. For example, if $v = 16$ and $k = 6$ the treatments are represented by elements of the form $a_0 + a_1x + a_2x^2 + a_3x^3$ where the coefficients a_i are either zero or one. Arithmetic is carried out mod 2. The representations of the treatments can be abbreviated to the 4-tuples $a_1a_2a_3a_4$. The initial block is 0000, 0001, 1000, 1111, 0101, 0011.

We shall see later that the use of difference sets may be extended to include designs involving several initial blocks. If there are t initial blocks, the design will have $b = tv$, and $r = tk$. We shall discuss Bose's two methods of using difference sets and his two fundamental theorems. After that we shall present some of the series of difference sets that have been discovered. Before proceeding to Bose's theorems, we shall mention two theorems about the symmetric designs.

13.5 Two Nonexistence Theorems

The following two theorems are important in establishing the nonexistence of symmetric balanced incomplete block designs. We state them without proof because the proofs depend upon results in the theory of numbers that fall outside the scope of this book. The first gives a necessary condition for the existence of a design with a particular set of parameters v, k, and λ which satisfy the equality $\lambda(v - 1) = k(k - 1)$. The second theorem gives a condition for the existence of a difference set, and hence a cyclic design, with a given set of parameters, when the treatments are represented residue classes mod v.

The first theorem uses the Legendre symbol $(a|b)$, which is defined as follows: $(a|b) = 1$ if a is a quadratic residue of b, i.e., if there is an integer x such that $x^2 \equiv a \pmod{b}$; if a is not a quadratic residue of b, $(a|b) = -1$.

Theorem 1. *Chowla and Ryser (1950). Let v, k, and λ be integers for which a symmetric BIBD exists. The following conditions hold:*

(i) *If v is even, $k - \lambda$ is a square;*
(ii) *If $v \equiv 1 \pmod 4$, and if there exists an odd prime p such that p divides the square-free part of $k - \lambda$, then $(\lambda|p) = 1$;*
(iii) *If $v \equiv 3 \pmod 4$, and if there exists an odd prime p, such that p divides the square-free part of $k - \lambda$, then $(-\lambda|p) = 1$.*

We have already discussed condition (i). Conditions (ii) and (iii) may be combined into the single statement that a necessary condition for the design to exist when v is odd is that the Diophantine equation

$$x^2 = (k - \lambda)y^2 + (-1)^{(v-1)/2}\lambda z^2$$

shall have a solution in integers other than the trivial solution $x = y = z = 0$.

The following examples illustrate the use of the theorem. The square-free part of $k - \lambda$ is the part remaining when all square factors have been removed. Thus the square free part of 12 is $12/2^2 = 3$.

If $v = 29$, $k = 8$, and $\lambda = 2$, we have $v \equiv 1 \pmod 4$, $k - \lambda = 6$. Taking $p = 3$, we note that $1^2 \equiv 1$, $2^2 \equiv 1 \pmod 3$. Hence, $\lambda = 2$ is not a square, so that $(\lambda|p) = -1$, and the design does not exist.

If $v = 67$, $k = 12$, and $\lambda = 2$, we have $v \equiv 3 \pmod 4$, $k - \lambda = 10$, and $p = 5$. Then $(-\lambda|p) = (3|5) = -1$, and the design does not exist.

Theorem 2. *Hall and Ryser (1951). Let v be odd and let e be any positive divisor of v. A necessary condition for the existence of a symmetric design with parameters v, k, and λ, which is cyclic, is that the Diophantine equation*

$$x^2 = (k - \lambda)y^2 + (-1)^{(e-1)/2}ez^2$$

must possess a solution in integers that are not all zero.

Smith and Hartley (1948) showed that if, for any design with $v = b$ and $r = k$ (whether balanced or not), we write the design as a rectangular array of v rows and k columns with blocks as rows, it is possible to arrange the treatments within the rows in such a way that each treatment appears exactly once in each column. Such an arrangement is called a Youden square; in the analysis, a sum of squares can be taken for differences between columns if desired.

It does not follow, however, that every symmetric design is a cyclic design, or is isomorphic to one in the sense that it could be made cyclic by a suitable relabeling of the treatments. Bhattacharya (1944*b*) has obtained a symmetric *BIBD* for $v = 25$, $k = 9$, and $\lambda = 3$.

Bhattacharya's design is the following:

ABEFKLQTW *ABIJOPQUY* *ABMNGHQVX*
CDEFOPQVX *ADEHJKSVY* *ADILNOSTX*
BCMPJKSTX *BCEHNOSUW* *ACEGJLRUX*
BDIKFHRUX *DBMOJLRVW* *BDEGNPRTY*
EIMHLPWXY *GKOHLPQRS* *FJNHLPTUV*
CDIJGHQTW *CDMNKLQUY*
ADMPFGSUW *BCILFGSVY*
ACIKNPRVW *ACMOFHRTY*
EIMFJNQRS *EIMGKOTUV*
FJNGKOWXY *QRSTUVWXY*.

However, the equation

$$x^2 = 6y^2 + 5z^2$$

has no solution in integers, other than $x = y = z = 0$, and so there can be no cyclic design with these parameters.

13.6 The Fundamental Theorems of Bose

We now extend this concept of generating a design from a difference set to Bose's methods of symmetrically repeated differences. We consider, as before, an Abelian group G with m elements ($x_0 = 0, x_1, \ldots, x_{m-1}$); the group operation is addition.

Let $v = mn$ treatments be divided into m subsets of n treatments each. We associate one subset with each element of the group G, and denote by v_{ij} the jth of the treatments in the subset, which has been assigned to the element x_i; treatments with the same second subscript, j, are said to belong to the same class. The notation used by Bose is to denote each treatment by the numeral of its subset, i, with the numeral, j, of its class as a subscript; thus our v_{12} would be denoted by 1_2.

Suppose that a block contains k treatments, of which m_j belong to the jth class. There are two kinds of differences to be considered: pure differences $[j, j]$ and mixed differences $[j, j']$, $j \neq j'$. The pure differences $[j, j]$ are the $m_j(m_j - 1)$ differences of the type $(x_i - x_{i'})$ where x_i and $x_{i'}$ correspond to treatments v_{ij} and $v_{i'j}$ in the block; the mixed differences $[j, j']$ are the $m_j m_{j'}$ differences of the type $(x_i - x_{i'})$ where x_i, $x_{i'}$ correspond to treatments v_{ij} and $v_{i'j'}$ in the block (we may have $i = i'$). If $n = 1$ there are no mixed differences, and we have the situation considered in the previous sections.

For example, if G consists of the residue classes, mod 5, and $n = 2$, there are ten treatments; consider the block 0_1, 1_1, 3_1, 3_2, and 4_2. The differences [1, 1] are $0 - 1 = 4$, $0 - 3 = 2$, $1 - 3 = 3$, $1 - 0 = 1$, $3 - 0 = 3$, $3 - 1 = 2$. The differences [2, 2] are $3 - 4 = 4$ and $4 - 3 = 1$. The differences [1, 2] are $0 - 3 = 2$, $1 - 3 = 3$, $3 - 3 = 0$, $0 - 4 = 1$, $1 - 4 = 2$, $3 - 4 = 4$. The differences [2, 1] are $3 - 0 = 3$, $3 - 1 = 2$, $3 - 3 = 0$, $4 - 0 = 4$, $4 - 1 = 3$, $4 - 3 = 1$.

Consider now a set of t blocks satisfying the following conditions, where m_{js} denotes the value of m_j in the sth block:

(i) For each j the set of $\sum_s m_{js}(m_{js} - 1)$ pure differences $[j, j]$ contains each nonzero element of G exactly λ times;

(ii) For each pair $j, j' (j \neq j')$ the set of $\sum_s m_{js} m_{j's}$ mixed differences $[j, j']$ contains each element of G, including zero, exactly λ times.

We say that in these t blocks the differences are symmetrically repeated.

For example, if G is the set of residue classes mod 5, and $n = 2$, the following set of $t = 6$ blocks has the property with $\lambda = 2$: $0_2, 1_2, 2_2$; $1_1, 4_1, 0_2$; $2_1, 3_1, 0_2$; $1_1, 4_1, 2_2$; $2_1, 3_1, 2_2$; $0_1, 0_2, 2_2$.

Theorem 1. *The first fundamental theorem of differences:*
Let G be an Abelian group of m elements, and let $v = mn$ treatments be divided into m subsets of n treatments each, denoted by v_{ij}, as described previously. Suppose that we can find a set of t blocks $B_1, B_2, \ldots, B_t$ which satisfy the following conditions:

(i) *Each block has k plots and contains k different treatments;*

(ii) *Among the kt varieties occurring in the set of t blocks, exactly r belong to each class. (Some treatments may appear in more than one block, and each appearance is counted; thus $kt = nr$);*

(iii) *The differences are symmetrically repeated, λ times each, in the t blocks.*

Suppose further that these t blocks are used as initial blocks in the following manner. From each initial block B_s a set of m blocks is generated. If v_{hj} is a treatment in B_s, and $x_g = x_h + x_i$, then v_{gj} appears in B_{si}; the block B_{s0} is the initial block B_s.

The theorem states that the resulting set of mt blocks forms a balanced incomplete block design with $v = mn$, $b = mt$, r, k, λ.

PROOF. Let v_{hj} be a treatment in B_s. The images of v_{hj} in the set of blocks B_{si} consist of all treatments v_{ij} belonging to the jth class. Thus each treatment v_{ij} appears in the design exactly once for each appearance of j in the initial blocks. Hence, by virtue of condition (ii), each treatment v_{ij}, $i = 0, 1, \ldots, m-1$, $j = 1, 2, \ldots, n$, appears exactly r times in the set of mt blocks.

Consider two treatments: v_{ij} and $v_{i'j'}$, $j \neq j'$. They appear together in a block B_{sg} if, and only if, we can find two treatments v_{hj} and $v_{h'j'}$ in B_s and an element x_g in G such that $x_h + x_g = x_i$ and $x_{h'} + x_g = x_{i'}$. Subtracting, we have the equivalent condition $x_h - x_{h'} = x_i - x_{i'}$. This implies that the pair v_{ij}, $v_{i'j'}$ occurs in exactly one of the blocks generated by B_s whenever $i - i'$ is one of the differences $[j, j']$ in that block. However, the difference $i - i'$ occurs exactly λ times among the set of differences $[j, j']$ in all t blocks, and so the pair of treatments v_{ij}, $v_{i'j'}$, occurs together in exactly λ blocks of the design. A similar argument involving the nonzero differences $[j, j]$ shows that the pair v_{ij}, $v_{i'j}$ also occurs together exactly λ times. Thus the design is a balanced incomplete block design with the required parameters, and the theorem is proved. ■

It should be noted that the requirement that each of the initial blocks should contain exactly k plots has not been used. Even if the blocks are of different sizes, the resulting design, though not a *BIBD*, is still equireplicate, and each pair of treatments appears in the same block exactly λ times.

EXAMPLE 1. The six blocks with $k = 3$, $v = 10$ just given may be used as initial blocks for a design with parameters $v = 10$, $b = 30$, $r = 9$, $k = 3$, $\lambda = 2$. The complete design is

$$
\begin{array}{llllll}
0_2, 1_2, 2_2; & 1_1, 4_1, 0_2; & 2_1, 3_1, 0_2; & 1_1, 4_1, 2_2; & 2_1, 3_1, 2_2; & 0_1, 0_2, 2_2; \\
1_2, 2_2, 3_2; & 2_1, 0_1, 1_2; & 3_1, 4_1, 1_2; & 2_1, 0_1, 3_2; & 3_1, 4_1, 3_2; & 1_1, 1_2, 3_2; \\
2_2, 3_2, 4_2; & 3_1, 1_1, 2_2; & 4_1, 0_1, 2_2; & 3_1, 1_1, 4_2; & 4_1, 0_1, 4_2; & 2_1, 2_2, 4_2; \\
3_2, 4_2, 0_2; & 4_1, 2_1, 3_2; & 0_1, 1_1, 3_2; & 4_1, 2_1, 0_2; & 0_1, 1_1, 0_2; & 3_1, 3_2, 0_2; \\
4_2, 0_2, 1_2; & 0_1, 3_1, 4_2; & 1_1, 2_1, 4_2; & 0_1, 3_1, 1_2; & 1_1, 2_1, 1_2; & 4_1, 4_2, 1_2.
\end{array}
$$

EXAMPLE 2. $v = 9$, $b = 18$, $r = 8$, $k = 4$, and $\lambda = 3$. G is the set of residue classes mod 9, $m = 9$, $n = 1$. The initial blocks are 0, 1, 2, 4; 0, 3, 4, 7.

EXAMPLE 3. $v = 25$, $b = 50$, $r = 8$, $k = 4$, $\lambda = 1$. G is the set of residue classes of the double modulus system mod (5, 5), or, equivalently, G can be the Galois field $GF(5^2)$. The initial blocks are 00, 01, 41, 13; 00, 32, 21, 02.

We begin the discussion of Bose's second method by considering an example. We can obtain two symmetric designs for $v = 7$: a design with $k = 3$, $\lambda = 1$, and the complementary design with $k = 4$, $\lambda = 2$. If we take the smaller design and add to each block a fourth plot with a new (eighth) treatment, then all fourteen blocks comprise a balanced incomplete block design with $v = 8$, $b = 14$, $r = 7$, $k = 4$, and $\lambda = 3$.

This approach can be generalized. We need to find two *BIBD*, one with parameters $v, b_1, r_1, k, \lambda_1$, and the other with parameters $v, b_2, r_2, k - 1, \lambda_2$. To each block of the second *BIBD* we add an extra plot with a new treatment ∞. If the composite design is a *BIBD*, its parameters must be $v + 1$, $b = b_1 + b_2$, $r = r_1 + r_2$, k, $\lambda = \lambda_1 + \lambda_2$. In the latter design the treatment ∞ appears in b_2 blocks, and each of the old treatments appears in r_2 of these. We must, therefore, have $r_2 = \lambda$ and $r = b_2$, whence $b_1 k = b_2(v - k + 1)$.

Theorem 2. *The second fundamental theorem of differences. Let G be an Abelian group of m elements under addition and, as in Theorem 1, let there be n treatments corresponding to each element x_i of G and denoted by v_{ij}. Furthermore, let there be an additional treatment denoted by ∞. Suppose that it is possible to find a set of $t + u$ blocks $B_1, \ldots, B_t, B_1^*, \ldots, B_u^*$ satisfying the following conditions:*

(i) *Each block contains k plots; each block B_s contains k of the mn treatments v_{ij}, each block B_s^* contains ∞ and $k - 1$ of the other treatments;*
(ii) *Among the kt treatments in the blocks B_s, exactly $(mu - \lambda)$ treatments belong to each class and among the $u(k - 1)$ treatments other than ∞ occurring in the blocks B^*, exactly λ belong to each class. (This implies that $kt = n(mu - \lambda)$ and $(k - 1)u = n\lambda$.);*
(iii) *The differences obtained from the $t + u$ blocks, ignoring ∞, are symmetrically repeated.*

From each of the initial blocks B_s we generate a set of m blocks in the same manner as in Theorem 1. From each of the initial blocks B_s^* we generate a set of m blocks by the same procedure, except that ∞ does not change as we go from block to block.

The theorem states that the resulting set of $m(t + u)$ blocks is a balanced incomplete block design with parameters $v = mn + 1$, $r = mu$, k, and λ.

The proof follows the same lines as the proof of the first fundamental theorem (Theorem 1) and will not be given.

EXAMPLE 4. Let G be the set of residue classes mod 11. The following three initial blocks generate a *BIBD* with $v = 12$, $b = 33$, $r = 11$, $k = 4$, and $\lambda = 3$; 0, 1, 3, 7; 0, 1, 3, 9; ∞, 0, 1, 5.

13.7 Designs Derived from Symmetric Designs

Lemma. *Any two blocks of a symmetric balanced incomplete block design have exactly λ treatments in common.*

PROOF. Consider a particular block, which we may take without loss of generality, to be labeled as the first. Let n_j denote the number of treatments common to the first block and the jth block with $j \neq 1$; it will suffice to show that $n_j = \lambda$ for all j.

The k treatments in the first block form $k(k-1)/2$ pairs. Each pair appears again in exactly $\lambda - 1$ of the remaining blocks; each of the treatments in the first block appears in exactly $k - 1$ of the remaining blocks. We thus have the equations

$$\sum_j n_j = k(k-1) = \lambda(v-1),$$

$$\sum_j n_j (n_j - 1)/2 = (\lambda - 1)k(k-1)/2 = \lambda(\lambda - 1)(v-1)/2.$$

Then $\bar{n}_j = \lambda$, $\sum n_j^2 = \lambda^2(v-1)$, and $\sum (n_j - \lambda)^2 = 0$, so that $n_j = \lambda$ for all j.

There are two methods of obtaining other balanced incomplete block designs from a symmetric design. The first is the method of block section (sometimes called residuation). The second is the method of block intersection (sometimes called derivation). Both are due to Bose (1939).

The method of block section consists of cutting one of the blocks out of the design, and dropping all the treatments contained in it from the other blocks. This gives a new design with parameters $v^* = v - k$, $b^* = b - 1$, r, $k - \lambda$, λ.

The simplest illustrations of this method occur in the orthogonal series of Yates. We have seen how a symmetric design with $b = v = s^2 + s + 1$, $r = k = s + 1$, and $\lambda = 1$ is obtained by augmenting the orthogonal (balanced lattice) design with $v = s^2$, $b = s(s+1)$, and $k = s$, $r = s + 1$. The method of block section takes us from the symmetric design to the lattice.

For another example, we may take the symmetric design with $v = b = 11$, $r = k = 5$, and $\lambda = 2$, and by block section obtain a design for six treatments in ten blocks of three plots each.

A partial converse result has been obtained by Hall and Connor (1953). They show that the existence of a balanced incomplete block design with parameters $v^* = v - k$, $b^* = v - 1$, $k^* = k - \lambda$, $r^* = k$, and $\lambda^* = \lambda$ implies the existence of the symmetric design with parameters v, k, and λ when λ is 1 or 2; i.e., when one is given the smaller design it is possible to augment it to obtain the larger symmetric design. Their result is not true for $\lambda = 3$. Connor (1952*a*) mentions the following counterexample: Bhattacharya gives a design for $v = 16$, $b = 24$, $r = 9$, $k = 6$, and $\lambda = 3$ which contains two blocks that have four treatments in common. If it were possible to augment this design to a symmetric design for $v = 25$, $k = 9$, and $\lambda = 3$, any pair of blocks of the new design would have exactly three treatments in common; with two of the blocks already having four treatments in common, this is clearly impossible. Bhattacharya's design for $v = 16$ was first given in reference (1944*a*). He repeated it (1944*b*), together with the symmetric design for $v = 25$ and $k = 9$ mentioned earlier and a second design for $v = 16$ obtainable by block section from it.

Bhattacharya's design for $v = 16$, $b = 24$, $r = 9$, $k = 6$, and $\lambda = 3$ is

ABGHNO	*CEHILN*	*CDGJLP*	*BDIJKM*	*ADGHKP*	*AFHJLM**
ADEMNP	*DFHIKO*	*CEGHKM*	*AFGILM**	*CDFMNO*	*CFGJKN*
BDHJLN	*ABCKLO*	*BEFKLP*	*AEIJKN*	*BCHIMP*	*BEGJMO*
DEGILO	*ABCDEF***	*EFHJOP*	*BFGINP*	*ACIJOP*	*KLMNOP***.

The two blocks marked with a single asterisk have four treatments in common; the two blocks marked with two asterisks have no treatments in common; the other blocks all have two treatments in common.

The method of block intersection is of less interest. It consists of omitting one of the blocks completely and retaining in the other blocks only those treatments that occur in the omitted block. This leads to a balanced incomplete block design with $v^* = k$, $b^* = v - 1$, $r^* = k - 1$, $k^* = \lambda$, and $\lambda^* = \lambda - 1$. For example, we may take the symmetric design for nineteen treatments in blocks of nine plots and obtain from it by block intersection a design for $v = 9$, $b = 18$, $r = 8$, $k = 4$, and $\lambda = 3$.

13.8 Designs with $r \leq 10$

In his 1939 paper Bose left unsolved these twelve cases of sets of parameters with $r \leq 10$ which had been listed in the Fisher and Yates tables. The reference numbers of the twelve cases are those given in the Fisher and Yates tables.

Reference Number	v	b	r	k	λ	Reference Number	v	b	r	k	λ
(8)	15	21	7	5	2	(24)	46	69	9	6	1
(10)	22	22	7	7	2	(26)	21	30	10	7	3
(12)	21	28	8	6	2	(27)	31	31	10	10	3
(14)	29	29	8	8	2	(28)	36	45	10	8	2
(17)	16	24	9	6	3	(30)	46	46	10	10	2
(20)	25	25	9	9	3	(31)	51	85	10	6	1

Since that time ten of the cases have been settled and only two, numbers (24) and (31), remain unsolved. We have already mentioned Bhattacharya's solutions to (17) and (20). In 1946 he obtained a solution to (27), and, hence, by block section, to (26). The impossibility of the symmetric designs (10) and (30) follows from the fact that v is even and $r - \lambda$ is not a square. The symmetric design (14) was shown to be impossible as the example following our presentation of the theorem of Chowla and Ryser (1950). The impossibility of (8), (12), and (28) then follows directly from the result of Hall and Connor (1953) because their existence would imply the existence of (10), (14), and (30) respectively. The nonexistence of (8) was established by Nandi (1945). An interesting alternate derivation of the nonexistence of (8) and (28) is given by Atiqullah (1958).

13.9 Series of Designs Based on Difference Sets

There are numerous series of designs based on difference sets. Most of them are to be found in four papers by Bose (1939) and (1942*a*) and Sprott (1954) and (1956*b*). Bose's series denoted by Latin letters are given in his 1939 paper, and his series denoted by Greek letters are given in reference (1942*a*). Sprott's series *A*, *B*, *C*, *D* are in his 1954 paper, and his numerical series are in his 1956 paper. Almost all these series involve a Galois field of p^α elements where p is a prime; most often there is the requirement that $v = p^\alpha$. We shall let x denote a primitive element of such a field. The standard procedure is to derive a set of initial blocks and then to generate the design by the method of one of the two fundamental theorems.

From the equations $\lambda(v - 1) = r(k - 1)$, $rv = bk$, we obtain the following necessary conditions for the existence of a *BIB* design:

$$\lambda(v - 1) \equiv 0 \quad \bmod (k - 1), \qquad \lambda v(v - 1) \equiv 0 \quad \bmod k(k - 1).$$

Hanani (1961) has shown that these two conditions are sufficient for $k = 3$ and 4 and all λ and for $k = 5$ with $\lambda = 1, 4$, or 20. (In his 1961 paper the design $v = 141$, $k = 5$, and $\lambda = 1$ was left open as a possible exception, but in 1965 Hanani gave a solution for that case.) Bose gave methods of obtaining many of these designs in his 1939 paper.

In particular, for $k = 3$ and $\lambda = 1$ we must have $v = 6t + 3$ or $v = 6t + 1$. Bose showed how to obtain designs when $v = 6t + 3$ for any t, and when $v = 6t + 1$ if either $v = p^\alpha$ or t is odd. The first of these is his T_1 series. The blocks in these designs are known as Steiner triplets after J. Steiner, who investigated them in 1853.

(a) Bose's T_1 Series

Let $v = 6t + 3$, $k = 3$, and $\lambda = 1$. We take the set of residue classes mod $2t + 1$ and associate three varieties with each of the elements. Consider the pairs $1, 2t; 2, 2t - 1; \ldots, i, 2t - i + 1; \ldots, t, t + 1$. The differences from the ith pair are $2t + 1 - 2i$ and $2i$, and so the set of differences covers all the nonzero elements of the set of residue classes.

The design is obtained from the following set of initial blocks:

$$\{1_1, 2t_1, 0_2\} \cdots \{i_1, (2t + 1 - i)_1, 0_2\} \cdots \{t_1, (t + 1)_1, 0_2\},$$
$$\{1_2, 2t_2, 0_3\} \cdots \{i_2, (2t + 1 - i)_2, 0_3\} \cdots \{t_2, (t + 1)_2, 0_3\},$$
$$\{1_3, 2t_3, 0_1\} \cdots \{i_3, (2t + 1 - i)_3, 0_1\} \cdots \{t_3, (t + 1)_3, 0_1\},$$
$$\{0_1, 0_2, 0_3\}.$$

The three smallest designs of the series are

$$t = 1, \quad v = 9, \quad b = 12, \quad r = 4, \quad k = 3, \quad \lambda = 1,$$
$$t = 2, \quad v = 15, \quad b = 35, \quad r = 7, \quad k = 3, \quad \lambda = 1,$$
$$t = 3, \quad v = 21, \quad b = 70, \quad r = 10, \quad k = 3, \quad \lambda = 1.$$

The design for $t = 1$ follows. The four blocks in the first row are the initial blocks.

$$\{1_1\ 2_1\ 0_2\} \quad \{1_2\ 2_2\ 0_3\} \quad \{1_3\ 2_3\ 0_1\} \quad \{0_1\ 0_2\ 0_3\}$$
$$\{2_1\ 0_1\ 1_2\} \quad \{2_2\ 0_2\ 1_3\} \quad \{2_3\ 0_3\ 1_1\} \quad \{1_1\ 1_2\ 1_3\}$$
$$\{0_1\ 1_1\ 2_2\} \quad \{0_2\ 1_2\ 2_3\} \quad \{0_3\ 1_3\ 2_1\} \quad \{2_1\ 2_2\ 2_3\}.$$

This design is isomorphic to the $OS1$ design for nine varieties.

(b) Sprott's A, B, C, D series

These four series require that $v = p^\alpha$. The treatments are represented by the elements of $GF(v)$, and x is a primitive element of the field. Series D has an additional requirement. The parameters of the designs that can be constructed with the series are

	v	k	r	λ
A	$tk + 1$	k	tk	$k - 1$
B	$2t(2\lambda + 1) + 1$	$2\lambda + 1$	$t(2\lambda + 1)$	λ
C	$2t(2\lambda - 1) + 1$	2λ	$2t\lambda$	λ
D	$4t(4\lambda + 1) + 1$	$4\lambda + 1$	$t(4\lambda + 1)$	λ.

In each of the designs there are t initial blocks, $0 \leq i \leq t - 1$:

$$\begin{array}{ll} A & x^i, x^{i+t}, x^{i+2t}, \ldots, x^{i+(k-1)t}; \\ B & x^i, x^{i+2t}, x^{i+4t}, \ldots, x^{i+4\lambda t}; \\ C & 0, x^i, x^{i+2t}, \ldots, x^{i+4(\lambda-1)t}; \\ D & x^{2i}, x^{2i+4t}, x^{2i+8t}, \ldots, x^{2i+16\lambda t}. \end{array}$$

In series A all the differences in the initial blocks may be written

$$x^{i+(s+u)t} - x^{i+ut} = x^{i+ut}(x^{st} - 1),$$

where $x^{st} - 1$ is an element of the field; $s = 1, \ldots, k - 1$; $u = 0, 1, \ldots, k - 1$. For any s, the number of such differences is kt and they are all distinct. If two such differences were identical, we should have

$$i + ut \equiv i' + u't \qquad (\text{mod } kt),$$

so that

$$i - i' \equiv (u - u')t \qquad (\text{mod } kt),$$

and

$$i - i' \equiv 0 \qquad (\text{mod } t).$$

But i, i' are both less than t, and so $i = i'$, $u = u'$. It follows that as s takes all its values, $0 < s \leq k - 1$, the differences are symmetrically distributed with $\lambda = k - 1$, and the required design is obtained by the method of the first fundamental theorem. The validity of the other series is established in the same way.

The additional requirement for series D is that among the 2λ quantities $x^{4ts} - 1$, which are themselves elements of the field, there should be exactly λ even powers of x and λ odd powers for each $s = 1, 2, \ldots, 2\lambda$. Sprott showed that if this requirement is violated by any primitive element, x, it is violated by all primitive elements, and so only one x need be checked. In the case $\lambda = 1$, this requirement simplifies to the requirement that $x^{4t} + 1$ is an odd power of x.

The numbered series 1, 2, and 3 of Sprott (1956*b*) are extensions of his letter series which also have restrictions in addition to the requirement that $GF(v)$ exists.

Bose's T_2 series for $v = 6t + 1 = p^{\alpha}$, $k = 3$, and $\lambda = 1$ is a special case of Sprott's B series with initial blocks $(x^i, x^{i+2t}, x^{i+4t})$. The smallest example occurs with $t = 1$ and $v = 7$. The elements of the field are the residue classes mod 7; $x = 3$ is a primitive element, and a difference set is (x^0, x^2, x^4), or $(1, 2, 4)$. It should be noted that Bose (1939) also gave a series for $v = 12t + 7$, $k = 3$, and $\lambda = 1$, which does not depend on v being a power of a prime.

(c) Other Series

Bose also gave series E_1 and E_2 for $k = 3$, $\lambda = 2$, and v even, series F_1 and F_2 for $k = 4$, $\lambda = 1$, and G_1, G_2 for $k = 5$, $\lambda = 1$. In his paper, the existence of designs in the E_2, F_2, and G_2 series was said to be dependent upon the solution of auxiliary problems. In F_2 and G_2 the problem was to find an integer θ such that $(x^{\theta} + 1)/(x^{\theta} - 1)$ is an odd power of x. Bose showed later (1942*b*) that this can always be done. The auxiliary problem for the E_2 series was solved by Bhattacharya (1943). Some of these designs may also be obtained as special cases of Sprott's letter series. The F_1 series is a particular case of Sprott's series 3.

Bose's designs α_1, α_2, α_3, and α_4 are special cases of Sprott's A, B, and C with $k = 4$ and 5. His series β_1, β_2, and γ have the following parameters:

	v	k	λ	Requirements
β_1	$2(3t + 2)$	4	2	$2t + 1 = p^\alpha, \quad t \geq 2,$
β_2	$10t + 5$	5	2	$2t + 1 = p^\alpha, \quad t \geq 2,$
γ	$4(3t + 2)$	4	3	$12t + 7 = p^\alpha.$

Bose's α_3 series is obtained from Sprott's series B by putting $\lambda = 2$. The other parameters are $v = 10t + 1, k = 5, r = 5t$. If we put $t = 1$, we have Bose's S_1 series. This a series of symmetric designs with $v = 4\lambda + 3$, $k = 2\lambda + 1$. A single initial block consists of all the nonzero elements of $GF(v)$ which are squares. The values of v for which this series is of practical interest are $v = 11$, 19, 23, 27, and 31. The missing value in the sequence, 15, is not a power of a prime, and this method does not apply in that case; we have, however, presented earlier in this chapter a difference set for the symmetric design with $v = 15$ and $k = 7$, using a double modulus system. The difference sets for the other values of v are

$v = 11$: 1 3 4 5 9,
$v = 19$: 1 4 5 6 7 9 11 16 17,
$v = 23$: 1 2 3 4 6 8 9 12 13 16 18,
$v = 27$: 001 211 020 021 022 100 111 110 120 121 102 202 221,
$v = 31$: 1 2 4 5 7 8 9 10 14 16 18 19 20 25 28.

For $v = 27$ we use $GF(3^3)$ with $x^3 = x + 2$ and arithmetic mod 3. The treatments are represented by triples $a_0a_1a_2$ corresponding to $a_0 + a_1x + a_2x^2$.

13.10 Resolvable Designs

A design is said to be resolvable [Bose (1942c)] if it can be split up into groups of blocks in such a way that each group contains each treatment exactly once. It was shown in Chapter 11 that for such a design $b \geq v + r - 1$.

Applying the method of block section to the symmetric designs with $v = 4\lambda + 3 = p^\alpha$, $k = 2\lambda + 1$, λ gives Bose's B_1 series of designs with $v = 2\lambda + 2$, $b = 4\lambda + 2$, $r = 2\lambda + 1$, and $k = \lambda + 1$. Bose (1947b) showed that when $2\lambda + 1$ is a prime, λ is odd and $\lambda > 1$, there are two nonisomorphic designs with the parameters $v = 2\lambda + 2$, $k = \lambda + 1$, $r = 2\lambda + 1$; one of these is resolvable and the other is not. When λ is even, there is a single solution which is not resolvable.

The resolvable design for odd λ is obtained from the initial blocks $(\infty, x, x^3, \ldots, x^{2\lambda-1})$, $(0, x^0, x^2, \ldots, x^{2\lambda-1})$ where x is a primitive element of $GF(2\lambda + 1)$. The nonresolvable design comes from the initial blocks $(\infty, x^0, x^2, \ldots)$, $(0, x, x^3, \ldots)$. The resolvable design is the same as Sprott's series 4 design, and it has the additional property that all triples of treatments occur the same number of times. The design with $\lambda = 3$ has $v = 8$, $b = 14$, $r = 7$, and $k = 4$. It is

$$\begin{array}{ccccccc} \infty\,3\,5\,6, & \infty\,4\,6\,0, & \infty\,5\,0\,1, & \infty\,6\,1\,2, & \infty\,0\,2\,3, & \infty\,1\,3\,4, & \infty\,2\,4\,5, \\ 0\,1\,2\,4, & 1\,2\,3\,5, & 2\,3\,4\,6, & 3\,4\,5\,0, & 4\,5\,6\,1, & 5\,6\,0\,2, & 6\,0\,1\,3. \end{array}$$

Sprott's series 5 designs are also resolvable designs with the property that every triple occurs the same number of times. They have $v = 2k$, $b = 4(2k - 1)$, $r = 2(2k - 1)$, $\lambda = 2(k - 1)$, and it is necessary that $2k - 1 = p^\alpha$. The design is obtained from the four initial blocks

$$(0, x^i, x^{i+2}, \ldots, x^{i+2k-4}), \quad (\infty, x^{i+1}, x^{i+3}, \ldots, x^{i+2k-3}), \qquad i = 0, 1.$$

The doubly balanced designs of Calvin (1954) are special cases of these series.

Exercises

1. Show that the complement of a *BIBD* is also a *BIBD*.
2. Show that the existence of a design with $v = n^2$, $k = n$, and $\lambda = 1$ implies the existence of a complete set of orthogonal Latin squares of side n. Hence there exists no design with $v = 36$, $k = 6$, and $\lambda = 1$.
3. Show that the equation

$$x^2 = 6y^2 + 5z^2$$

   has no solution in integers other than the trivial solution $x = y = z = 0$.
4. If v is a prime and $v = 4t + 3$, show that the quadratic residues of v form a perfect difference set.
5. (Continuation). Show that if $v = 27$, and if the varieties are represented by elements of $GF(3^3)$, the quadratic residues also form a difference set. That is the example given in the text.
6. We stated in the discussion of Sprott's series D that if the necessary requirement is violated by any primitive x, it is violated by all primitive elements. Prove this.
7. Show that the differences are symmetrically repeated in Bose's resolvable B_1 designs.
8. Find counterexamples to show the following:

   (i) The equality $b = v + r - 1$ does not ensure that $v = nk$ where n is an integer;
   (ii) $v = nk$ is not sufficient to ensure resolvability.

9. Show that for an affine resolvable design (i.e., a design with $b = v + r - 1$) the parameters can be written as

$$v = n^2((n-1)t + 1), \qquad b = n(n^2t + n + 1), \qquad k = n((n-1)t + 1),$$
$$r = n^2t + n + 1, \qquad \lambda = nt + 1.$$

CHAPTER 14

The Existence and Construction of Partially Balanced Designs

In Chapter 12 we initiated our discussion of partially balanced association schemes and partially balanced designs. These were introduced in 1939 by Bose and Nair because of the restrictions on the existence of balanced incomplete block designs, especially in the case where $v - 1$ is a prime. We have postponed until now the discussion of the existence and construction of designs with the several association schemes. This chapter, like Chapter 13, is of more mathematical interest than the others. The reader who does not wish to work his way through it may find a design to fit his requirements by using the tables of designs assembled by Bose, Clatworthy, and Shrikhande (1954), and by Clatworthy (1956).

We shall present in the next section some elementary designs that can be obtained from any association scheme. In the subsequent sections we shall, for the most part, confine our attentions to partially balanced designs with only two associate classes, and present methods of obtaining designs with the group divisible, triangular, and Latin square association schemes. Even for these schemes there are still some sets of parameters with $k \leq 10$ and $r \leq 10$ for which it is not known whether or not a design exists. The eighteen unsolved cases for the triangular scheme and the five unsolved cases for the L_2 scheme are listed in the relevant sections of the chapter. More problems of interest remain to be solved for $PBIB(m)$ designs with $m > 2$.

The reader will wish to refer to Chapter 12 for the introductory remarks about partially balanced designs. He will then recall that we begin with a partially balanced association scheme. A *PBIB* design with that scheme is then an equireplicate, proper, binary design, i.e., a design in which the treatments each appear r times in blocks of k plots each, with no treatment appearing more than once in any block. Each pair of ith associates appears together in exactly λ_i blocks. The λ_i need not all be different.

We have already given a proof in Chapter 11 of Kempthorne's conjecture that, of all proper, binary, equireplicate designs with a given set of parameters (b, v, r, k), the balanced incomplete block design, if it exists, is the most efficient. The experimenter will usually not, therefore, choose a partially balanced design in preference to a balanced design, nor, if the blocks are large enough, will he choose an incomplete block design over the randomized complete block design. The efficiency of an incomplete block design has been shown (Chapter 11) to be given by $N^{-1}v\delta$ where δ is the harmonic mean of the nonzero latent roots of the matrix **A** of the reduced intrablock equations.

Let **N** be the incidence matrix of a partially balanced design. We shall show in Chapter 15 that if θ_i is a latent root of $\mathbf{NN}'$ with multiplicity α_i, then

$$\psi_i = (rk - \theta_i)/k$$

is a root of **A** with the same multiplicity α_i. For every connected design $\mathbf{NN}'$ has a root $\theta_0 = rk$ with $\alpha_0 = 1$. It is only meaningful to talk about efficiency for connected designs, in which case

$$E = (v - 1)/(r \sum \alpha_i \psi_i^{-1}).$$

The largest value of E that can be hoped for with v varieties in blocks of size k is

$$E = (\lambda v)/(rk)$$

where $\lambda(v - 1) = r(k - 1)$, i.e.,

$$E_{\max} = \frac{v(k - 1)}{k(v - 1)} = 1 - \frac{v - k}{k(v - 1)}.$$

For any design the parameters r, k, n_i, and λ_i must satisfy the equation

$$r(k - 1) = \sum_i \lambda_i n_i.$$

Furthermore, since $\mathbf{NN}'$ is a nonnegative definite matrix, its roots θ_i must all be nonnegative. We shall see in Chapter 15 that for a $PBIB(m)$ design there are, other than the simple root $\theta_0 = rk$, only m distinct roots θ_i; their values and the values of the multiplicities α_i are readily obtained by the methods to be given in that chapter. This provides us with a convenient method of establishing the nonexistence of designs with some sets of parameters. In addition, for a symmetric design the determinant $|\mathbf{NN}'|$ is a square and so $rk\theta_1^{\alpha_1}\theta_2^{\alpha_2}\ldots$ is a square. Thus, for example, if all the roots are positive and if α_i is even for all i, $i > 1$, but α_1 is odd, then θ_1 must be a perfect square. This is, of course, an extension of the method that we used earlier to show that in a symmetric balanced incomplete block design with v even, $(r - \lambda)$ must be a square.

For some association schemes, other conditions for the nonexistence of symmetric designs have been derived. These conditions make use of the Hilbert symbol $(a, b)_p$ and depend upon the Hasse–Minkowski invariant. Because the calculation of these invariants is usually quite tedious, we shall content ourselves by giving references where needed to the appropriate papers. The latent roots of $\mathbf{NN}'$ are also of importance in linked block designs.

14.1 Duality and Linked Block Designs

The concept of *duality* comes from projective geometry in which it is possible to obtain from one theorem a second one by interchanging in the statement such words as *line* and *point*. The statements "There is one straight line common to any two points" and "There is one point common to any two lines" are duals. In the present context, the dual design is obtained by interchanging blocks and varieties.

Consider the unreduced *BIBD* with $v^* = 4$, $k^* = 2$. The parameters are $v^* = 4$, $b^* = 6$, $k^* = 2$, $r^* = 3$, and $\lambda^* = 1$; the design is *AB, AC, AD, BC, BD, CD*.

In the dual the varieties are *a, b, c, d, e, f*, corresponding to the six original blocks. The statement "Variety *A* appears in blocks 1, 2, and 3" becomes "Block 1 contains the varieties *a, b*, and *c*." This gives the new design *abc*; *ade*; *bdf*; *cef* with parameters $\tilde{v} = b^* = 6$, $\tilde{b} = v^* = 4$, $\tilde{k} = r^* = 3$, and $\tilde{r} = k^* = 2$. The dual design is actually a triangular *PBIB*(2) design. It does not, however, follow that the dual of a balanced incomplete block design is always a *PBIB*(2) design. This will be discussed further in Chapter 14.

Youden (1951) defined linked block designs as designs with the property that each pair of blocks has the same number (μ) of varieties in common. It follows that the dual has the property that each pair of varieties has the same number (μ) of blocks in common, i.e., the dual is a balanced incomplete block design. If $\mu = 1$, we have singly linked blocks; if $\mu = 2$, we have doubly linked blocks. The main properties of linked block designs are given by J. Roy and Laha (1957*a,b*).

Let $\mathbf{NN}'$ be the interblock matrix for a linked block design; $\mathbf{N}'\mathbf{N}$ is the interblock matrix for the dual (*BIBD*). We recall that if $\mathbf{A}$ and $\mathbf{B}$ are matrices such that $\mathbf{AB}$ and $\mathbf{BA}$ both exist, then $\mathbf{AB}$ and $\mathbf{BA}$ have the same nonzero latent roots. Thus $\mathbf{NN}'$ has the same nonzero latent roots as $\mathbf{N}'\mathbf{N}$, namely r^*k^* and $r^*(v^* - k^*)/(v^* - 1) = r^* - \lambda^*$. Furthermore, $\mathbf{NN}'$ and $\mathbf{N}'\mathbf{N}$ have the same rank and the same trace. This leads to the following result, due to J. Roy and Laha (1957*b*).

A necessary and sufficient condition that a design be a linked block design is that $\mathbf{NN}'$ has only one nonzero latent root other than rk, namely,

$$\theta = \frac{k(b - r)}{b - 1}$$

and the multiplicity of θ is $\alpha = b - 1$.

Shrikhande (1952) gave two cases in which the duals of balanced incomplete block designs are $PBIB(2)$ designs. The first is the dual of a balanced incomplete block design with parameters

$$v^* = rk - k + 1, \qquad r^* = k, \qquad k^* = r, \qquad \lambda^* = 1.$$

It is a $PBIB(2)$ design with parameters

$$v = k(kr - k + 1)/r, \qquad b = rk - k + 1, \qquad r, k,$$

$$n_1 = r(k - 1), \qquad n_2 = (k - r)(r - 1)(k - 1)/r, \qquad \lambda_1 = 1, \lambda_2 = 0.$$

The **P** matrices are

$$\mathbf{P}_1 = \begin{bmatrix} (k-2) + (r-1)^2 & (r-1)(k-r) \\ (r-1)(k-r) & (r-1)(k-r)(k-r-1)/r \end{bmatrix}$$

$$\mathbf{P}_2 = \begin{bmatrix} r^2 & r(k-r-1) \\ r(k-r-1) & n_2 - r(k-r-1) - 1 \end{bmatrix}$$

This is called the singly linked block (SLB) association scheme [Bose (1963)].

If we take the balanced design with $v^* = 9$, $b^* = 12$, $r^* = 4$, $k^* = 3$, and $\lambda^* = 1$ used in Chapter 11 and number the blocks 1 through 12, the dual is the design

1, 4, 7, 10; 1, 5, 8, 11; 1, 6, 9, 12; 2, 4, 9, 11; 2, 5, 7, 12;
2, 6, 8, 10; 3, 4, 8, 12; 3, 5, 9, 10; 3, 6, 7, 11.

The second case is the triangular doubly linked block design, which is discussed later in this chapter.

14.2 The Elementary Designs

Corresponding to each association scheme are designs that we shall call the *elementary designs.* The first of these designs are the elementary symmetric designs denoted by $ES(i)$ and $E'S(i)$. In the design $ES(i)$ where $i = 1, 2, \ldots, m$, the hth block consists of all those varieties which are ith associates of the hth variety. For the design $ES(i)$, we have $r = k = n_i$ and $\lambda_t = p_{ii}^t$, $t = 1, 2, \ldots, m$.

In the design $E'S(i)$, the hth block consists of the hth treatment together with all its ith associates, and so $r = k = n_i + 1$, $\lambda_i = p_{ii}^i + 2$, and $\lambda_t = p_{ii}^t$ $(t \neq i)$. For the group divisible association scheme, the designs $ES(1)$ and $E'S(1)$ are both unconnected. If there are only two associate classes, the $ES(1)$ design and the $E'S(2)$ design are complementary, as are the $E'S(1)$ and $ES(2)$ designs.

For breeding experiments, designs with $k = 2$ can be obtained by crossing each variety with its ith associates. We gave an example in Chapter 12.

We can also make up elementary designs by combining a collection of v *BIB* designs. The hth of these balanced designs is constructed from the set of ith associates of the hth variety and has parameters b^*, $v^* = n_i$, k^*, r^*, and λ^*. Then we have for the combined design $b = vb^*$, $k = k^*$, $r = n_i r^*$, and $\lambda_t = p_{ii}^t \lambda^*$. If we now add the hth variety itself to each of the blocks in the hth *BIBD*, we have a design with $b = vb^*$, $k = k^* + 1$, $r = n_i r^* + b^*$, $\lambda_i = 2r^* + p_{ii}^i \lambda^*$, and $\lambda_t = p_{ii}^t \lambda^*$ $(t \neq i)$.

14.3 Group Divisible Designs

This was the first association scheme to be investigated extensively. Contributions have been made by Bose and Connor (1952), Connor (1952*b*), Bose et al. (1953), and Freeman (1957). The matrix $\mathbf{NN}'$ has each diagonal element r; the off-diagonal elements are either λ_1 or λ_2. For the *GD* design, if we order the varieties by groups so that the first n varieties are in the first group, and so on, $\mathbf{NN}'$ consists of submatrices $(r - \lambda_1)\mathbf{I}_n + \lambda_1 \mathbf{J}_n$ along the main diagonal and submatrices $\lambda_2 \mathbf{J}_n$ off the diagonal. We can evaluate the determinant as $|\mathbf{NN}'| = (r + (n - 1)\lambda_1 + (m - 1)n\lambda_2)(r - \lambda_1 + n(\lambda_1 - \lambda_2))^{m-1}(r - \lambda_1)^{m(n-1)}$. Substituting $r - \theta$ for r we see that the latent roots and their multiplicities are

$$\theta_0 = r + (n - 1)\lambda_1 + (m - 1)n\lambda_2 = rk, \qquad \alpha_0 = 1;$$

$$\theta_1 = r + \lambda_1(n - 1) - \lambda_2 n = rk - \lambda_2 v, \qquad \alpha_1 = m - 1;$$

$$\theta_2 = r - \lambda_1, \qquad \alpha_2 = m(n - 1)$$

The latent roots and their multiplicities will be obtained more easily later. For binary designs we cannot have $\lambda_1 > r$, and so the condition $\theta_2 \geq 0$ is trivial. However, the condition $\theta_1 \geq 0$ can be used to establish the nonexistence of some designs, as the following example shows.

Let $m = 6$ and $n = 4$. The condition $r(k - 1) = \sum_i n_i \lambda_i$ becomes

$$r(k - 1) = (n - 1)\lambda_1 + n(m - 1)\lambda_2.$$

A design with $\lambda_1 = 0$, $\lambda_2 = 11$, $r = 20$, and $k = 12$ would satisfy this requirement and also the requirement that b be an integer. However, $\theta_1 = 20 - 44 = -24$, and so no such design exists.

For symmetric designs, we note that the determinant

$$|\mathbf{NN}'| = r^2(r^2 - v\lambda_2)^{m-1}(r - \lambda_1)^{m(n-1)}$$

must be a perfect square. If neither θ_1 nor θ_2 is zero, the following conditions are necessary:

(i) If m is odd and n is even, then $r - \lambda_1$ must be a square;
(ii) If m is even, then $r^2 - v\lambda_2$ is a square.

Thus, for example, condition (i) rules out the design for $v = b = 22$, $r = k = 5$, $m = 11$, $n = 2$, $\lambda_1 = 0$, and $\lambda_2 = 1$; condition (ii) rules out $v = b = 44$, $r = k = 7$, $m = 22$, $n = 2$, $\lambda_1 = 0$, and $\lambda_2 = 1$. Bose and Connor (1952) have extended these results, using the Hasse–Minkowski invariant. Bose and Connor (1952) divided the *GD* designs into three classes which are mutually exclusive:

(i) Singular designs, which are designs for which $\theta_2 = 0$, i.e., $r = \lambda_1$;
(ii) Semiregular designs which have $\theta_1 = 0$, $\theta_2 > 0$;
(iii) Regular designs which have $\theta_1 > 0$, $\theta_2 > 0$.

If both θ_1 and θ_2 are zero, we have the complete block design with $b = r$ and $k = v$.

14.4 Singular Designs

For a design with $r = \lambda_1$, the varieties have to occur by groups. Whenever a given variety appears in a block, all varieties in the same group must also appear in that block. The simplest example is the design with $b = m$, $k = n$, and $r = 1$, in which each block consists of a single group of varieties. The $E'S(1)$ design repeats this design n times. The $ES(2)$ design is also singular.

Singular designs with $\lambda_2 > 0$ are obtained from balanced incomplete block designs. If we take a balanced design with parameters $v^* = m$, b^*, r^*, k^*, and λ^*, and replace each treatment by a group of n varieties, we obtain a singular *GD* design with $v = mn$, $b = b^*$, $r = \lambda_1 = r^*$, $k = nk^*$, and $\lambda_2 = \lambda^*$. Indeed, every singular *GD* design with $\lambda_1 = 0$ can be obtained by this method. It is possible to have $b < v$, as we saw in the preceding paragraph, because the rank of $\mathbf{NN}'$ is $1 + \alpha_1 = m$. The inequality $b \geq v$, which holds for balanced incomplete block designs, is thus replaced by $b \geq m$.

Zelen (1954) points out that this method of replacing each variety by a group of n new varieties can be used to derive a *PBIB*($m + 1$) design from any *PBIB*(m) design, and P. M. Roy (1953) has incorporated this idea into his hierarchical *GD* designs.

14.5 Semiregular Designs

Bose and Connor (1952) prove the following theorem.

Theorem. *A necessary and sufficient condition for a group divisible design to be semiregular is that $k = cm$ where c is an integer, and that each block contain exactly c varieties from each of the groups.*

PROOF. Suppose that we have a *GD* design with parameters $v = mn$, r, k, λ_1, and λ_2.

Sufficiency: Consider a variety in the jth group. It appears r times, and on each occasion there are, in the same block, c varieties from the hth group, $(h \neq j)$. Then

$$rc = \lambda_2 n.$$

We also have $k = cm$ so that

$$rk = rcm = mn\lambda_2 = \lambda_2 v,$$

and the design is semiregular.

Necessity: Suppose that the design is semiregular and that the jth block contains e_j varieties belonging to the first group. Then $\sum_{j=1}^{b} e_j = nr$.

The jth block contains $\binom{e_j}{2}$ pairs of varieties from the first group. Each of these pairs of varieties occurs together in λ_1 blocks, and so

$$\sum_{j=1}^{b} e_j(e_j - 1) = \lambda_1 n(n - 1).$$

It follows that

$$\sum e_j^2 = n(n - 1)\lambda_1 + nr = n(rk - n(m - 1)\lambda_2) = n^2\lambda_2.$$

Let $\bar{e} = nr/b = vr/(mb) = k/m = c$. Then $b \sum (e_j - \bar{e})^2 = n^2 b\lambda_2 - n^2r^2 = 0$, and $e_1 = e_2 = \cdots = e_m = \bar{e} = c$. But e_j is an integer and so c is an integer. The same argument applies for each group of varieties, and so each block contains exactly c varieties from each group. ■

It is easily shown that for a semiregular *GD* design $b \geq v - m + 1$. From any semiregular design with $v = mn$ treatments, a design with parameters $v^* = m_1 n$, b, r, $k^* = cm_1$, λ_1, λ_2 may be obtained by cutting out $(m - m_1)$ of the groups of treatments.

When Bose (1942*c*) established the inequality

$$b \geq v + r - 1$$

for resolvable balanced incomplete block designs, he called affine resolvable those designs for which the equality $b = v + r - 1$ held. He showed that for an affine resolvable design with $v = nk$, each block in a given replication has exactly k/n varieties in common with each of the blocks in every other replication; k/n is an integer.

It follows that the dual of an affine resolvable balanced incomplete block design is a semiregular *GD* design. When we take the dual, blocks that were in the same replication become varieties in the same group, and, since the blocks in any replication are disjoint, we have $\lambda_1 = 0$. Since blocks in different replications become varieties in different groups, it follows from the property of affine resolvable designs just mentioned that $\lambda_2 = k/n$. We summarize this in the following statement. Let D be an affine resolvable balanced incomplete block design with parameters

$$v^* = nk^*, \qquad r^* = m, \qquad k^*, \qquad \lambda^*;$$

then the dual of D is a semiregular GD design with

$$v = mn, \qquad k = m \quad (c = 1), \qquad r = k^*, \qquad \lambda_1 = 0, \qquad \lambda_2 = k/n.$$

The duals of the designs in Yates' orthogonal series $OS1$ are semiregular designs with parameters

$$v = s^2 + s, \qquad b = s^2, \qquad r = s, \qquad k = s + 1,$$
$$m = s + 1, \qquad n = s, \qquad \lambda_1 = 0, \qquad \lambda_2 = 1.$$

If we take only the first p replications, we have a design with

$$v = ps, \qquad b = s^2, \qquad k = p, \qquad m = p, \qquad n = s, \qquad \lambda_1 = 0, \qquad \lambda_2 = 1.$$

14.6 Regular Designs

For regular designs $\theta_1 > 0$ and $\theta_2 > 0$; $\mathbf{NN}'$ is of full rank, and $b \geq v$. Several methods of obtaining regular designs have been given by Bose et al. (1953), and by Freeman (1957). We will present some of these methods.

(a) Section from a Balanced Design

From a balanced design with parameters v^*, r^*, k^*, and $\lambda = 1$, we may delete all the blocks containing a particular variety. The resulting design will be a GD design with $v = v^* - 1$, $m = r^*$, $n = k^* - 1$, $r = r^* - 1$, $k = k^*$, $\lambda_1 = 0$, $\lambda_2 = 1$. The groups of varieties correspond to the discarded blocks. The latent roots of $\mathbf{NN}'$ are $\theta_1 = r^* - k^*$ and $\theta_2 = r^* - 1$. If the balanced design is symmetric, we have $\theta_1 = 0$, and the design is semiregular.

(b) Complete and Incomplete Groups

These are designs in which each block consists of u complete groups of varieties together with h varieties from another group. An example is the $E'S(2)$ design, which has $u = m - 1$, $h = 1$. The $ES(1)$ design has $u = 0$, $h = n - 1$. There are several methods of obtaining designs of this type. In an earlier section we mentioned unconnected designs in which a balanced incomplete block design with $v^* = n$, b^*, k, $\lambda = \lambda_1$ is taken for each group. The combined design has $b = mb^*$, $\lambda_1 = \lambda$, $\lambda_2 = 0$. Its complement is a design of this type with $h = n - k$ and $u = m - 1$.

Other designs may be obtained by using two balanced incomplete block designs. In this context, either or both of them may be degenerate designs with $r = k = 1$, $\lambda = 0$. Call the designs D_1, D_2; D_1 is a design for n treatments; D_2 is a design for $(m - 1)$ treatments. We begin by taking the design D_1 and using the varieties in the first group as the treatments. We then substitute for the $m - 1$ treatments in D_2 the other $m - 1$ complete groups, and combine each block of D_1 with each block of D_2. The procedure is then repeated using the varieties of the second group for D_1 and so on. If the parameters of D_i are $b_i, k_i, r_i, \lambda_i^*$, the overall design has

$$b = mb_1b_2, \qquad r = r_1b_2 + (m - 1)r_2b_1, \qquad k = k_1 + nk_2,$$

$$\lambda_1 = (m - 1)b_1r_2 + b_2\lambda_1^*, \qquad \lambda_2 = (m - 2)b_1\lambda_2 + 2r_1r_2.$$

Suppose, for example, that there are twelve varieties divided into three groups of four: *A, B, C, D*; *E, F, G, H*; *I, J, K, L*.

EXAMPLE 1. The $E'S(2)$ design is

AEFGHIJKL,	*BEFGHIJKL*,	*CEFGHIJKL*,	*DEFGHIJKL*,
ABCDEIJKL,	*ABCDFIJKL*,	*ABCDGIJKL*,	*ABCDHIJKL*,
ABCDEFGHI,	*ABCDEFGHJ*,	*ABCDEFGHK*,	*ABCDEFGHL*.

$$v = b = 12, \qquad r = k = 9, \qquad \lambda_1 = 8, \qquad \lambda_2 = 6.$$

EXAMPLE 2. Take for the incomplete group blocks from the balanced design: 12, 13, 14, 23, 24, 34. Augment each block with the varieties in the remaining two groups. This gives

AB EFGH IJKL,	*AC EFGH IJKL*,	*AD EFGH IJKL*,
BC EFGH IJKL,	*BD EFGH IJKL*,	*CD EFGH IJKL*,
ABCD EF IJKL,	*ABCD EG IJKL*,	*ABCD FG IJKL*, and so on.

$$b = 18, \qquad k = 10, \qquad r = 15, \qquad \lambda_1 = 13, \qquad \lambda_2 = 12.$$

EXAMPLE 3. Taking $u = h = 1$ gives a design with $b = 24$, $k = 5$, $r = 10$, $\lambda_1 = 8$, and $\lambda_2 = 2$. (Each of the two balanced designs used is degenerate with $r = k = 1$ and $\lambda = 0$.)

A IJKL,	*B IJKL*,	*C IJKL*,	*D IJKL*,
E ABCD,	*F ABCD*,	*G ABCD*,	*H ABCD*,
I EFGH,	*J EFGH*,	*K EFGH*,	*L EFGH*,
A EFGH,	*B EFGH*,	*C EFGH*,	*D EFGH*,
E IJKL,	*F IJKL*,	*G IJKL*,	*H IJKL*,
I ABCD,	*J ABCD*,	*K ABCD*,	*L ABCD*.

It is not, however, always necessary to take the complete design. In Example 3, either the first twelve blocks or the last twelve blocks constitute a half design which is partially balanced. Indeed, whenever m is odd such half designs are available. They are obtained by taking, for example, those blocks in which the hth group is complete and the $(h + 1)$th group is incomplete, and omitting the corresponding blocks in which the $(h + 1)$th group is complete and the hth group is incomplete ($h + 1$ being reduced mod m).

In analogous fashion, designs may be obtained in which each block consists of a subset from each of two or more groups, the subsets again being chosen as blocks of balanced designs with $v = m$. Thus, with six varieties in two groups $A, B, C;\ D, E, F$, we may take all possible pairs from the first group with each member of the second group, and vice versa, to obtain a design with $b = 18$. $k = 3$, $\lambda_1 = 3$, and $\lambda_2 = 4$:

ABD,	*ABE*,	*ABF*,	*ACD*,	*ACE*,	*ACF*,
BCD,	*BCE*,	*BCF*,	*ADE*,	*BDE*,	*CDE*,
ADF,	*BDF*,	*CDF*,	*AEF*,	*BEF*,	*CEF*.

Again, fractional designs exist. The complete design for $m = n = 3$, with each block containing two treatments from one group and one from another, calls for fifty-four blocks. There is a partially balanced half design with $v = 9$, $b = 27$, $k = 3$, $r = 9$, $\lambda_1 = 3$, and $\lambda_2 = 2$. For $m = n = 4$, with each block containing two varieties from one group and one from another, 288 blocks are needed for the complete design, but there is a one-sixth fraction with $v = 16$, $b = 48$, $k = 3$, $r = 9$, $\lambda_1 = 2$, and $\lambda_2 = 1$.

Freeman (1957) also discusses variations of these methods in which $v = 2n^2$ varieties are split into $m = 2n$ groups, and the groups are divided into two sets of n groups each. Designs can then be found with $k = 2n$, $b = n(3n - 1)$, $r = (3n - 1)$, $\lambda_1 = 2n - 1$, and $\lambda_2 = 2$. It is necessary that there exist $(n - 1)$ orthogonal Latin squares of side n. There are two types of block in the design. The first $n(n - 1)$ blocks consist of all possible pairs of groups from the same set. The remaining $2n^2$ blocks each consist of a group from one set and a single variety from each group of the other set; each group appears complete in n of these blocks, and the assignment of the varieties from the other set uses the orthogonal Latin squares.

The design for $n = 2$ has $v = 8$, $b = 10$, $k = 4$, $r = 5$, $\lambda_1 = 3$, and $\lambda_2 = 2$. If the groups are $A, B;\ C, D;\ E, F;\ G, H$, we have

ABCD, *EFGH*, *ABEG*, *ABFH*, *CDEH*, *CDFG*, *ACEF*, *BDEF*,
ADGH, *BCGH*.

Freeman gives a design for $n = 3$.

(c) Adding Designs

Suppose that there exists a balanced design with parameters $v = mn$, $k = n$, b^*, r^*, λ^*. A disconnected GD design with m blocks of size k can be obtained by letting each of the m blocks consist of a different group of n varieties. The design obtained by repeating the balanced design s times and then repeating the disconnected design t times is a regular GD design with parameters

$$v = mn, \qquad b = b^*s + mt, \qquad k = n, \qquad r = r^*s + t,$$
$$\lambda_1 = \lambda^*s + t, \qquad \lambda_2 = \lambda^*s.$$

A design for $v = 9$, $m = 3$, $n = 3$, $b = 27$, $k = 3$, $\lambda_1 = 3$, and $\lambda_2 = 2$ has just been mentioned; each block contained two varieties from one group and one from another. Bose, Clatworthy, and Shrikhande (1954) give a different (nonisomorphic) design with the same parameters. They repeat the orthogonal series design with $v = k^2 = 9$ and $b = 12$ twice, and then repeat one of the squares a third time, which amounts to taking $s = 2$ and $t = 1$.

In the previous example of adding a disconnected design to a balanced design it happened that the balanced design was resolvable and that the disconnected design comprised one of the replications in it. Bose et al. (1953) use a method of repeating a resolvable balanced incomplete block design t times and further repeating one of the replicates an additional a times where a may be positive (addition) or negative (subtraction).

Other designs may be obtained by adding partially balanced designs to balanced designs. We have mentioned Freeman's design for $v = 16$, $m = n = 4$, $b = 48$, $k = 3$, $\lambda_1 = 2$, and $\lambda_2 = 1$. A nonisomorphic design is found in the Bose, Clatworthy, and Shrikhande tables, design $R38$ (1954). It may be constructed from other designs in the following way. We first take for each group the $BIBD$ for $v = b = 4$, $r = k = 3$, and $\lambda = 2$. These sixteen blocks are a design with $\lambda_1 = 2$ and $\lambda_2 = 0$, and we add to them design $R35$ of Bose et al. (1954), which has $v = 16$, $m = n = 4$, $k = 3$, $\lambda_1 = 0$, and $\lambda_2 = 1$. Other examples of this method are given by Bose et al. (1953).

This method of adding designs is not confined in its application to group divisible schemes. If in the general case we can find designs D_i of block size k with the property that $\lambda_i = 1$ and $\lambda_h = 0$ for all $h \neq i$, we can easily construct a design of block size k for any set of parameters $\lambda_1, \lambda_2, \ldots$ by repeating each basic design D_i exactly λ_i times. An example will be found in the section on Latin square designs.

(d) Extension of Designs

It will be recalled that the complete lattice design with $v = k^2 = 9$, $b = 12$, $r = 4$, and $\lambda = 1$ is resolvable and can be made into a balanced design for $v = b = 13$ and $r = k = 4$ in the following way. A new block is added containing the four new varieties, a plot with the first new variety is added to each block in the first replicate, the second new variety is added to each block in the second replicate, and so on. This method of adding r additional varieties to make a design with block size $k + 1$ has its parallel in the construction of GD designs. Suppose that there exists a resolvable GD design with parameters $v = mn = ks$, $b = rs$, r, k, λ_1, $\lambda_2 = 1$, and also that there exists another GD design with parameters $v' = r = m'n$, b', $r' = r - s$, $k' = k + 1$, $\lambda_1' = \lambda_1$, and $\lambda_2' = 1$ where m' is an integer; then we may allocate one new variety to each replicate in the first design, adding it to each block in the replicate, and obtain a new *GD* design with parameters

$$\tilde{v} = v + v' = v + r, \quad \tilde{b} = b + b', \quad \tilde{r} = r, \quad \tilde{k} = k + 1,$$
$$\tilde{m} = m + m', \quad \tilde{n} = n, \quad \tilde{\lambda}_1 = \lambda_1, \quad \tilde{\lambda}_2 = 1.$$

There is a paucity of examples of moderate size. The following one is found in the paper by Bose et al. (1953).

EXAMPLE. A resolvable *GD* design for $v = 6$ and $k = 2$ is obtained by taking the unreduced design with $b = \binom{6}{2} = 15$ and repeating one of the replicates. This gives a design with $v = 6$, $b = 18$, $r = 6$, $k = 2$, $m = 3$, $n = 2$, $\lambda_1 = 2$, and $\lambda_2 = 1$:

$$AB, CD, EF; \quad AB, CD, EF; \quad AC, BE, DF; \quad AD, BF, CE; \quad AE, BD, CF;$$
$$AF, BC, DE.$$

A design for $v' = r = 6$, $k' = k + 1 = 3$, $m = 3$, $n = 2$, $\lambda_1 = 2$, and $\lambda_2 = 1$ is

$$UWX, \quad VWX, \quad WYZ, \quad XYZ, \quad UVY, \quad UVZ.$$

The combined design with $\tilde{v} = 12$, $\tilde{b} = 24$, $\tilde{r} = 6$, $\tilde{k} = 3$, $\tilde{m} = 6$, $\tilde{n} = 2$, $\tilde{\lambda}_1 = 2$, and $\tilde{\lambda}_2 = 1$ is

$$\begin{array}{llllllll} ABU, & CDU, & EFU, & ABV, & CDV, & EFV, & ACW, & BEW, \\ DFW, & ADX, & BFX, & CEX, & AEY, & BDY, & CFY, & AFZ, \\ BCZ, & DEZ, & UWX, & VWX, & WYZ, & XYZ, & UVY, & UVZ. \end{array}$$

(e) Difference Sets

We quote without proof the following theorem of Bose et al. (1953), which is similar to Bose's fundamental theorems for balanced incomplete block designs.

Theorem. *Let M be a module with m elements and let n treatments correspond to each element. Suppose that we can find a set of t initial blocks $B_1, B_2, \ldots, B_t$ each of size k, and an initial group G of n elements, such that*

(i) *The $n(n - 1)$ differences (reduced mod m) arising from G are all different, and*

(ii) *Among the $k(k - 1)t$ differences occurring in the initial blocks, each difference occurs λ_2 times, except those that arise in G, which each occur λ_1 times.*

Then a GD design is obtained by cycling (mod m) from the initial blocks B_1, $B_2, \ldots, B_t$.

EXAMPLE. $v = b = 14$, $m = 7$, $n = 2$, $\lambda_1 = 0$, and $\lambda_2 = 1$. There are two initial blocks $(1_1, 2_1, 4_1, 0_2)$, $(1_2, 2_2, 4_2, 0_1)$. We denote the jth variety in the ith group by i_j, $0 \leq i \leq m - 1$, $0 \leq j \leq n - 1$. Arithmetic is mod 7. The initial group is $0_1, 0_2$. In the differences [1, 1] each value $1, 2, \ldots, 6$ occurs once; the same is true for the differences [2, 2]. In the differences [1, 2] and again in the differences [2, 1], each value $0, 1, 2, \ldots, 6$ occurs once, except for zero, which does not appear. In the initial group there are no differences [1, 1] or [2, 2]; the only difference [1, 2] or [2, 1] is zero. Thus, the conditions of the theorem are satisfied and we have the following design:

$$\begin{array}{llll}
1_1\ 2_1\ 4_1\ 0_2, & 2_1\ 3_1\ 5_1\ 1_2, & 3_1\ 4_1\ 6_1\ 2_2, & 4_1\ 5_1\ 0_1\ 3_2, \\
5_1\ 6_1\ 1_1\ 4_2, & 6_1\ 0_1\ 2_1\ 5_2, & 0_1\ 1_1\ 3_1\ 6_2, & \\
1_2\ 2_2\ 4_2\ 0_1, & 2_2\ 3_2\ 5_2\ 1_1, & 3_2\ 4_2\ 6_2\ 2_1, & 4_2\ 5_2\ 0_2\ 3_1, \\
5_2\ 6_2\ 1_2\ 4_1, & 6_2\ 0_2\ 2_2\ 5_1, & 0_2\ 1_2\ 3_2\ 6_1. &
\end{array}$$

Again, it is sometimes possible to find fractional or partial designs, and the previous theorem can be generalized.

Let M be a module with cm elements and let n/c (an integer) varieties correspond to each element of M. Suppose that we can find a set of t initial blocks of k treatments each and an initial group G of n treatments such that

(i) The differences arising from G take $n(n - 1)/c$ different values, each of which is repeated c times, the values being equally spaced;
(ii) Among the $kt(k - 1)$ differences in the initial blocks, each value occurs λ_2 times, except for those occurring in G, which occur λ_1 times each.

Then, developing the initial blocks gives a *GD* design with parameters $v = mn$, $b = mct$, $r = kct/n$, m, n, λ_1, λ_2.

EXAMPLE. $v = b = 12$, $r = k = 4$, $m = 6$, $n = 2$, $\lambda_1 = 2$, $\lambda_2 = 1$. Initial block 0, 1, 4, 6; initial group (0, 6); $c = 2$. The design is

$$\begin{array}{llll}
0\ 1\ 4\ 6, & 3\ 4\ 7\ 9, & 6\ 7\ 10\ 0, & 9\ 10\ 1\ 3, \\
1\ 2\ 5\ 7, & 4\ 5\ 8\ 10, & 7\ 8\ 11\ 1, & 10\ 11\ 2\ 4, \\
2\ 3\ 6\ 8, & 5\ 6\ 9\ 11, & 8\ 9\ 0\ 2, & 11\ 0\ 3\ 5.
\end{array}$$

(f) Designs Based on Projective Geometries

Some designs based on finite geometries are given in the original paper by Bose and Nair (1939). We give an illustrative example here.

In the Euclidean geometry with $N = 2$, consider only lines that do not pass through the origin. Then lines are blocks, points are treatments, and we have a design with $v = b = s^2 - 1$, $r = k = s$. If points on the same line are second associates, $\lambda_2 = 1$ and $\lambda_1 = 0$. This is the design that is obtained by starting with the orthogonal series balanced design $v = k^2 = s^2$, $r = k + 1$, and $\lambda = 1$, and cutting out all the blocks that contain a particular treatment, namely the treatment represented by (0, 0). It is a particular case of the method (a) presented in this section.

14.7 Designs with the Triangular Scheme

We recall that in the triangular scheme $v = n(n-1)/2$, $n_1 = 2(n-2)$, $n_2 = (n-2)(n-3)/2$, $p_{11}^1 = n - 2$, and $p_{11}^2 = 4$. The varieties are arranged above the diagonal of a square array of side n, the diagonal positions in the array are left blank, and the varieties are repeated below the diagonal so as to make the array symmetric. Two varieties are first associates if they appear in the same row (or column) of the array; otherwise they are second associates. The latent roots of $\mathbf{NN}'$ and their multiplicities are

$$\theta_1 = r + (n-4)\lambda_1 - (n-3)\lambda_2, \qquad \alpha_1 = n - 1;$$

$$\theta_2 = r - 2\lambda_1 + \lambda_2, \qquad \alpha_2 = n(n-3)/2.$$

Connor (1958) showed that for $n \geq 9$ the triangular scheme is unique in the sense that if a *PBIB*(2) scheme has the values just given for the parameters of the first kind, then it must be a triangular scheme. Shrikhande (1959*a*) proved uniqueness for $n \leq 6$. Hoffman (1960) showed that the scheme was unique for $n = 7$ but not for $n = 8$. Chang (1959) also showed that the scheme was not unique for $n = 8$ and later (1960) showed that for the parameters $v = 28$, $n_1 = 12$, $n_2 = 15$, $p_{11}^2 = 4$, and so on, there are exactly three schemes possible in addition to the triangular scheme.

Chang et al. (1965) found that there are exactly 225 combinations of values, all integers, of n, r, k, λ_1, and λ_2, with $r \leq 10$ and $k \leq 10$, satisfying the following necessary conditions:

$$2v = n(n-1), \qquad b = rv/k \text{ is an integer},$$

$$2(n-2)\lambda_1 + (n-2)(n-3)\lambda_2/2 = r(k-1), \qquad 0 \leq \lambda_1 < r, \quad 0 \leq \lambda_2 < r,$$

and, since the complement is also a triangular design,

$$2r - b < \lambda_1, \qquad 2r - b < \lambda_2.$$

For seventy-six of these combinations, designs are given by Clatworthy (1956); some of them are also included in the tables of Bose et al. (1954). Chang et al.

provide designs for twenty-one more combinations and show that no designs exist in a hundred and ten other cases. Eighteen remain unsolved: their parameters are given in Table 14.1. Number 7 has been solved by K. R. Aggarwal. See *Ann. Math. Statist.* (1972), **43**, p. 371.

TABLE 14.1

Reference Number	n	v	k	r	b	λ_1	λ_2
1	6	15	5	9	27	3	2
2				10	30	2	4
3	7	21	5	10	42	1	3
4						3	1
5			6	6	21	2	1
6				8	28	3	1
7				10	35	2	3
8						3	2
9			7	10	30	2	4
10						4	2
11						5	1
12			10	10	21	4	5
13	9	36	4	7	63	0	1
14	10	45	5	7	63	0	1
15			9	9	45	1	2
16	11	55	5	9	99	0	1
17			10	10	55	3	1
18	12	66	6	9	99	0	1

There are four methods of obtaining triangular designs in addition to the general methods for all schemes which were mentioned before.

(i) The dual of a balanced incomplete block design with parameters

$$v^* = n, \qquad b^* = n(n-1)/2, \qquad r^* = n-1, \qquad k^* = 2, \qquad \lambda^* = 1,$$

i.e., the unreduced design for n treatments in blocks of two, is a triangular design with

$$b = n, \qquad r = 2, \qquad k = n-1, \qquad \lambda_1 = 1, \qquad \lambda_2 = 0.$$

This design is a special case of the first of the two dual methods given by Shrikhande (1952). Each block is one of the rows of the array in the association scheme. These are called the triangular singly linked blocked

designs because each pair of blocks has a single treatment in common.
For $n = 5$ the array is

$$\begin{matrix} * & 1 & 2 & 3 & 4 \\ 1 & * & 5 & 6 & 7 \\ 2 & 5 & * & 8 & 9 \\ 3 & 6 & 8 & * & 0 \\ 4 & 7 & 9 & 0 & *, \end{matrix}$$

and this design has the five blocks 1234, 1567, 2589, 3680, 4790.

(ii) Shrikhande's second method gives the triangular doubly linked block designs. These are the duals of balanced designs with parameters

$$v^* = \binom{n-1}{2}, \qquad b^* = \binom{n}{2}, \qquad r^* = n, \qquad k^* = n - 2, \qquad \lambda^* = 2.$$

The triangular designs have parameters

$$v = \binom{n}{2}, \quad b = \binom{n-1}{2}, \quad k = n, \quad r = n - 2, \quad \lambda_1 = 1, \quad \lambda_2 = 2.$$

For $n = 5$ we may take the balanced design

ABC, *BDE*, *AEF*, *CDF*, *CEF*, *BDF*, *ADE*, *ACD*, *ABF*, *BCE*,

and, letting the first block correspond to treatment 1, the second to treatment 2, and so on, obtain the triangular design

13789, 12690, 14580, 24678, 23570, 34569.

Conversely, a triangular design with $r = n - 2$, $\lambda_1 = 1$, and $\lambda_2 = 2$ has for the latent roots of $\mathbf{NN}'$: $\theta_1 = 0$, $\theta_2 = r = n - 2$, and $\alpha_2 = b - 1$. Hence, its dual has only one latent root other than $\theta_0 = rk$, namely $n - 2$ with multiplicity $\alpha_2 = v^* - 1$. Thus the dual is a balanced design.

Raghavarao (1960*b*) proved that if for a triangular design $\theta_1 = 0$, or, equivalently,

$$rk - v\lambda_1 = n(r - \lambda_1)/2,$$

which is true for the doubly linked block designs, then $2k$ is divisible by n and every block of the design contains exactly $2k/n$ treatments from each of the n rows of the scheme. This result is analogous to the similar result for semiregular *GD* designs and the proof follows the same lines.

The complement of the design of method (ii) for $n = 6$ is a linked block design of triangular type with $v = 15$, $b = 10$, $r = 6$, $k = 9$, $n_1 = 8$, $n_2 = 6$, $\lambda_1 = 3$, and $\lambda_2 = 4$; each block contains three treatments from each row of the scheme. It is the dual of the balanced design with parameters

$$v^* = 10, \qquad b^* = 15, \qquad r^* = 9 = n(n-3)/2,$$
$$k^* = 6 = (n-2)(n-3)/2, \qquad \lambda^* = 5 = (n-1)(n-4)/2.$$

It follows from the foregoing discussion that all triangular designs with $r = n - 2$, $\lambda_1 = 1$, and $\lambda_2 = 2$ are doubly linked block designs and that the existence of such a design for any n implies the existence of the corresponding balanced design. For $n = 7, 8, 10$ the corresponding designs would be balanced designs with parameters

$$v = 15,\ k = 5,\ \lambda = 2; \quad v = 21,\ k = 6,\ \lambda = 2; \quad v = 36,\ k = 8,\ \lambda = 2.$$

However, Hall and Connor (1953) showed that these three designs are impossible. This implies that triangular designs with $r = n - 2$, $\lambda_1 = 1$, and $\lambda_2 = 2$ do not exist for $n = 7, 8$, or 10.

(iii) Each of the rows in the scheme constitutes a set of $(n - 1)$ varieties with the property that any pair of them are first associates. If there exists a balanced incomplete block design with parameters $v^* = n - 1$, b^*, r^*, $k^* = k$, λ^*, we can take such a design for each row of the scheme and combine them into a triangular design with

$$v = n(n-1)/2, \quad b = nb^*, \quad r = 2r^*, \quad k, \quad \lambda_1 = \lambda^*, \quad \lambda_2 = 0.$$

For $n = 5$, $v = 10$, and $k = 3$, we note that each row of the scheme has $n - 1 = 4$ varieties and use the balanced design with $v = b = 4$, $r = k = 3$ to obtain the design

1 2 3,	1 5 6,	2 5 8,	3 6 8,	4 7 9,
1 2 4,	1 5 7,	2 5 9,	3 6 0,	4 7 0,
1 3 4,	1 6 7,	2 8 9,	3 8 0,	4 9 0,
2 3 4,	5 6 7,	5 8 9,	6 8 0,	7 9 0.

For $k \geq 4$, this is the only way of obtaining a design with $\lambda_2 = 0$. Hence, a triangular design with $4 \leq k \leq n - 1$, r, λ_1, and $\lambda_2 = 0$ exists if, and only if, a balanced incomplete block design with $v^* = n - 1$, $k^* = k \geq 4$, $r^* = r/2$, and $\lambda^* = \lambda_1$ exists.

(iv) The method of triads is due to Clatworthy. When $k = 3$, method (iii) is not the only method of obtaining a design with $\lambda_1 > 0$ and $\lambda_2 = 0$. Another method is to obtain the blocks as follows. Take a pair of varieties that appear in the same row. The first variety appears again in another row. For the third member of the block take the variety that appears in that row and in the same column as the second member. Thus, in the case $n = 5$ the blocks would be

$$\begin{array}{ccccc} 1\,2\,5, & 1\,3\,6, & 1\,4\,7, & 2\,3\,8, & 2\,4\,9, \\ 3\,4\,0, & 5\,6\,8, & 5\,7\,9, & 6\,7\,0, & 8\,9\,0. \end{array}$$

In the general case, $r = n - 2$, $6b = n(n - 1)(n - 2)$.

14.8 The Nonexistence of Some Triangular Designs

The latent roots of $\mathbf{NN}'$ are not negative; thus it is necessary that

$$r + (n - 4)\lambda_1 - (n - 3)\lambda_2 \geq 0, \qquad r - 2\lambda_1 + \lambda_2 \geq 0.$$

This rules out, for example, designs with parameters

$$n = 5, v = 10, r = 9, k = 5, b = 18, \lambda_1 = 2, \lambda_2 = 8, (\theta_1 = -5);$$

$$n = 5, v = 10, r = 3, k = 5, b = 6, \ \lambda_1 = 2, \lambda_2 = 0, (\theta_2 = -1).$$

For symmetric designs it is necessary that

$$|\mathbf{NN}'| = r^2\theta_1^{n-1}\theta_2^{n(n-3)/2}$$

be a square. If neither θ_1 nor θ_2 is zero, it follows that

(i) If $n = 4t$, θ_1 must be a square;
(ii) If $n = 4t + 1$, θ_2 must be a square;
(iii) If $n = 4t + 2$, $\theta_1\theta_2$ must be a square.

Examples of designs that are found to be nonexistent by these criteria are

(a) $n = 8, v = b = 28, r = k = 10, \lambda_1 = 5, \lambda_2 = 2, (\theta_1 = 20)$;
(b) $n = 9, v = b = 36, r = k = 8, \ \lambda_1 = 1, \lambda_2 = 2, (\theta_2 = 8)$;
(c) $n = 6, v = b = 15, r = k = 5, \ \lambda_1 = 1, \lambda_2 = 2, (\theta_1\theta_2 = 5)$.

Further such conditions involving the Hasse–Minkowski invariant are given by Ogawa (1959).

14.9 Designs for the Latin Square Scheme

These designs are discussed by Clatworthy (1956) and (1967), and by Chang and Liu (1964). They include as a special case the designs obtained by taking some, but not all, of the replicates in the Yates orthogonal series designs for

$v = k^2$, $b = k(k + 1)$, $r = k + 1$, and $\lambda = 1$. Clatworthy (1956) gives a listing of 111 designs most of which are L_2 designs. Chang and Liu have calculated that there are 155 combinations of parameters that might lead to L_2 designs with $r \leq 10$ and $k \leq 10$. Of these, seventy-two are included in Clatworthy's listing; Chang and Liu give twenty-one others and show that fifty-seven of the designs do not exist. The five unsolved cases are

$$\begin{array}{lllll} v = 36, & k = 6, & r = 9, & \lambda_1 = 2, & \lambda_2 = 1, \\ v = 36, & k = 6, & r = 10, & \lambda_1 = 0, & \lambda_2 = 2, \\ v = 36, & k = 9, & r = 10, & \lambda_1 = 3, & \lambda_2 = 2, \\ v = 49, & k = 9, & r = 9, & \lambda_1 = 3, & \lambda_2 = 1, \\ v = 100, & k = 10, & r = 9, & \lambda_1 = 0, & \lambda_2 = 1. \end{array}$$

Shrikhande (1959*b*) has shown that the L_2 scheme is unique except for $n = 4$. In this case there are two schemes with $v = 16$, $n_1 = 6$, $n_2 = 9$, and $p_{11}^1 = 2$. The second scheme arises when, in one of the L_3 schemes, we interchange first and second associates. Suppose that the square array and the Latin square are

$$\begin{array}{rrrr} 1 & 2 & 3 & 4 \\ 5 & 6 & 7 & 8 \\ 9 & 10 & 11 & 12 \\ 13 & 14 & 15 & 16 \end{array} \qquad \begin{array}{cccc} A & B & C & D \\ B & C & D & A \\ C & D & A & B \\ D & A & B & C. \end{array}$$

Let two varieties be first associates if they do not appear in the same row or column or with the same letter; this gives a set of parameters $v = 16$, $n_1 = 6$, $n_2 = 9$, and $p_{11}^1 = 2$ as in the L_2 scheme, but consider the first associates of variety 1. They are 6, 7, 10, 12, 15, and 16. The first associates of 6 are 1, 4, 11, 12, 13, and 15, and so varieties 12 and 15 are first associates of both 1 and 6. If the scheme is an L_2 scheme, it follows that 1, 6, 12, and 15 must appear in the same column or row of a square array and, hence, 12 and 15 are first associates. Twelve and 15 both go with the letter B, and so they are second associates, in contradiction. This scheme was first mentioned by Mesner (1956) and (1967).

In the same paper Shrikhande gives conditions for the nonexistence of symmetric L_2 designs based on the Hasse–Minkowski invariant.

The latent roots of $\mathbf{NN}'$ and their multiplicities are

$$\theta_1 = r - (i - n)(\lambda_1 - \lambda_2) - \lambda_2, \qquad \alpha_1 = i(n - 1),$$

$$\theta_2 = r - i(\lambda_1 - \lambda_2) - \lambda_2, \qquad \alpha_2 = (n - 1)(n - i + 1).$$

For the L_2 case, α_1 is always even; α_2 is odd when n is odd, and then $\theta_2 = r - 2\lambda_1 + \lambda_2$ must be a square if the design is symmetric and $\theta_1 > 0$.

Raghavarao (1960*b*) gives a result for L_2 designs similar to those just mentioned for semiregular GD designs and triangular designs. If $rk - v\lambda_1 = n(r - \lambda_1)$, then k is divisible by n and every block of the design contains k/n varieties from each row of the association scheme. It should be noted, however, that this condition is not equivalent either to $\theta_1 = 0$ or to $\theta_2 = 0$.

Clatworthy (1967) has shown that most of the known L_i designs belong to four families. We shall discuss each of the families in turn. Our notation differs from that of Clatworthy in that he talks of $v = s^2$ varieties while we have $v = n^2$. We shall alter the definitions of families A and D slightly in order to include more designs. Clatworthy considers numerous special cases in each family; the reader is referred to his paper for these.

Family A: These are designs with $k \leq n$ and $\lambda_2 = 0$. Let D be a balanced incomplete block design with parameters $v^* = n, b^*, k^* = k, r^*, \lambda^* = \lambda_1$. The first b^* blocks of our design consist of the blocks of D using the varieties in the first row of the square array of the association scheme; D is then repeated using the varieties in the second row, and so on. After all the rows have been used, we repeat the procedure using the columns of varieties. If $i > 2$, the procedure is then repeated again for each of the $i - 2$ orthogonal squares. We then have a design with

$$b = inb^*, \qquad r = ir^*, \qquad k, \qquad \lambda_1, \qquad \lambda_2 = 0.$$

If the balanced design D is the unreduced design consisting of all $\binom{n}{k}$ subsets of k out of the n varieties, we have

$$b = ni\binom{n}{k}, \qquad \lambda_1 = \binom{n-2}{k-2}, \qquad \lambda_2 = 0.$$

If $k = n$, the design consists of ni blocks forming the first i replicates of the balanced lattice, if that design exists, with $\lambda_1 = 1$ and $\lambda_2 = 0$. We shall use this design later.

Family B: These are designs with $k = 2n$. The first $n(n - 1)/2$ blocks consist of all possible pairs of rows from the square array, the next $n(n - 1)/2$ blocks consist of pairs of columns, and so on. For the L_i design we have

$$b = in(n - 1)/2, r = i(n - 1), k = 2n, \lambda_1 = n + i - 2, \lambda_2 = i.$$

Family C: These designs have $k = n$, and they exist whenever n is a prime, or a power of a prime. We have already mentioned the design in family A with $k = n$, $\lambda_1 = 1$, and $\lambda_2 = 0$. If we proceed in the same way to set up a second design using the remaining Latin squares of the complete orthogonal set, we shall have a second design with $b = n(n - i + 1)$, $k = n$, $\lambda_1 = 0$, and $\lambda_2 = 1$. If we wish to construct a design with parameters λ_1, λ_2, we have now merely to repeat the first design λ_1 times and the second design λ_2 times. The reader will appreciate that if there exists a balanced design with $\lambda = 1$ for a value of $k \leq s$, we may use a similar procedure.

Family D: This family consists of the $ES(i)$ and $E'S(i)$ designs. The parameters for these symmetric designs are

$$\begin{aligned}
&ES(1): && k = i(n-1), \quad \lambda_1 = i^2 - 3i + n, \quad \lambda_2 = i(i-1),\\
&ES(2): && k = (n-1)(n-i+1), \quad \lambda_1 = (n-i)(n-i+1),\\
& && \lambda_2 = (n-i)^2 + i - 2,\\
&E'S(1): && k = i(n-1) + 1, \quad \lambda_1 = i^2 - 3i + n + 2, \quad \lambda_2 = i(i-1),\\
&E'S(2): && k = (n-1)(n-i+1) + 1, \quad \lambda_1 = (n-i)(n-i+1),\\
& && \lambda_2 = (n-i)^2 + i.
\end{aligned}$$

Exercises

1. Prove that if $\mathbf{A}$ and $\mathbf{B}$ are square matrices such that $\mathbf{AB}$ and $\mathbf{BA}$ both exist, then $\mathbf{AB}$ and $\mathbf{BA}$ both have the same nonzero latent roots.
2. Derive the $\mathbf{P}$ matrices for Shrikhande's two linked block designs.
3. Let $\mathbf{U}$ be a square matrix of order mn partitioned into m^2 square matrices U_{ij} of order n where $U_{ij} = a\mathbf{I}_n + b\mathbf{J}_n$ if $i = j$, and $c\mathbf{I}_n + d\mathbf{J}_n$ if $i \neq j$. Recalling that $|a\mathbf{I}_n + b\mathbf{J}_n| = a^{n-1}(a + nb)$, show that

$$|\mathbf{U}| = (a - c)^{(n-1)(m-1)}(a + nb - c - nd)^{m-1}(a + (m-1)c)^{n-1} \times \{a + nb + (m-1)(c + nd)\}.$$

4. (Continuation). Order the varieties of a GD scheme so that the ith group consists of varieties $(i-1)n + 1, \ldots, in$. Then evaluate directly the determinant $|\mathbf{NN}'|$.
5. (Continuation). Show that if the varieties in an L_2 scheme are suitably ordered

$$|\mathbf{NN}'| = rk(r + (n-2)\lambda_1 - (n-1)\lambda_2)^{2(n-1)}(r - 2\lambda_1 + \lambda_2)^{(n-1)^2}.$$

   Hence find the latent roots of $\mathbf{NN}'$ and their multiplicities.
6. Prove that if a triangular design has $\theta_1 = 0$ then $2k$ is divisible by n and each block contains exactly $2k/n$ treatments from each row of the scheme [Raghavarao (1960*b*)].
7. Prove that if, in a design with the L_2 scheme,

$$rk - v\lambda_1 = n(r - \lambda_1),$$

   then k is divisible by n and every block of the design contains k/n treatments from each row of the scheme [Raghavarao (1960*b*)].
8. Show that for any L_2 design the following inequalities hold for b. Let $\alpha = \min(n-1, [r/\lambda_1])$, $\beta = \min(n-1, [r/\lambda_2])$, where $[x]$ is the integer part of x. Then [Chang and Liu (1964)]

$$2b \geq (\alpha + 1)(2r - \alpha\lambda_1), \qquad 2b \geq (\beta + 1)(2r - \beta\lambda_2).$$

9. Consider a $PBIB(2)$ association scheme A in which there exists a set of u_i treatments that are all ith associates of each other. Let

$$w_i = \min(u_i - 1, [r/\lambda_i]), \qquad w_i' = \min(u_i - 1, [(b - r)/(b - 2r + \lambda_i)]).$$

Show that for any design having associate scheme A

$$2b \geq (w_i + 1)(2r - w_i\lambda_i), \qquad 2b \geq (w_i' + 1)(2b - 2r - w_i'(b - 2r + \lambda_i)).$$

Adapt this result to obtain a nonexistence theorem for triangular designs [Chang, et al. (1965)].

10. Show that for any $PBIB(2)$ design which is not of the GD type [Chang et al. (1965)],

$$\max(2\lambda_1 - \lambda_2, 2\lambda_2 - \lambda_1) \leq r.$$

CHAPTER 15

Additional Topics in Partially Balanced Designs

In this chapter we shall continue our investigation of partially balanced designs, with emphasis on designs having more than two associate classes. We begin, however, with a short introduction to the topic of partial geometries and their application to *PBIB* (2) designs.

We have already emphasized the usefulness of the latent roots of $\mathbf{NN}'$ in determining the nonexistence of some designs. We shall discuss the subject of the latent roots of $\mathbf{NN}'$ and their multiplicities in *PBIB*(m) schemes at some length. The discussion of that topic will conclude with a few words about cyclic and pseudocyclic schemes with two associate classes. In the succeeding section, we shall introduce several *PBIB*(3) schemes. The chapter will conclude with a discussion of cyclic designs.

15.1 Geometric Designs

In the previous chapter we dealt with designs having the three principal types of schemes with two associate classes: group divisible, triangular, and Latin square. There was only one reference to designs obtained from finite geometries, and that example duplicated a design given earlier. Finite geometries do, however, provide a series of simple partially balanced designs, i.e., designs having either $\lambda_1 = 1$, $\lambda_2 = 0$, or $\lambda_1 = 0$, $\lambda_2 = 1$, which do not belong to the three principal types.

Bose and Nair (1939) presented several examples of designs obtained from Euclidean or projective geometries. If we consider the set of all lines in a Euclidean geometry with s^N points and omit all the lines that pass through the origin O, we can let points denote varieties and lines denote blocks. The remaining points and lines form a GD design for $v = s^N - 1$ varieties in blocks of size $k = s$ with $\lambda_1 = 0$ and $\lambda_2 = 1$. Two varieties are first associates if the line joining the points passes through the origin. Then $n_1 = s - 1$.

Other GD designs are obtained if we consider m-dimensional subspaces $(m > 1)$ and again omit those subspaces that contain the origin O. Bose and Nair also considered omitting from a projective geometry, $PG(2, s = p^n)$, all the points on three lines that are not concurrent, and all the lines passing through the vertices of the triangle formed by those three lines.That procedure led them to some symmetric L_3 designs with $v = (s - 1)^2$, $k = (s - 2)$, $\lambda_2 = 1$, and $\lambda_1 = 0$.

More interesting, perhaps, are the designs obtained from partial geometries because every partial geometry is isomorphic to a simple $PBIB(2)$ design. A partial geometry is a collection of v points (varieties) and b lines (blocks) with the following axioms:

(i) There is only one line through any pair of points;
(ii) There are exactly k points on each line;
(iii) There are exactly r lines through each point;
(iv) If a point P does not lie on a line p, then there are exactly t lines through P that intersect p.

This system constitutes a partial geometry with parameters (r, k, t). It follows that

$$tv = k[(r - 1)(k - 1) + t], \; tb = r[(r - 1)(k - 1) + t].$$

To derive the latter relationship, we note that there are r lines passing through P, each with $k - 1$ points other than P. Thus, there are $r(k - 1)$ points connected to P. On the other hand, there are $b - r$ lines that do not pass through P. Each of these $b - r$ lines contains t points that are connected to P; each of the points is on $r - 1$ of the $b - r$ lines; thus, there are $t(b - r)/(r - 1)$ points connected to P. Hence,

$$r(k - 1) = t(b - r)/(r - 1) \qquad \text{and} \qquad b = r[(r - 1)(k - 1) + t]/t.$$

If we now let points denote varieties and lines denote blocks, we have a simple partially balanced design. Two varieties are first associates if their points are connected in the partial geometry; otherwise, they are second associates. We have $n_1 = r(k - 1)$, $n_2 = (r - 1)(k - 1)(k - t)/t$. To obtain p_{11}^1 we consider two first associates P and Q. There are $k - 2$ other points on PQ. There are $r - 1$ lines through Q other than PQ; each of these lines contains t first associates of P. Thus,

$$p_{11}^1 = (t - 1)(r - 1) + k - 2, \qquad p_{12}^1 = (r - 1)(k - t),$$
$$p_{11}^2 = rt, \qquad p_{12}^2 = r(k - t - 1).$$

Bose and Clatworthy (1955) show that every $PBIB(2)$ design with $\lambda_1 = 1$, $\lambda_2 = 0$, and $b < v$ is a partial geometry. It follows that the singly linked block designs are partial geometries. These are one of the series of linked block designs obtained by Shrikhande (1952), and are duals of the balanced incomplete block designs with $v^* = rk - k + 1$, $r^* = k$, $k^* = r$, and $\lambda^* = 1$. They were mentioned at the beginning of Chapter 14. When $r = 2$, the singly linked block design is a triangular design. Thus the triangular association scheme can be obtained from a partial geometry configuration.

For a further discussion of this topic the reader is referred to the review paper by Bose (1963) and the two monographs by Vajda (1967*a*,*b*). Further research in this area is being carried out by students of graph theory. The representation of association schemes as concordant graphs is mentioned by Bose and Mesner (1959).

Mesner (1967) has extended the L_i scheme to include cases where n and i are both negative, although all the design parameters are positive. This gives the NL family of schemes (negative Latin square). The $NL_i(n)$ scheme has $v = n^2$, and the other parameters are obtained by putting $-n$ for n and $-i$ for i in the corresponding formulae for the L_i scheme. For example, the $NL_3(8)$ scheme has $v = 64$, $n_1 = -3(-8 - 1) = 27$, $n_2 = (-8 + 3 + 1)(-8 - 1) = 36$. The $\mathbf{P}$ matrices are

$$\mathbf{P}_1 = \begin{bmatrix} 10 & 16 \\ 16 & 20 \end{bmatrix}, \qquad \mathbf{P}_2 = \begin{bmatrix} 12 & 15 \\ 15 & 20 \end{bmatrix}.$$

Numerous designs with NL schemes can be obtained from finite geometries, and Mesner gives some examples in the reference cited.

15.2 Latent Roots of the Intrablock and Interblock Matrices

The formula for the intrablock efficiency of an incomplete block design obtained at the end of Chapter 11 involves the nonzero latent roots of $\mathbf{A}$. We have seen that for a linked block design $\mathbf{A}$ has a single nonzero latent root with multiplicity $b - 1$. We now investigate further the latent roots of $\mathbf{A}$ and $\mathbf{NN}'$ for $PBIB$ designs.

In solving the intrablock equations for the case of two associate classes (Chapter 12) we obtained (12.2). Adding the quantity $\sum_{i=1}^{v} \lambda_1 n_1 \hat{\tau}_i$ gives

$$\begin{aligned} kS_1(Q_h) = {} & S_1(\hat{\tau}_h)\{r(k-1) - \lambda_1 p_{11}^1 - \lambda_2 p_{12}^1 + \lambda_1 n_1\} \\ & + S_2(\hat{\tau}_h)\{-\lambda_1 p_{11}^2 - \lambda_2 p_{12}^2 + \lambda_1 n_1\}. \end{aligned}$$

In the general case of m associate classes

$$S_i(Q_h) = \sum_{j=1}^{m} \varphi_{ij} S_j(\hat{\tau}_h),$$

where $k\varphi_{ij} = r(k-1)\delta_{ij} - \sum_{k=1}^{v} \lambda_k p_{ik}^j + \lambda_i n_i$, and δ_{ij} is the Kronecker delta.

Let $\mathbf{C}_h$ be a matrix of m rows and v columns defined as follows: $c_{ij} = 1$ if the jth variety is an ith associate of the hth variety; otherwise $c_{ij} = 0$. Then $\mathbf{C}_h\mathbf{Q}$ is the vector $(S_i(Q_h))$ and $\mathbf{C}_h\hat{\boldsymbol{\tau}}$ is the vector $(S_i(\hat{\tau}_h))$. Thus, writing $\boldsymbol{\Phi} = (\varphi_{ij})$ we have

$$\mathbf{C}_h\mathbf{Q} = \boldsymbol{\Phi}\mathbf{C}_h\hat{\boldsymbol{\tau}}.$$

If $\boldsymbol{\Phi}$ is not singular (i.e., if the design is connected),

$$\mathbf{C}_h\hat{\boldsymbol{\tau}} = \boldsymbol{\Phi}^{-1}\mathbf{C}_h\mathbf{Q};$$

but

$$\hat{\tau}_h = -\sum_{i=1}^{m} S_i(\hat{\tau}_h) = -\mathbf{1}'\mathbf{C}_h\hat{\boldsymbol{\tau}},$$

hence,

$$\hat{\tau}_h = -\mathbf{1}'\boldsymbol{\Phi}^{-1}\mathbf{C}_h\mathbf{Q}.$$

We now drop the subscript h, noting that for any choice of h,

$$\boldsymbol{\Phi}\mathbf{C} = \mathbf{C}\mathbf{A} + \mathbf{D}\mathbf{1}',$$

where $k\mathbf{D}' = (\lambda_1 n_1, \lambda_2 n_2, \lambda_3 n_3, \ldots)$.

$\mathbf{A}$ has a latent root $\psi_0 = 0$, with the corresponding latent vector $\mathbf{1}$; if the design is not connected, the zero latent root will be repeated.

We now show that the matrices $\mathbf{A}$ and $\boldsymbol{\Phi}$ have the same nonzero roots.

Let ψ be a nonzero latent root of $\mathbf{A}$ and $\mathbf{x}$ a corresponding vector such that $\mathbf{x}'\mathbf{1} = 0$. Then

$$\mathbf{A}\mathbf{x} = \psi\mathbf{x} \quad \text{and} \quad \mathbf{C}\mathbf{A}\mathbf{x} = \psi\mathbf{C}\mathbf{x},$$

but

$$\mathbf{C}\mathbf{A}\mathbf{x} = (\mathbf{C}\mathbf{A} + \mathbf{D}\mathbf{1}')\mathbf{x} = \boldsymbol{\Phi}\mathbf{C}\mathbf{x},$$

and so ψ is a latent root of $\boldsymbol{\Phi}$ with vector $\mathbf{C}\mathbf{x}$.

Conversely, let θ be a nonzero root of $\boldsymbol{\Phi}'$ with corresponding vector $\mathbf{x}$. Then

$$\mathbf{x}'\boldsymbol{\Phi} = \theta\mathbf{x}',$$

and

$$\mathbf{x}'\boldsymbol{\Phi}\mathbf{C} = \theta\mathbf{x}'\mathbf{C}.$$

Then

$$\mathbf{x}'\mathbf{C}\mathbf{A} + \mathbf{x}'\mathbf{D}\mathbf{1}' = \theta\mathbf{x}'\mathbf{C},$$

$$\mathbf{x}'(\mathbf{C} - \theta^{-1}\mathbf{D}\mathbf{1}')\mathbf{A} = \theta\mathbf{x}'(\mathbf{C} - \theta^{-1}\mathbf{D}\mathbf{1}'),$$

and θ is a latent root of $\mathbf{A}$ with $[\mathbf{C} - \theta^{-1}\mathbf{D}\mathbf{1}]'\mathbf{x}$ as vector.

Let $\mathbf{\Pi}^* = (p^*_{ij})$ where

$$p^*_{ij} = r\delta_{ij} + \sum_{k=1}^{m} \lambda_k p^j_{ik} - n_i\lambda_i,$$

then

$$k\mathbf{\Phi} = kr\mathbf{I} - \mathbf{\Pi}^*.$$

Since $k\mathbf{A} = rk\mathbf{I} - \mathbf{NN}'$, it follows that if θ_i is a latent root of $\mathbf{NN}'$ with multiplicity α_i, then $\psi_i = (rk - \theta_i)/k$ is a root of $\mathbf{A}$ with the same multiplicity α_i. It also follows, from the similar relationship between $\mathbf{\Phi}$ and $\mathbf{\Pi}^*$, that because the nonzero roots of $\mathbf{\Phi}$ are the nonzero roots of $\mathbf{A}$, the roots of $\mathbf{NN}'$ are, with the exception of rk, the same as the roots of $\mathbf{\Pi}^*$. This result was first shown by Connor and Clatworthy (1954). Bose and Mesner (1959) derived it by considering the algebra of the association matrices; in their paper they also proved that the multiplicities α_i are independent of $r, \lambda_1, \lambda_2, \ldots, \lambda_m$. Since $\mathbf{NN}'$ is nonnegative definite, none of its roots are negative.

The ith association matrix $\mathbf{B}_i$ is defined as follows: the element β_{ab} in the ath row and bth column, $0 \leq a \leq m$, $0 \leq b \leq m$, is 1 if a and b are ith associates and zero otherwise.

The equation $|\mathbf{\Pi}^*| = 0$ is a polynomial of the mth degree in r. If the determinant vanishes when r takes the values $s_1, s_2, \ldots, s_m$, the corresponding latent roots of $\mathbf{\Pi}^*$, and hence of $\mathbf{NN}'$, are $\theta_i = r - s_i$. The zeros s_h are linear functions of the λ_i, and so we may write

$$\boldsymbol{\theta} = r\mathbf{1} + \mathbf{Z}^*\boldsymbol{\lambda},$$

where $\boldsymbol{\theta}' = (\theta_1, \theta_2, \ldots, \theta_m)$, $\boldsymbol{\lambda}' = (\lambda_1, \lambda_2, \ldots, \lambda_m)$.

Adding the root $\theta_0 = rk$, we have

$$\text{tr}(\mathbf{NN}') = rv = rk + \sum_{i=1}^{m} \alpha_i\theta_i,$$

or, writing $\boldsymbol{\alpha}' = (\alpha_1, \alpha_2, \ldots, \alpha_m)$, $\mathbf{n}' = (n_1, n_2, \ldots, n_m)$,

$$r(v - k) = \boldsymbol{\alpha}'\boldsymbol{\theta} = r\boldsymbol{\alpha}'\mathbf{1} + \boldsymbol{\alpha}'\mathbf{Z}^*\boldsymbol{\lambda}.$$

Then, since $\boldsymbol{\alpha}'\mathbf{1} = v - 1$,

$$\boldsymbol{\alpha}'\mathbf{Z}^*\boldsymbol{\lambda} = -r(k - 1) = -\mathbf{n}'\boldsymbol{\lambda},$$

which is, since the α_i are independent of the λ_i, an identity in the λ_i. This leads to m equations to be solved for the m unknowns $\alpha_1, \ldots, \alpha_m$, $(\mathbf{Z}^*)'\boldsymbol{\alpha} = -\mathbf{n}$.

For two associate classes det $\mathbf{\Pi}^* = 0$ is a quadratic in θ. Connor and Clatworthy (1954) have shown that after some simplification it may be written

$$(r - \theta)^2 + [(\lambda_1 - \lambda_2)(p^2_{12} - p^1_{12}) - (\lambda_1 + \lambda_2)](r - \theta)$$
$$+ [(\lambda_1 - \lambda_2)(\lambda_2 p^1_{12} - \lambda_1 p^2_{12}) + \lambda_1\lambda_2] = 0,$$

whence, putting $\gamma = p_{12}^2 - p_{12}^1$, $\beta = p_{12}^1 + p_{12}^2$, and $\Delta = \gamma^2 + 2\beta + 1$, it follows that

$$\theta_1 = r + \tfrac{1}{2}(\sqrt{\Delta} + \gamma - 1)\lambda_1 - \tfrac{1}{2}(\sqrt{\Delta} + \gamma + 1)\lambda_2,$$

$$\theta_2 = r - \tfrac{1}{2}(\sqrt{\Delta} - \gamma + 1)\lambda_1 + \tfrac{1}{2}(\sqrt{\Delta} - \gamma - 1)\lambda_2.$$

Furthermore,

$$\alpha_1 = \frac{n_1 + n_2}{2} - \frac{(n_1 - n_2) + \gamma(n_1 + n_2)}{2\sqrt{\Delta}},$$

and

$$\alpha_2 = \frac{n_1 + n_2}{2} + \frac{(n_1 - n_2) + \gamma(n_1 + n_2)}{2\sqrt{\Delta}}.$$

It is necessary that these multiplicities be positive integers. This imposes some restrictions upon the parameters for the existence of designs, especially if Δ is not a perfect square.

We calculate now the latent roots and their multiplicities for a group divisible design. In this case

$$|\mathbf{\Pi}^*| = \begin{vmatrix} r - \lambda_1 & (\lambda_2 - \lambda_1)(n - 1) \\ 0 & r + \lambda_1(n - 1) - \lambda_2 n_2 \end{vmatrix},$$

whence $\theta_1 = r + \lambda_1(n - 1) - \lambda_2 n = rk - v\lambda_2$ and $\theta_2 = r - \lambda_1$. We have $\gamma = \beta = n - 1$ and $\Delta = n^2$. Hence,

$$\alpha_1 = m - 1, \qquad \alpha_2 = m(n - 1).$$

Bose and Connor (1952) obtained this result directly by evaluating $|\mathbf{NN}'| = rk(rk - v\lambda_2)^{m-1}(r - \lambda_1)^{m(n-1)}$. Alternatively,

$$\mathbf{Z}^* = \begin{bmatrix} n - 1 & -n \\ -1 & 0 \end{bmatrix}, \qquad (\mathbf{Z}^*)^{-1} = -\frac{1}{n}\begin{bmatrix} 0 & n \\ 1 & n - 1 \end{bmatrix},$$

$$\boldsymbol{\alpha}' = -\mathbf{n}'(\mathbf{Z}^*)^{-1} = \frac{1}{n}[(n - 1), n(m - 1)]\begin{bmatrix} 0 & n \\ 1 & n - 1 \end{bmatrix} = [(m - 1), m(n - 1)].$$

15.3 Cyclic Association Schemes

We have already noted that the requirements that the latent roots θ_i be non-negative and that for a symmetric design the determinant $|\mathbf{NN}'|$ be a square enable us to establish that designs with certain sets of parameters do not exist. In all the standard association schemes for *PBIB*(m) designs, the formulae for α_i automatically give only integer values, and so the requirement that α_i be an integer is not very helpful. It is of interest, however, in the schemes for two associate classes when we consider the results of Connor and Clatworthy that have just been obtained. If Δ is not a square, $\sqrt{\Delta}$ is irrational and the term which has $\sqrt{\Delta}$ in the denominator must vanish. Since γ is either zero or a positive integer, it follows that if $\sqrt{\Delta}$ is irrational we must have $n_1 = n_2$ and $\gamma = 0$, which leads to the following result [Mesner (1965)].

If, in a *PBIB*(2) scheme, $\sqrt{\Delta}$ is irrational, there is an integer t, such that

$$v = \Delta = 4t + 1, \qquad p_{12}^1 = p_{12}^2 = t, \qquad n_1 = n_2 = \alpha_1 = \alpha_2 = 2t.$$

Mesner also notes that the only known cyclic association schemes as defined by Bose and Shimamoto (1952) have parameters that satisfy these conditions. He therefore calls any *PBIB*(2) scheme with $\sqrt{\Delta}$ irrational *pseudocyclic*, and proves that a two-class association scheme with v prime must be of this type. It follows that there are no two-class association schemes, and thus no designs, for which v is a prime of the form $4t + 3$.

We mention in passing that the only cyclic association schemes listed by Bose, Clatworthy, and Shrikhande (1954) have $v = 13, 17, 29$, and 37. The varieties can be represented by integers $1, \ldots, v - 1$, and 0. For $v = 13$ and 17, their scheme gives as the first associates of the vth variety the set of nonquadratic residues of v. For $v = 29$ and 37, they have the quadratic residues of v. In all cases, either set of residues will suffice.

Mesner (1967) also uses the term *pseudo-Latin square* to denote any scheme that has the same parameters as an $L_i(n)$ scheme (the *NL* schemes are excluded). He points out that there are such schemes which are not L_i schemes, and gives as an example the complement of an $L_3(6)$ design. This has the same parameters as an $L_4(6)$ scheme would. However, no $L_4(6)$ scheme exists because there is no Graeco-Latin square of side six.

15.4 Partially Balanced Association Schemes with Three Associate Classes

Any *PBIB*(2) design can be made into a *PBIB*(3) design with $v' = sv$ by replacing each treatment by a set of s new treatments. These designs are said to be singular, in the sense that the matrix $\mathbf{NN}'$ of the new design is singular. In this section we shall present several partially balanced association schemes with $m = 3$. In most cases the authors of the papers cited have followed the procedure of giving the $\mathbf{P}$ matrices and calculating the latent roots of $\mathbf{NN}'$ and their multiplicities; they have then given some examples in which designs are found to be nonexistent because their existence would lead to a negative latent root or, in the case of a symmetric design, to a violation of the requirement that

$$\theta_1^{\alpha_1}\theta_2^{\alpha_2}\theta_3^{\alpha_3}$$

be a perfect square.

The intrablock and interblock equations for *PBIB*(3) designs can be solved by extending Rao's method for two-associate class designs. This is discussed in Rao's paper (1947*a*) and also by Nair (1952).

For certain sets of parameter values, these schemes degenerate into two-class schemes, or, more often, certain designs are two-class designs. If, in a $PBIB(m)$ design all the λ_i are equal, then the design is a balanced incomplete block design. It is not, however, true that if in a $PBIB(3)$ design two of the λ_i are equal, the design is automatically a two-class design and that there are only two variances; whether or not that is true depends upon the number of distinct latent roots.

We again have for each scheme the two series of elementary symmetric designs $ES(i)$ and $E'S(i)$. In the $ES(i)$ design, the jth block consists of the n_i ith associates of the jth variety; $\lambda_h = p_{ii}^h$. In the $E'S(i)$ design, the jth block also contains the jth variety itself; then $\lambda_i = p_{ii}^i + 2$ and $\lambda_h = p_{ii}^h$ where $h \neq i$.

15.5 The Hierarchic Group Divisible Scheme

This is due to P. M. Roy (1953). He considered the case of m classes; we shall confine ourselves to the case $m = 3$. Suppose that $v = N_1N_2N_3$. We divide the treatments into N_1 groups of N_2N_3 treatments each at the first stage. Treatments in different groups are first associates; $n_1 = (N_1 - 1)N_2N_3$. Then we divide each group into N_2 subgroups of N_3 treatments each. Treatments in different subgroups of the same group are second associates; $n_2 = (N_2 - 1)N_3$. Treatments in the same subgroup are third associates; $n_3 = N_3 - 1$. Raghavarao (1960a) considers the same scheme, but he reverses Roy's definitions. What Roy calls first associates, Raghavarao calls third associates, and vice versa.

The $\mathbf{P}$ matrices are

$$\mathbf{P}_1 = \begin{bmatrix} (N_1 - 2)N_2N_3 & (N_2 - 1)N_3 & N_3 - 1 \\ (N_2 - 1)N_3 & 0 & 0 \\ N_3 - 1 & 0 & 0 \end{bmatrix},$$

$$\mathbf{P}_2 = \begin{bmatrix} (N_1 - 1)N_2N_3 & 0 & 0 \\ 0 & (N_2 - 2)N_3 & N_3 - 1 \\ 0 & N_3 - 1 & 0 \end{bmatrix},$$

$$\mathbf{P}_3 = \begin{bmatrix} (N_1 - 1)N_2N_3 & 0 & 0 \\ 0 & (N_2 - 1)N_3 & 0 \\ 0 & 0 & N_3 - 2 \end{bmatrix}.$$

The latent roots are

$$\begin{aligned} \theta_1 &= r - \lambda_1N_2N_3 + \lambda_2(N_2 - 1)N_3 + \lambda_3(N_3 - 1) = r - \lambda_1(n_2 + n_3 + 1) \\ &\quad + \lambda_2n_2 + \lambda_3n_3, \\ \theta_2 &= r - \lambda_2N_3 + \lambda_3(N_3 - 1) = r - \lambda_2(n_3 + 1) + \lambda_3n_3, \\ \theta_3 &= r - \lambda_3. \end{aligned}$$

To find the multiplicities we solve

$$\begin{bmatrix} -(n_2 + n_3 + 1) & 0 & 0 \\ n_2 & -(n_3 + 1) & 0 \\ n_3 & n_3 & -1 \end{bmatrix} \begin{bmatrix} \alpha_1 \\ \alpha_2 \\ \alpha_3 \end{bmatrix} = - \begin{bmatrix} n_1 \\ n_2 \\ n_3 \end{bmatrix},$$

and obtain

$$\alpha_1 = N_1 - 1, \qquad \alpha_2 = N_1(N_2 - 1), \qquad \alpha_3 = N_1N_2(N_3 - 1).$$

Roy (1953) gives the following example of a resolvable design with parameters

$$v = 8, \quad b = 10, \quad r = 5, \quad k = 4, \quad N_1 = N_2 = N_3 = 2,$$
$$n_1 = 4, \quad n_2 = 2, \quad n_3 = 1, \quad \lambda_1 = 2, \quad \lambda_2 = 3, \quad \lambda_3 = 1:$$

1234, 5678; 1357, 2468; 1368, 2457; 1458, 2367; 1467, 2358.

The subgroups are 12, 34; 56, 78.

If $\lambda_2 = \lambda_3$, the design is merely a *GD* design with N_1 groups, each of which has N_2N_3 treatments. In this case, $\theta_2 = \theta_3 = r - \lambda_3$, and there are only two distinct roots.

15.6 The Rectangular Scheme

Vartak (1959) discusses designs with this scheme and calls them series A. Suppose that we arrange $v = mn$ treatments in a rectangular array of m rows and n columns. Treatments are first associates if they appear in the same row, and second associates if they appear in the same column; otherwise they are third associates. Then $n_1 = n - 1$, $n_2 = m - 1$, and $n_3 = (m - 1)(n - 1)$,

$$\mathbf{P}_1 = \begin{bmatrix} n - 2 & 0 & 0 \\ 0 & 0 & m - 1 \\ 0 & m - 1 & (m - 1)(n - 2) \end{bmatrix},$$

$$\mathbf{P}_2 = \begin{bmatrix} 0 & 0 & n - 1 \\ 0 & m - 2 & 0 \\ n - 1 & 0 & (m - 2)(n - 1) \end{bmatrix},$$

$$\mathbf{P}_3 = \begin{bmatrix} 0 & 1 & n - 2 \\ 1 & 0 & m - 2 \\ n - 2 & m - 2 & (m - 2)(n - 2) \end{bmatrix}.$$

$$\theta_1 = r - \lambda_1 + (m - 1)(\lambda_2 - \lambda_3), \qquad \alpha_1 = n - 1 = n_1,$$

$$\theta_2 = r + (n - 1)(\lambda_1 - \lambda_3) - \lambda_2, \qquad \alpha_2 = m - 1 = n_2,$$

$$\theta_3 = r - \lambda_1 - \lambda_2 + \lambda_3, \qquad \alpha_3 = (m - 1)(n - 1) = n_3.$$

Vartak also proves that the scheme is unique in the sense that if a *PBIB*(3) association scheme has **P** matrices identical with those above, then the scheme must be a rectangular scheme.

If $m = n$ and also $\lambda_1 = \lambda_2$, we have designs with the L_2 scheme. In this case $\theta_1 = \theta_2$. Note, however, that unless the array is square, putting $\lambda_1 = \lambda_2$ will not make the two roots equal, and the number of associate classes is still three.

15.7 The Cubic Scheme

This is due to Raghavarao and Chandrasekhararao (1964). There are $v = s^3$ treatments, each of which is denoted by a different set of three coordinates (x, y, z) where x, y, and z are all integers such that $1 \leq x \leq s$, $1 \leq y \leq s$, $1 \leq z \leq s$. Two treatments are defined to be ith associates if they have exactly $(3 - i)$ coordinates in common. For convenience we write (x, y, z) as xyz. Thus, for $s = 3$, $v = 27$, the treatment 000 has six first associates: 001, 002, 010, 020, 100, and 200. It has eight third associates denoted by abc, where each of the coordinates a, b, and c takes the values 1 or 2. In the general case, $n_1 = 3(s - 1)$, $n_2 = 3(s - 1)^2$, and $n_3 = (s - 1)^3$.

$$\mathbf{P}_1 = \begin{bmatrix} s-2 & 2(s-1) & 0 \\ 2(s-1) & 2(s-1)(s-2) & (s-1)^2 \\ 0 & (s-1)^2 & (s-1)^2(s-2) \end{bmatrix},$$

$$\mathbf{P}_2 = \begin{bmatrix} 2 & 2(s-2) & (s-1) \\ 2(s-2) & 2(s-1)+(s-2)^2 & 2(s-1)(s-2) \\ (s-1) & 2(s-1)(s-2) & (s-1)(s-2)^2 \end{bmatrix},$$

$$\mathbf{P}_3 = \begin{bmatrix} 0 & 3 & 3(s-2) \\ 3 & 6(s-2) & 3(s-2)^2 \\ 3(s-2) & 3(s-2)^2 & (s-2)^3 \end{bmatrix}.$$

The latent roots of $\mathbf{NN}'$ and their multiplicities are

$$\theta_0 = rk, \quad \theta_1 = r + (2s - 3)\lambda_1 + (s - 1)(s - 3)\lambda_2 - (s - 1)^2\lambda_3,$$

$$\theta_2 = r + (s - 3)\lambda_1 - (2s - 3)\lambda_2 + (s - 1)\lambda_3, \quad \theta_3 = r - 3\lambda_1 + 3\lambda_2 - \lambda_3,$$

$$\alpha_0 = 1, \quad \alpha_1 = 3(s - 1) = n_1, \quad \alpha_2 = 3(s - 1)^2 = n_2, \quad \alpha_3 = (s - 1)^3 = n_3.$$

The $E'S(2)$ design for $v = 64$ is of particular interest. We have $s = 4$, $\lambda_1 = p^1_{22} = 2(s-1)(s-2) = 12$, $\lambda_2 = p^2_{22} + 2 = 2(s-1) + (s-2)^2 + 2 = 12$, and $\lambda_3 = 6(s-2) = 12$. The design is thus a balanced incomplete block design with $v = b = 64$, $r = k = 28$, and $\lambda = 12$.

15.8 The Triangular Scheme for $m = 3$

We defined the triangular scheme for $m = 2$ in two equivalent ways. The definition involving the square array is the one most commonly used; the second definition, in which the treatments are denoted by pairs of integers (x, y), $1 \leq x < y \leq n$, may be extended to a scheme for three-associate classes. In this scheme [John, (1966*c*)] we have $6v = (s+2)(s+3)(s+4)$. Each treatment is represented by three integers (x, y, z) where $1 \leq x < y < z \leq s + 4$. Two treatments are ith associates if there are exactly $(3 - i)$ integers which appear in both representations. If $s = 1$, there are ten treatments. The first associates of (1, 2, 3) are (1, 2, 4), (1, 2, 5), (1, 3, 4), (1, 3, 5), (2, 3, 4), (2, 3, 5); the second associates are (1, 4, 5), (2, 4, 5), (3, 4, 5), and the scheme reduces to a triangular scheme for $m = 2$. If $s = 2$, treatment (1, 2, 3) has a single third associate (4, 5, 6). In general, $n_1 = 3(s+1)$, $n_2 = 3s(s+1)/2$, $n_3 = (s-1)s(s+1)/6$.

$$\mathbf{P}_1 = \begin{bmatrix} s+2 & 2s & 0 \\ 2s & s^2 & s(s-1)/2 \\ 0 & s(s-1)/2 & s(s-1)(s-2)/6 \end{bmatrix},$$

$$\mathbf{P}_2 = \begin{bmatrix} 4 & 2s & (s-1) \\ 2s & (s-1)(s+6)/2 & (s-1)(s-2) \\ (s-1) & (s-1)(s-2) & (s-1)(s-2)(s-3)/6 \end{bmatrix},$$

$$\mathbf{P}_3 = \begin{bmatrix} 0 & 9 & 3(s-2) \\ 9 & 9(s-2) & 3(s-2)(s-3)/2 \\ 3(s-2) & 3(s-2)(s-3)/2 & (s-2)(s-3)(s-4)/6 \end{bmatrix}.$$

$$\theta_1 = r + (2s-1)\lambda_1 + \tfrac{1}{2}s(s-5)\lambda_2 - \tfrac{1}{2}s(s-1)\lambda_3, \qquad \alpha_1 = s+3,$$

$$\theta_2 = r + (s-3)\lambda_1 - (2s-3)\lambda_2 + (s-1)\lambda_3, \qquad \alpha_2 = (s+1)(s+4)/2,$$

$$\theta_3 = r - 3\lambda_1 + 3\lambda_2 - \lambda_3, \qquad \alpha_3 = (s-1)(s+3)(s+4)/6.$$

This scheme should not be confused with the $T_n(m)$ scheme given by Singh and Singh (1964); they obtain $PBIB(3)$ designs by replacing each treatment in a design which has the triangular scheme for two associate classes, and $v = n(n-1)/2$ by a set of m new treatments. These singular designs are discussed at some length in their paper.

The dual of the balanced incomplete block design with $v^* = 6$, $b^* = 20$, $k^* = 3$ is a *PBIB*(3) design with this scheme. We have $s = 2$, $v = 20$, $b = 6$, $r = 3$, $k = 10$, $\lambda_1 = 2$, $\lambda_2 = 1$, and $\lambda_3 = 0$. The latent roots of $\mathbf{NN}'$ are

$$\theta_1 = 3 + 3\lambda_1 - 3\lambda_2 - \lambda_3 = 6, \qquad \alpha_1 = 5;$$
$$\theta_2 = 3 - \lambda_1 - \lambda_2 + \lambda_3 = 0, \qquad \alpha_2 = 9;$$
$$\theta_3 = 3 - 3\lambda_1 + 3\lambda_2 - \lambda_3 = 0, \qquad \alpha_3 = 5.$$

In the dual, the block that contained the old varieties x, y, and z is transformed into the treatment (x, y, z). This method is general. An unreduced *BIBD* with $v^* = n$, $b^* = \binom{n}{3}$, and $k = 3$ has for its dual a triangular *PBIB*(3) design with $s = n - 4$, $\lambda_1 = 2$, $\lambda_2 = 1$, and $\lambda_3 = 0$.

15.9 The *LS*3 Association Scheme

This is an extension of the two-class L_2 scheme for $v = n^2$ treatments [Johnson (1967)]. The $v = n^2$ treatments are arranged in a square array and a Latin square is superimposed. Two treatments are said to be first associates if they are in the same row or in the same column of the array. They are second associates if they correspond to the same letter in the Latin square. Otherwise they are third associates. Then $n_1 = 2(n - 1)$, $n_2 = (n - 1)$, $n_3 = (n - 1)(n - 2)$.

$$\mathbf{P}_1 = \begin{bmatrix} n-2 & 1 & n-2 \\ 1 & 0 & n-2 \\ n-2 & n-2 & (n-2)(n-3) \end{bmatrix},$$

$$\mathbf{P}_2 = \begin{bmatrix} 2 & 0 & 2(n-2) \\ 0 & n-2 & 0 \\ 2(n-2) & 0 & (n-2)(n-3) \end{bmatrix},$$

$$\mathbf{P}_3 = \begin{bmatrix} 2 & 2 & 2(n-3) \\ 2 & 0 & n-3 \\ 2(n-3) & n-3 & n^2-6n+10 \end{bmatrix}.$$

The nonzero latent roots are

$$\theta_1 = r + (n-2)\lambda_1 - \lambda_2 - (n-2)\lambda_3, \qquad \alpha_1 = 2(n-1),$$
$$\theta_2 = r - 2\lambda_1 + (n-1)\lambda_2 - (n-2)\lambda_3, \qquad \alpha_2 = n-1,$$
$$\theta_3 = r - 2\lambda_1 - \lambda_2 + 2\lambda_3, \qquad \alpha_3 = (n-1)(n-2).$$

If $\lambda_2 = \lambda_3$, the design becomes an L_2 design. For $n = 6$, the $E'S(3)$ design is a balanced incomplete block design with $b = v = 36$, $r = k = 21$, and $\lambda = 12$; its complement has $r = k = 15$ and $\lambda = 6$.

15.10 Cyclic Designs

We have seen in Chapters 13 and 14 how designs can be obtained by taking one or more initial blocks and, in the manner of Bose, developing other blocks by cycling. From any initial block we obtain another block by adding the same number to each element of the initial block. It is a popular method of obtaining balanced designs, and we have referred to its use in obtaining GD designs in Chapter 14.

Kempthorne (1953) and David (1963) investigated cyclic designs with only two plots per block. Kempthorne's motivation derived from breeding experiments, especially diallel crosses. Later, David and Wolock (1965) and J. A. John (1966) discussed cyclic designs with larger blocks. All these authors confined themselves to representing the varieties by the residue classes $0, 1, \ldots, v - 1 \pmod v$, and for the present we shall keep this restriction. These designs are all balanced or partially balanced, but they may have considerably more than two classes. The schemes have a cyclic structure in the sense that if x, y, and z are integers and x is an ith associate of y, then $z + x$ is an ith associate of $z + y$. Furthermore, if x is an ith associate of zero, then $v - x$ is also an ith associate of zero.

A consequence of this is that the determinant $|\mathbf{NN}'|$ is a circulant. If a circulant has for its first row the elements $a_0, a_1, \ldots, a_{n-1}$, the latent roots are given by

$$\theta_j = a_0 + a_1\omega_j + a_2\omega_j^2 + \cdots + a_{n-1}\omega_j^{n-1},$$

where $\omega_j = \cos 2\pi j/n + i \sin 2\pi j/n$, and $i^2 = -1, j = 1, 2, \ldots, n$.

In this particular case, $\mathbf{NN}'$ is symmetric and $n = v$ so that $a_1 = a_{v-1}$, $a_2 = a_{v-2}, \ldots$, and the imaginary parts vanish. Hence,

$$\theta_j = a_0 + 2a_1 \cos 2\pi j/v + 2a_2 \cos 4\pi j/v + \cdots :$$

i.e.,

$$\theta_j = r + 2 \sum_{h=1}^{(v-1)/2} a_h \cos 2\pi hj/v,$$

if v is odd;

$$\theta_j = r + 2 \sum_{h=1}^{(v-2)/2} a_h \cos 2\pi hj/v + a_{v/2} \cos \pi,$$

if v is even.

As an example we may consider the design with the cyclic association scheme given in Chapter 12:

$$012,\ 123,\ 234,\ 340,\ 401.$$

The first associates of zero are 1 and 4, $\lambda_1 = 2$, $\lambda_2 = 1$, and $r = 3$.

Applying the results of Connor and Clatworthy, we have $\gamma = 0$, $\beta = 2$, $\Delta = 5$, $\alpha_1 = \alpha_2 = 2$, $\theta_i = (3 \pm \sqrt{5})/2$.

The matrix $\mathbf{NN}'$ is

$$\begin{bmatrix} 3 & 2 & 1 & 1 & 2 \\ 2 & 3 & 2 & 1 & 1 \\ 1 & 2 & 3 & 2 & 1 \\ 1 & 1 & 2 & 3 & 2 \\ 2 & 1 & 1 & 2 & 3 \end{bmatrix}.$$

Hence, $\theta_j = 3 + 4 \cos 2\pi j/5 + 2 \cos 4\pi j/5$.

Recalling that $\cos 2\pi/5 = (-1 + \sqrt{5})/4$, $\cos 4\pi/5 = (-1 - \sqrt{5})/4$, we have

$$\theta_1 = 3 + 4 \cos 2\pi/5 + 2 \cos 4\pi/5 = (3 + \sqrt{5})/2 = \theta_4,$$

$$\theta_2 = 3 + 4 \cos 4\pi/5 + 2 \cos 8\pi/5 = 3(-\sqrt{5})/2 = \theta_3.$$

Cyclic designs can be obtained with *GD* schemes by appropriately labeling the treatments. We denote the jth treatment in the ith group by $(i - 1) + (j - 1)m$, $1 \le i \le m$, $1 \le j \le n$. In essence, this is what Bose et al. (1953) did in the second example that we presented in Chapter 14. The design was for $v = 12$, $m = 6$, and $n = 2$. The treatments in the first group were zero and 6, in the second group 1 and 7, and so on. The initial block was 0 1 4 6 and the design consisted of twelve blocks. The differences in the initial block are

$$1 - 0 = 1, 4 - 0 = 4, 6 - 0 = 6, 4 - 1 = 3, 6 - 4 = 2,$$

and their negatives

$$0 - 1 = 11, 8, 6, 9, 10.$$

We note that the set contains each nonzero difference except 6 once, while 6 itself appears twice. We thus have a design with $\lambda_1 = 2$ and $\lambda_2 = 1$.

J. A. John (1966) suggests using 0 1 3 7 as the initial block, and we may ask whether the two designs are equivalent. It should first be pointed out that Bose and his colleagues would have obtained the same design if they had started with any of the other three blocks which contain zero, namely, 6 7 10 0, 8 9 0 2, or 11 0 3 5. Similarly, John could have suggested 11 0 2 6, 9 10 0 4, or 5 6 8 0. If we multiply each of the members of an initial block by a number prime to v, in this case 5, 7, or 11, and reduce mod v, we shall get an equivalent design [David

(1963)], inasmuch as our action will amount only to a relabeling of the treatments. In each case, 6 will remain unchanged. Similarly, if we replaced each treatment by its negative (mod v), we should get an equivalent design. Multiplying 0 1 4 6 by 5 gives 0 5 8 6, which is the fourth of the possible initial blocks in John's design. Thus the two designs are equivalent.

Cyclic designs with the hierarchic group divisible scheme can be obtained by labeling the jth treatment in the ith subgroup of the hth group as

$$h - 1 + (i - 1)N_1 + (j - 1)N_1N_2.$$

With this labeling, a design with $N_1 = N_2 = N_3 = 2$ would have varieties 0, 2 and 4, 6 in the subgroups of the first group and 1, 3 and 5, 7 in the other group. The $E'S(1)$ design is now the cyclic design generated by 0 1 3 5 7. The $ES(1)$ design is also cyclic and consists of the two blocks 0 2 4 6 and 1 3 5 7, each repeated four times. The pair of blocks is called a partial cycle.

15.11 Generalized Cyclic Designs

We may generalize the concept of a cyclic design by allowing the varieties to be elements of an Abelian group under addition. As before, it is convenient to take the initial block to contain the identity zero. The jth block is then obtained by adding the jth element of the group to each member of the initial block. Again, several initial blocks may be used, and partial cycles may be included.

For group divisible designs we may use a double modulus system, representing the jth treatment in the ith group by the ordered pair ij; addition is then defined by $gh + ij = xy$ where $x \equiv g + i \pmod{m}$ and $y \equiv h + j \pmod{n}$, and the same system can be used for L_2 designs. Consider, for example, the array of nine treatments

$$\begin{matrix} 00 & 01 & 02 \\ 10 & 11 & 12 \\ 20 & 21 & 22, \end{matrix}$$

with arithmetic for both coordinates being mod 3.

We can consider the array as a GD scheme with rows as groups. Then the initial block 00 01 02 11 generates a GD design with $\lambda_1 = 3$ and $\lambda_2 = 1$. Considering the array as an L_2 scheme, the initial block 00 01 12 22 generates the $ES(1)$ design with $\lambda_1 = 1$ and $\lambda_2 = 2$.

We can also use this labeling of treatments for the rectangular scheme. Suppose we take a scheme with $m = 2$ and $n = 3$.

$$\begin{matrix} 00 & 01 & 02 \\ 10 & 11 & 12. \end{matrix}$$

Arithmetic for the first digit is mod 2, and for the second digit mod 3. Then the initial block 00 11 12 gives the $E'S(3)$ design with $\lambda_1 = 1$, $\lambda_2 = 0$, and $\lambda_3 = 2$.

In the cubic scheme, the coordinate system is immediately adaptable to cyclic designs. We give two examples for the schemes with $v = 27$ and $v = 64$.

In the scheme for $v = 27$, we denote the treatments by 000, 001, and so on, with arithmetic mod 3. Consider the design generated by the initial block 000, 001, 002. The six differences are 001 and 002, each repeated three times. The twenty-seven blocks consist of a partial cycle of nine blocks repeated three times. Suppose now that we take the set of three initial blocks, 000, 001, 002; 000, 010, 020; 000, 100, 200, and take one partial cycle from each set of blocks. The resulting design will have $b = v = 27$, $r = k = 3$, $\lambda_1 = 1$, $\lambda_2 = 0$, and $\lambda_3 = 0$.

In a similar way, partial cycles from the six initial blocks 000, 012, 021; 000, 011, 022; 000, 102, 201; 000, 101, 202; 000, 120, 210; 000, 110, 220 give $b = 54$, $r = 6$, $k = 3$, $\lambda_1 = 0$, $\lambda_2 = 1$, and $\lambda_3 = 0$. The initial blocks 000, 111, 222; 000, 112, 221; 000, 121, 212; 000, 211, 112 likewise give a design with $b = 36$, $r = 4$, $\lambda_1 = \lambda_2 = 0$, and $\lambda_3 = 1$, and all three of these *PBIB*(3) designs combined form a *BIBD* with $b = 117$, $k = 3$, $r = 13$, $\lambda = 1$.

For $v = 64$ we can use for the coordinates the elements of the Galois field $GF(2^2)$, namely, 0, 1, x, and $1 + x$, which we write as y. We can obtain a similar set of designs with $k = 4$ and other parameters:

(i)	$b = 48$,	$r = 3$,	$\lambda_1 = 1$,	$\lambda_2 = 0$,	$\lambda_3 = 0$;
(ii)	$b = 144$,	$r = 9$,	$\lambda_1 = 0$,	$\lambda_2 = 1$,	$\lambda_3 = 0$;
(iii)	$b = 144$,	$r = 9$,	$\lambda_1 = 0$,	$\lambda_2 = 0$,	$\lambda_3 = 1$.

In design (iii), one of the initial blocks used is 000, 111, xxx, yyy. The differences are 111, xxx, yyy, $xxx - 111 = yyy$, $yyy - 111 = xxx$, $yyy - xxx = 111$ and their negatives; but since all arithmetic is mod 2, taking the negatives amounts to repeating each difference again. The set of twelve differences is thus 111, xxx, and yyy, each appearing four times. The actual design consists of partial cycles (sixteen blocks) on each of the following nine initial blocks:

000, 111, xxx, yyy; 000, $11x$, xxy, $yy1$; 000, $11y$, yyx, $xx1$;

000, $1x1$, xyx, $y1y$; 000, $1y1$, yxy, $x1x$; 000, $x11$, $1yy$, yxx;

000, $y11$, xyy, $1xx$; 000, $1xy$, $xy1$, $yx1$; 000, $1yx$, $yx1$, $x1y$.

15.12 Designs for Eight Treatments in Blocks of Three

As an illustration of the techniques that have been developed so far, we consider designs with $v = 8$ and $k = 3$. In particular, we shall look at designs with two or three associate classes. We have the following possible schemes:

(i) *GD* with $m = 2$, $n = 4$, and ordering 0 2 4 6, 1 3 5 7;
(ii) *GD* with $m = 4$, $n = 2$, and ordering 04, 15, 26, 37;
(iii) Hierarchic *GD* with ordering 04, 26, 15, 37;
(iv) Rectangular;
(v) Cubic.

For $v = 8$, the last two schemes are equivalent.

We turn first to the ordinary cyclic designs with a single initial block. Each initial block consists of zero and two other treatments. There are thus $\binom{7}{2} = 21$ possible initial blocks, but they occur in equivalence sets.

Consider the block 0, 1, 2. When we cycle on it, we obtain 0 1 7, 0 6 7, and so these three blocks are equivalent in the sense that they generate the same design. Furthermore, multiplying by 3 gives 0 3 6, 0 3 5, 0 2 5, and we have one equivalence set with six triples. The initial block 0 1 3 also gives us 0 2 7, 0 5 6, with differences $1 - 0$, $3 - 0$, $3 - 1$, and their negatives. We could as well have considered the negative of the block, replacing 1 by 7 and 3 by 5 to give 0 5 7, and, similarly, 0 1 6, 0 2 3, for another set of six. There are two more equivalence sets, 0 1 4, 0 3 7, 0 4 5, 0 4 7, 0 1 5, 0 3 4, and the set of three triples, 0 2 4, 0 2 6, 0 4 6. For designs with a single initial block, we need therefore consider only the four cases:

(i) 0 1 2, with differences 1, 1, 2, 6, 7, 7 [or, as David (1963) writes, $(1^2, 2)$]; this gives four associate classes, with the associates of zero being 1, 7; 2, 6; 3, 5; 4, and $\lambda_1 = 2$, $\lambda_2 = 1$, $\lambda_3 = \lambda_4 = 0$;

(ii) 0 1 3, which gives a *GD* design with groups 0, 4; 1, 5; 2, 6; 3, 7 and $\lambda_1 = 0$, $\lambda_2 = 1$;

(iii) 0 1 4, which gives a hierarchic *GD* design with $\lambda_1 = 1$, $\lambda_2 = 0$, $\lambda_3 = 2$; and

(iv) 0 2 4, which gives a disconnected *GD* design with $m = 2$, $n = 4$, and $\lambda_1 = 2$, $\lambda_2 = 0$.

J. A. John (1966) also considers combinations of several initial blocks. The set (0 1 2), (0 1 3), (0 1 4) gives a design for four classes; $\lambda_1 = 4$, $\lambda_2 = \lambda_3 = \lambda_4 = 2$. There are three other designs with three associate classes, all of which have the hierarchic scheme. They are generated by

(0 1 3), (0 2 4) $\quad \lambda_1 = 1, \quad \lambda_2 = 3, \quad \lambda_3 = 2;$

(0 1 3), (0 1 4) $\quad \lambda_1 = 2, \quad \lambda_2 = 1, \quad \lambda_3 = 2;$

(0 1 3), (0 1 4), (0 2 4) $\quad \lambda_1 = 2, \quad \lambda_2 = 3, \quad \lambda_3 = 4.$

We can obtain three more designs by using the cubic scheme for $v = 8$. Denoting the treatments by 000, 001, 010, and so on, with arithmetic mod 2, the initial blocks and values of λ_i are

(000, 001, 010), (000, 010, 100), (000, 100, 001) $\quad \lambda_1 = 4, \quad \lambda_2 = 2, \quad \lambda_3 = 0;$

(000, 001, 110), (000, 010, 101), (000, 100, 011) $\quad \lambda_1 = \lambda_2 = 2, \quad \lambda_3 = 6;$

(000, 011, 110) $\quad \lambda_1 = \lambda_3 = 0, \quad \lambda_2 = 2.$

15.13 Conclusion

The construction of designs and the discovery of new schemes remains a fertile field for research, and in this chapter we have given only a limited survey of the field. It might, however, be appropriate to recall the last pages of Chapter 12 in which it was suggested that from a practical standpoint there is sometimes little to choose between the various alternative designs that can be used in some problems. We mentioned at that time the nineteen designs that J. A. John listed for $v = 12$ and $k = 4$, with efficiencies ranging from 0.8007 to 0.8166, the maximum being attained by a *GD* design. The mathematical problems discussed in the last three chapters are very interesting from a theoretical point of view and have a fascination for pure mathematicians as well as for statisticians.

In spite of the beauty and elegance of the mathematical results, the practicing statistician should still bear in mind that the more complex the design, the easier it is for something to go wrong. We conclude with the maxim that has already appeared at the end of Chapter 12: When you design an experiment, keep it as simple as you can.

Exercises

1. Find the latent roots of $\mathbf{NN}'$ and their multiplicities for the triangular designs.
2. Find the latent roots of $\mathbf{NN}'$ and their multiplicities for designs with the $L_i(n)$ scheme.
3. Verify the results given for the latent roots of $\mathbf{NN}'$ and their multiplicities for the several three-associate class schemes.
4. The hierarchic *GD* design is obtained from a *GD* design by replacing each treatment by a set of N_3 new treatments. We may similarly take an $L_i(n)$ scheme and replace each treatment by a group of m new treatments. Two treatments are said to be first associates if they belong to the same group, and second associates if their groups correspond to first associates in the $L_i(n)$ scheme. This is the $L_i(n, m)$ scheme of Singh and Singh (1962). Find the $\mathbf{P}$ matrices, the latent roots of $\mathbf{NN}'$, and their multiplicities for this scheme.
5. (Continuation) They also defined in the same paper the $T_n(m)$ scheme, in which they take a triangular scheme and replace each treatment by a group of m new treatments. The three associate classes are defined as for the $L_i(n, m)$ scheme. Derive the $\mathbf{P}$ matrices, the latent roots of $\mathbf{NN}'$, and their multiplicities for this scheme.
6. Show that if in a *PBIB*(2) design Δ is not a perfect square then

$$p_{21}^1 = p_{12}^2 = p_{11}^2 = t, \qquad n_1 = n_2 = \alpha_1 = \alpha_2 = 2t, \qquad v = 4t + 1.$$

7. Show that if $v = 4t + 1$ and v is a prime we may obtain a cyclic association scheme by taking as the first associates of zero either the set of quadratic residues of v or its complement; the first associates of i are then obtained by adding i to each of the first associates of zero.

8. (Continuation) Obtain a similar result when $v = 9$ or 25 by allowing the treatments to be the elements of the Galois field of v elements.

9. Show that for a $PBIB(2)$ scheme $vn_1n_2 = \Delta\alpha_1\alpha_2$ [Mesner (1965)].

10. Show that a necessary and sufficient condition for a $PBIB(2)$ scheme to have α_1 and α_2 equal in some order to n_1 and n_2 is that the scheme be of the pseudo-cyclic, the pseudo-Latin square or the NL types [Mesner (1967)].

11. Prove the following results for designs with the rectangular scheme:

(i) If $\theta_1 = 0$, k is divisible by n and each block contains exactly k/n treatments from each column of the array;

(ii) If $\theta_2 = 0$, k is divisible by m and each block contains k/m treatments from each row;

(iii) If $\theta_1 = \theta_2 = 0$, there are $k(r - 1)/(b - 1)$ treatments common to any two blocks of the design [Vartak (1959)].

12. Show that there are only five nonisomorphic cyclic designs with $v = b = 8$ and $r = k = 4$ and a single initial block.

13. Show that although the cyclic design for $v = 8$ generated by (0, 1, 2, 4) has only two distinct values of λ_i, nevertheless there are four associate classes.

14. Prove that $\cos(2\pi/5) = (-1 + \sqrt{5})/4$.

Appendix

A.1 Matrices

We review in this appendix some properties of matrices, vector spaces, and quadratic forms. The elements of all the matrices will be real numbers, and some of the results given are not true without this restriction. Some of the results will be proved as theorems; others will be stated without proof, and some of these will appear later as exercises. Although it is reasonable to assume that most students entering a graduate course in the mathematical theory of design of experiments will have had some exposure to elementary matrix theory, many undergraduate courses in linear algebra do not progress as far as the discussion of quadratic forms, which are essential to the development of our subject. This does not attempt to be a complete presentation, and the reader is referred for further details to the standard texts. Some basic results about determinants will be assumed.

A matrix, in the context in which we shall be considering them, is a rectangular array of elements, which in our case will be real numbers (or, at a later stage, smaller matrices of real numbers). We talk of a matrix **A** having m rows and n columns, sometimes written $(m \times n)$, and denote by a_{ij} the element in the ith row and the jth column. We shall use boldface type, **A**, to denote matrices and we shall write $\mathbf{A} = (a_{ij})$. A square matrix with m rows and m columns is said to be of order m. If $a_{ij} = a_{ji}$, **A** is called symmetric. **O** will denote a matrix of zero elements. $\mathbf{I}_n$ is the identity matrix, the square matrix of order n with ones along the main diagonal and zeros elsewhere, i.e., $a_{ii} = 1$, $a_{ij} = 0$, $i \neq j$. If there is no danger of confusion, we shall omit the subscript n and write **I**.

A matrix $\mathbf{A}$ can be multiplied by a scalar (a real number) c. The product is defined as $c\mathbf{A} = \mathbf{B}$ where $b_{ij} = ca_{ij}$.

Two matrices $\mathbf{A}$ and $\mathbf{B}$, each with m rows and n columns, can be added. Addition is defined by $\mathbf{A} + \mathbf{B} = \mathbf{C}$ where $c_{ij} = a_{ij} + b_{ij}$.

The product $\mathbf{AB}$ of two matrices is defined only if the number of columns of $\mathbf{A}$ is the same as the number of rows of $\mathbf{B}$, in which case we define $\mathbf{AB} = \mathbf{C}$ where $c_{ij} = \sum_k a_{ik}b_{kj}$. In general it is not true that $\mathbf{AB} = \mathbf{BA}$; indeed, one product may exist and the other not. If $\mathbf{AB} = \mathbf{BA}$, the two matrices are said to commute.

The transpose of a matrix $\mathbf{A}$ with m rows and n columns will be written as $\mathbf{A}'$. Some authors write $\mathbf{A}^T$. It is defined as $\mathbf{A}' = \mathbf{C}$ where $\mathbf{C}$ has n rows and m columns and $c_{ij} = a_{ji}$. If $\mathbf{A}$ and $\mathbf{B}$ are two matrices such that $\mathbf{AB}$ exists, then $(\mathbf{AB})' = \mathbf{B}'\mathbf{A}'$.

So far as we are concerned, a vector of n elements may be considered as a matrix with n rows and a single column. Its transpose is sometimes called a row vector. We shall use $\mathbf{1}_n$ to denote a vector of n unit elements. $\mathbf{E}_{m,n}$ will denote a matrix of m rows and n columns with every element unity: $\mathbf{E}_{m,n} = \mathbf{1}_m\mathbf{1}_n'$. We shall usually write $\mathbf{J}_n$ for $\mathbf{E}_{n,n}$. The length of a vector $\mathbf{y}$ is defined as d where $d^2 = \mathbf{y}'\mathbf{y} = \sum y_i^2$. Two vectors, $\mathbf{x}$ and $\mathbf{y}$, are said to be orthogonal if $\mathbf{x}'\mathbf{y} = \mathbf{y}'\mathbf{x} = 0$; this implies that $\mathbf{x}$ and $\mathbf{y}$ have the same number of elements.

Consider a set of vectors $\mathbf{y}_1, \ldots, \mathbf{y}_m$, each with n elements. They are said to be linearly independent if there exists no set of scalars $c_1, \ldots, c_m$ (except the set $c_1 = \cdots = c_m = 0$) for which $\sum c_i\mathbf{y}_i = \mathbf{0}$ where $\mathbf{0}$ is a vector of zero elements. Otherwise, we have a linearly dependent set. The zero vector $\mathbf{0}$ is itself a linearly dependent set. The collection of all linear combinations $\sum c_i\mathbf{y}_i$ is called the vector space, V, spanned by the set of vectors $\mathbf{y}_1, \ldots, \mathbf{y}_m$. If the set $\mathbf{y}_1, \ldots, \mathbf{y}_m$ is a linearly dependent set, V is also spanned by a subset of $\mathbf{y}_1, \ldots, \mathbf{y}_m$. This leads us to the idea of a basis for a vector space. We define a basis (or a Hamel basis) for a vector space V to be a set of linearly independent vectors that span V. It can be shown that every vector space has a basis and that all bases for a given vector space contain the same number of vectors. This number is called the dimension of the space. It is closely connected with the next concept, that of the rank of a matrix. We may consider a matrix $\mathbf{A}$, $(m \times n)$, as a collection of n columns or vectors.

The rank of $\mathbf{A}$, written $r(\mathbf{A})$, is the number of linearly independent columns in $\mathbf{A}$, which is the same as the dimension of the vector space spanned by the columns of $\mathbf{A}$. It can be shown that this is the same as the number of linearly independent rows of $\mathbf{A}$, and that $r(\mathbf{A})$ is equal to the order of the largest nonvanishing determinant that can be formed from the columns and the rows of $\mathbf{A}$. It follows that $r(\mathbf{A}) \leq \min(m, n)$, and that $r(\mathbf{0}) = 0$. We conclude these preliminaries by defining the inverse of a matrix and proving some theorems about the ranks of matrices.

Let $\mathbf{A}$ be a square matrix, and let $|\mathbf{A}|$ be the determinant with elements a_{ij}. If $|\mathbf{A}| = 0$, $\mathbf{A}$ is said to be singular; if $|\mathbf{A}| \neq 0$, $\mathbf{A}$ is nonsingular. Let A_{ij} denote the cofactor of the element a_{ij} in $|\mathbf{A}|$; A_{ij} is the value of the determinant obtained by striking out the ith row and the jth column of $\mathbf{A}$, multiplied by $(-1)^{i+j}$.

Suppose now that $\mathbf{A}$ is nonsingular, and consider the matrix $\mathbf{C} = (c_{ij})$ where $c_{ij} = A_{ji}/|\mathbf{A}|$. It follows from elementary properties of determinants, namely $\sum_j a_{ij}A_{ij} = |\mathbf{A}|$ and $\sum_j a_{kj}A_{ij} = 0$ where $i \neq k$, that $\mathbf{AC} = \mathbf{CA} = \mathbf{I}$. $\mathbf{C}$ is called the inverse of $\mathbf{A}$ and is written as $\mathbf{A}^{-1}$. $\mathbf{C}$ is unique, and, if $\mathbf{A}$ and $\mathbf{B}$ are both nonsingular, $(\mathbf{AB})^{-1} = \mathbf{B}^{-1}\mathbf{A}^{-1}$. Every square nonsingular matrix $\mathbf{A}$ has an inverse. If $r(\mathbf{A})$ is less than the order of $\mathbf{A}$, $|\mathbf{A}| = 0$, and $\mathbf{A}$ has no inverse.

Theorem. *Let* $\mathbf{A}$ *and* $\mathbf{B}$ *be* $(m \times n)$ *matrices. Then* $r(\mathbf{A} + \mathbf{B}) \leq r(\mathbf{A}) + r(\mathbf{B})$.

PROOF. Let α be a set of $r(\mathbf{A})$ linearly independent columns of $\mathbf{A}$, and β a set of $r(\mathbf{B})$ linearly independent columns of $\mathbf{B}$; let γ be the union of the two sets α and β. Every column of $\mathbf{C} = \mathbf{A} + \mathbf{B}$ is a linear combination of the columns of γ. Then $r(\mathbf{C})$ is not greater than the number of columns in γ, and so $r(\mathbf{A} + \mathbf{B}) \leq r(\mathbf{A}) + r(\mathbf{B})$. ■

Theorem. *Let* $\mathbf{A}$ *and* $\mathbf{B}$ *be matrices such that* $\mathbf{AB}$ *exists. Then* $r(\mathbf{AB}) \leq r(\mathbf{A})$ and $r(\mathbf{AB}) \leq r(\mathbf{B})$.

PROOF. The columns of $\mathbf{AB}$ are linear combinations of the columns of $\mathbf{A}$. Thus they belong to the space spanned by the columns of $\mathbf{A}$, and so $r(\mathbf{AB}) \leq r(\mathbf{A})$. Similarly, $r(\mathbf{B}'\mathbf{A}') \leq r(\mathbf{B}')$. But $r(\mathbf{B}'\mathbf{A}') = r(\mathbf{AB})$ and $r(\mathbf{B}') = r(\mathbf{B})$ so that $r(\mathbf{AB}) \leq r(\mathbf{B})$. ■

Corollary 1. *If* $\mathbf{A}$ *is nonsingular and* $\mathbf{AB}$ *exists,* $r(\mathbf{AB}) = r(\mathbf{B})$.

PROOF. By the theorem $r(\mathbf{AB}) \leq r(\mathbf{B})$; but $\mathbf{B} = (\mathbf{A}^{-1})(\mathbf{AB})$, so that $r(\mathbf{B}) \leq r(\mathbf{AB})$. Hence, $r(\mathbf{AB}) \leq r(\mathbf{B}) \leq r(\mathbf{AB})$, and so $r(\mathbf{AB}) = r(\mathbf{B})$. We may similarly prove that, if $\mathbf{CA}$ exists, $r(\mathbf{CA}) = r(\mathbf{C})$. ■

Corollary 2. *If* $\mathbf{A}$ *and* $\mathbf{C}$ *are nonsingular* $r(\mathbf{ABC}) = r(\mathbf{B})$, *i.e., the rank of a matrix is not changed by multiplying on the left and the right by nonsingular matrices.*

Corollary 3. *If* $\mathbf{A}$ *and* $\mathbf{B}$ *are square matrices and* $\mathbf{AB} = \mathbf{0}$, *then* $\mathbf{A} = \mathbf{0}$, *or* $\mathbf{B} = \mathbf{0}$, *or else* $\mathbf{A}$ *and* $\mathbf{B}$ *are both singular.*

PROOF. Suppose, in contradiction, that $\mathbf{A}$ is nonsingular and $\mathbf{B} \neq \mathbf{0}$. Then $r(\mathbf{AB}) = r(\mathbf{B}) > 0$, and $\mathbf{AB} \neq \mathbf{0}$. Similarly, if $\mathbf{B}$ is nonsingular and $\mathbf{A} \neq \mathbf{0}$, $\mathbf{AB} \neq \mathbf{0}$. ■

A.2 Orthogonality

We have mentioned that two vectors, $\mathbf{x}$ and $\mathbf{y}$, are said to be orthogonal if the product $\mathbf{x}'\mathbf{y} = \mathbf{y}'\mathbf{x} = 0$. A set of vectors of unit length which are mutually orthogonal is called an *orthonormal set*. A set of n mutually orthogonal vectors of

n elements each is a linearly independent set; it is not possible to find a $(n + 1)$th vector other than $\mathbf{0}$ which is orthogonal to all the others. A matrix $\mathbf{P}$ $(n \times n)$ which has for its columns a set of orthonormal vectors is called an orthogonal matrix. If $\mathbf{P}$ is orthogonal, $\mathbf{P}'\mathbf{P} = \mathbf{I}$ and $\mathbf{P}' = \mathbf{P}^{-1}$.

If $\mathbf{x}$ is a vector of coordinates in Euclidean n-dimensional space and $\mathbf{P}$ is an orthogonal matrix, the transformation into new coordinates $\mathbf{y}$ defined by $\mathbf{y} = \mathbf{Px}$ is called an orthogonal transformation and corresponds to a rotation of the axes. A simple example occurs in two-dimensional analytic geometry when the coordinate axes are rotated through an angle θ. With respect to the new axes, the coordinates of the point (x_1, x_2) become (y_1, y_2), where $y_1 = x_1 \cos\theta - x_2 \sin\theta$, $y_2 = x_1 \sin\theta + x_2 \cos\theta$, or

$$\begin{bmatrix} y_1 \\ y_2 \end{bmatrix} = \begin{bmatrix} \cos\theta & -\sin\theta \\ \sin\theta & \cos\theta \end{bmatrix} \begin{bmatrix} x_1 \\ x_2 \end{bmatrix}.$$

A.3 Quadratic Forms

Let $\mathbf{Y}$ be a vector of n real elements $y_1, \ldots, y_n$. A homogeneous second-degree multinomial in these elements, $\sum_i b_{ii} y_i^2 + \sum_i \sum_j b_{ij} y_i y_j$, where $i < j$, is called a quadratic form. We shall write such a form as $\mathbf{Y}'\mathbf{AY}$ where $\mathbf{A}$ is a symmetric matrix having $a_{ii} = b_{ii}$ and $a_{ij} = a_{ji} = b_{ij}/2$; $\mathbf{A}$ is called the matrix of the form. We shall consider only forms for which $\mathbf{A}$ is a symmetric matrix. A quadratic form $\mathbf{Q}$ is said to be positive definite if $\mathbf{Q}$ is positive for all values of $\mathbf{Y}$ except $\mathbf{Y} = \mathbf{0}$. The form $Q = 2y_1^2 + 2y_1y_2 + y_2^2$ is positive definite; it may be written as the sum of two squares, $Q = y_1^2 + (y_1 + y_2)^2$. If $\mathbf{Q}$ is never negative, but is zero for some values of $\mathbf{Y}$ other than $\mathbf{0}$, it is said to be positive semidefinite. $Q = (y_1 - y_2)^2$ is an example; it is never negative, but it takes the value zero whenever $y_1 = y_2$. The term nonnegative definite is used to cover both positive definite forms and positive semidefinite forms. The same terms are applied to the corresponding matrices $\mathbf{A}$.

Suppose that $a_{11} = 0$. Then the form $a_{11}y_1^2 = 0$ for all y_1, and $\mathbf{A}$ cannot be positive definite. If $a_{11} < 0$, $a_{11}y_1^2 < 0$ for $y_1 \neq 0$. If $a_{11} = 0$ and a_{12} is positive, $Q = a_{11}y_1^2 + 2a_{12}y_1y_2$ is negative whenever y_1y_2 is negative. It follows that a positive definite matrix has all its diagonal elements, a_{ii}, positive. In a positive semidefinite matrix all the diagonal elements are nonnegative, and, if $a_{ii} = 0$, then $a_{ij} = a_{ji} = 0$ for all j; i.e., y_i does not appear in the form.

A.4 Latent Roots and Latent Vectors

If $\mathbf{A}$ is a square matrix of order n, and λ is a root of the equation $|\mathbf{A} - \lambda\mathbf{I}| = 0$, there exists a vector $\mathbf{x}$ other than $\mathbf{0}$ such that $[\mathbf{A} - \lambda\mathbf{I}]\mathbf{x} = \mathbf{0}$, and,

$$\mathbf{Ax} = \lambda\mathbf{x};$$

λ is called a latent root (or characteristic root or proper value or eigenvalue) of **A**, and **x** is a latent (characteristic, proper or eigen-) vector corresponding to λ. Counting multiplicities, there are exactly n roots; if **A** is symmetric, the number of nonzero roots of **A** is equal to the rank of **A**. We now confine ourselves to symmetric matrices **A**.

Theorem 1. *If λ_i, λ_j are two latent roots of* **A** *such that $\lambda_i \neq \lambda_j$, and $\mathbf{x}_i$, $\mathbf{x}_j$ are two corresponding latent vectors, then $\mathbf{x}_i$ and $\mathbf{x}_j$ are orthogonal.*

PROOF. We have $\mathbf{Ax}_i = \lambda_i\mathbf{x}_i$, $\mathbf{Ax}_j = \lambda_j\mathbf{x}_j$, so that $\mathbf{x}_j'\mathbf{Ax}_i = \lambda_i\mathbf{x}_j'\mathbf{x}_i$, $\mathbf{x}_i'\mathbf{Ax}_j = \lambda_j\mathbf{x}_i'\mathbf{x}_j$. Then, since **A** is symmetric, taking the transpose of the latter expression gives $\mathbf{x}_j'\mathbf{Ax}_i = \lambda_j\mathbf{x}_j'\mathbf{x}_i$. Thus $\lambda_i\mathbf{x}_j'\mathbf{x}_i = \lambda_j\mathbf{x}_j'\mathbf{x}_i$, and since $\lambda_i \neq \lambda_j$, $\mathbf{x}_j'\mathbf{x}_i = 0$. ■

It can also be shown that if λ is a latent root with multiplicity m, there is a set of exactly m mutually orthogonal latent vectors corresponding to λ. We may take the lengths of the latent vectors to be unity, and thus obtain for **A** a set of n orthonormal vectors. Let **P** denote the matrix whose columns are these orthonormal vectors; **P** is orthogonal and

$$\mathbf{AP} = \mathbf{P\Lambda},$$

where $\mathbf{\Lambda}$ is a diagonal matrix (i.e., a matrix with all its off-diagonal elements zero), having as the diagonal elements the latent roots, each appearing as many times as its multiplicity. Multiplying on the left by $\mathbf{P}'$ gives $\mathbf{P'AP} = \mathbf{P'P\Lambda} = \mathbf{\Lambda}$. We state this result as a theorem.

Theorem 2. *Let* **A** *be a square symmetric matrix, and* $\mathbf{\Lambda}$ *be the diagonal matrix just defined. There exists an orthogonal matrix* **P** *such that* $\mathbf{P'AP} = \mathbf{\Lambda}$. *Conversely, if there exists an orthogonal matrix* **P** *such that* $\mathbf{P'AP} = \mathbf{D}$ *where* **D** *is diagonal, the diagonal elements are the latent roots of* **A**.

Since **P** is nonsingular, $r(\mathbf{A}) = r(\mathbf{P'AP}) = r(\mathbf{\Lambda})$, and we see that the rank of **A** is equal to the number of nonzero latent roots, counting their multiplicities. Denote the ith diagonal element of $\mathbf{\Lambda}$ by λ_i. Let a diagonal matrix **D** be constructed as follows. The ith diagonal element d_{ii} of $\mathbf{D} = 1$ if $\lambda_i = 0$ and $|\lambda_i|^{-1/2}$ if $\lambda_i \neq 0$. Let $\mathbf{Q} = \mathbf{PD}$. **Q** is not singular. Then the transformation $\mathbf{Y} = \mathbf{QZ}$ changes $\mathbf{Y'AY}$ into $\sum \pm z_i^2$ where the number of positive signs is equal to the number of positive latent roots, and the number of negative signs is equal to the number of negative latent roots.

If **B** is a nonsingular matrix we have

$$\mathbf{B}^{-1}\mathbf{Ax} = \mathbf{B}^{-1}\mathbf{ABB}^{-1}\mathbf{x} = \mathbf{B}^{-1}\lambda\mathbf{x} = \lambda\mathbf{B}^{-1}\mathbf{x},$$

so that λ is also a latent root of $\mathbf{B}^{-1}\mathbf{AB}$ with latent vector $\mathbf{B}^{-1}\mathbf{x}$. The two matrices **A** and $\mathbf{B}^{-1}\mathbf{AB}$ are said to be similar.

Theorem 3. **A** *is positive definite if, and only if, all its latent roots are positive.*

PROOF. Let $\mathbf{Z} = \mathbf{P}'\mathbf{Y}$, where $\mathbf{P}$ is an orthogonal matrix such that $\mathbf{P}'\mathbf{A}\mathbf{P} = \mathbf{\Lambda}$. Then $\mathbf{Y}'\mathbf{A}\mathbf{Y} = \mathbf{Y}'\mathbf{P}\mathbf{P}'\mathbf{A}\mathbf{P}\mathbf{P}'\mathbf{Y} = \mathbf{Z}'\mathbf{\Lambda}\mathbf{Z}$, and, since $\mathbf{Z} = \mathbf{0}$ if, and only if, $\mathbf{Y} = \mathbf{0}$, the form $\mathbf{Y}'\mathbf{A}\mathbf{Y}$ is positive definite if, and only if, $\mathbf{Z}'\mathbf{\Lambda}\mathbf{Z}$ is positive definite. Thus $\mathbf{A}$ is positive definite if, and only if, $\mathbf{\Lambda}$ is positive definite, i.e., since $\mathbf{\Lambda}$ is diagonal, if, and only if, all the diagonal elements of $\mathbf{\Lambda}$ are positive. ■

Similarly, $\mathbf{A}$ is positive semidefinite if, and only if, $\mathbf{A}$ has no negative latent roots and has at least one zero latent root, which implies that $\mathbf{A}$ is singular.

The trace of a square matrix $\mathbf{A}$ is defined as the sum of the diagonal elements: $\text{tr}(\mathbf{A}) = \sum a_{ii}$. It is easily shown by performing the multiplications that, if $\mathbf{A}$ and $\mathbf{B}$ are both square matrices, $\text{tr}(\mathbf{AB}) = \text{tr}(\mathbf{BA})$.

Theorem 4. *The sum of the latent roots of* $\mathbf{A}$ *is equal to the trace of* $\mathbf{A}$.

PROOF. $\sum \lambda_i = \text{tr}(\mathbf{\Lambda}) = \text{tr}(\mathbf{P}'\mathbf{A}\mathbf{P}) = \text{tr}(\mathbf{A}\mathbf{P}\mathbf{P}') = \text{tr}(\mathbf{A}\mathbf{I}) = \text{tr}(\mathbf{A})$. ■

Theorem 5. *The latent roots of a real symmetric matrix are real.*

PROOF. Suppose that we have a complex root $\lambda + i\mu$ with a complex vector $\mathbf{x} + i\mathbf{y}$ where λ, μ, $\mathbf{x}$, and $\mathbf{y}$ are all real. Then

$$\mathbf{A}(\mathbf{x} + i\mathbf{y}) = (\lambda + i\mu)(\mathbf{x} + i\mathbf{y})$$

and, equating the real and imaginary parts,

$$\mathbf{A}\mathbf{x} = \lambda\mathbf{x} - \mu\mathbf{y} \tag{A.1}$$

$$\mathbf{A}\mathbf{y} = \lambda\mathbf{y} + \mu\mathbf{x}. \tag{A.2}$$

Multiply Equation A.1 on the left by $\mathbf{y}'$. Multiply Equation A.2 on the left by $\mathbf{x}'$ and transpose it. Then

$$\mathbf{y}'\mathbf{A}\mathbf{x} = \lambda\mathbf{y}'\mathbf{x} - \mu\mathbf{y}'\mathbf{y} = \lambda\mathbf{y}'\mathbf{x} + \mu\mathbf{x}'\mathbf{x}, \tag{A.3}$$

so that $\mu(\mathbf{x}'\mathbf{x} + \mathbf{y}'\mathbf{y}) = 0$. But $\mathbf{x}'\mathbf{x} + \mathbf{y}'\mathbf{y} > 0$, and so $\mu = 0$. ■

Theorem 6. *The latent roots of* $\mathbf{A}^2$ *are the squares of the latent roots of* $\mathbf{A}$.

PROOF. We have $\mathbf{P}'\mathbf{A}\mathbf{P} = \mathbf{\Lambda}$. Then

$$\mathbf{P}'\mathbf{A}\mathbf{P}\mathbf{P}'\mathbf{A}\mathbf{P} = \mathbf{P}'\mathbf{A}^2\mathbf{P} = \mathbf{\Lambda}^2,$$

so that the orthogonal matrix $\mathbf{P}$ also diagonalizes $\mathbf{A}^2$, and the diagonal elements of the diagonalized matrix are $d_i = \lambda_i^2$. Thus the latent roots of $\mathbf{A}^2$ are λ_i^2. Furthermore, multiplying $\mathbf{A}\mathbf{x} = \lambda\mathbf{x}$ on the left by $\mathbf{A}$ gives

$$\mathbf{A}^2\mathbf{x} = \lambda\mathbf{A}\mathbf{x} = \lambda^2\mathbf{x},$$

and we see that each vector $\mathbf{x}$ corresponding to a root λ of $\mathbf{A}$ also corresponds to a root λ^2 of $\mathbf{A}^2$. ■

The following theorem is of particular use to us.

Theorem 7. *Let* $\mathbf{X}$ *be a real matrix with* m *rows and* n *columns. Then* $\mathbf{X}'\mathbf{X}$ *is nonnegative definite and* $r(\mathbf{X}'\mathbf{X}) = r(\mathbf{X})$.

PROOF. Since $\mathbf{X}'\mathbf{X}$ is symmetric, there is an orthogonal $\mathbf{P}$ such that $\mathbf{P}'\mathbf{X}'\mathbf{X}\mathbf{P} = \mathbf{\Lambda}$. Let $\mathbf{B} = (b_{ij}) = \mathbf{XP}$. Then $\mathbf{B}'\mathbf{B} = \mathbf{\Lambda}$, and we have, writing the columns of $\mathbf{B}$ as vectors, $\mathbf{b}_i$, $\mathbf{b}_i'\mathbf{b}_i = \lambda_i$ and $\mathbf{b}_i'\mathbf{b}_j = 0$ where $i \neq j$. It follows that $\lambda_i = 0$ if and only if $\mathbf{b}_i = \mathbf{0}$; otherwise, $\lambda_i > 0$. Thus the latent roots are nonnegative and $\mathbf{X}'\mathbf{X}$ is nonnegative definite.

The vectors $\mathbf{b}_i$ are mutually orthogonal. The rank of $\mathbf{X}'\mathbf{X}$ is equal to the number of positive diagonal elements of $\mathbf{\Lambda}$, which, in turn, is the number of $\mathbf{b}_i$ that are not $\mathbf{0}$. Thus, $r(\mathbf{X}'\mathbf{X}) = r(\mathbf{B}) = r(\mathbf{XP}) = r(\mathbf{X})$. ■

We state without proof the following theorem, which is to be found in many of the standard textbooks.

Theorem 8. *A necessary and sufficient condition for* $\mathbf{A}$ *to be positive definite is that every principal minor of* $|\mathbf{A}|$ *shall be positive. A principal minor of* $|\mathbf{A}|$ *is the determinant of a submatrix of* $\mathbf{A}$ *obtained by striking out some rows of* $\mathbf{A}$ *and the corresponding columns so that the diagonal elements of the submatrix are also diagonal elements of* $\mathbf{A}$.

A square symmetric matrix $\mathbf{A}$ is said to be idempotent if $\mathbf{A}^2 = \mathbf{A}$. The following properties of idempotent matrices are readily established:

(i) A necessary and sufficient condition for $\mathbf{A}$ to be idempotent is that the latent roots of $\mathbf{A}$ are either zero or one;
(ii) If $\mathbf{A}$ is idempotent, $r(\mathbf{A}) = \mathrm{tr}(\mathbf{A})$;
(iii) If $\mathbf{A}$ is idempotent and $\mathbf{Q}$ is orthogonal, $\mathbf{Q}'\mathbf{A}\mathbf{Q}$ is idempotent;
(iv) If $\mathbf{A}$ is idempotent and $a_{ii} = 0$, then $a_{ij} = a_{ji} = 0$ for all j;
(v) If $\mathbf{A}$ is idempotent, all its diagonal elements are nonnegative.

A.5 Simultaneous Linear Equations

We conclude this discussion of matrices by recalling some results about the solution of sets of simultaneous linear equations.

Consider the set of linear equations $\mathbf{Ax} = \mathbf{b}$ where $\mathbf{A}$ has m rows and n columns, and suppose that we wish to solve for $\mathbf{x}$. If $m = n$, and $r(\mathbf{A}) = n$, the solution vector is $x = \mathbf{A}^{-1}\mathbf{b}$. Otherwise, there are three possibilities to be considered:

(i) There is no solution;
(ii) There is an infinite number of solution vectors;
(iii) There is a unique solution vector.

For example, we may have the following four equations in two unknowns:

$$x_1 + x_2 = 3 \tag{A.4}$$

$$x_1 + 2x_2 = 4 \tag{A.5}$$

$$2x_1 + 3x_2 = 7 \tag{A.6}$$

$$2x_1 + 3x_2 = 8. \tag{A.7}$$

Equation A.4 alone admits an infinity of solutions; we may take any number t that we wish and obtain the solution vector $\mathbf{x}' = (t, 3 - t)$.

The pair of equations A.4 and A.5 has the unique solution vector $\mathbf{x}' = (2, 1)$. The addition of equation A.6 makes no difference because it is actually just the sum of Equations A.4 and A.5. When, however, Equation A.7 is included we have an inconsistent set; subtracting Equation A.6 from Equation A.7 would give $0 = 1$.

We state the following results without proof. **B** is the augmented matrix (**A**, **b**), i.e., the matrix **A** with the column **b** added to it. A set of equations is said to be consistent if it has at least one solution.

A necessary and sufficient condition for the equations to be consistent is that $r(\mathbf{A}) = r(\mathbf{B})$. If $r(\mathbf{A}) = r(\mathbf{B}) = n$, there is a unique solution vector; if $r(\mathbf{A}) = r(\mathbf{B}) < n$, then there is an infinity of solutions and $n - r(\mathbf{A})$ of the x_i may be assigned arbitrarily.

Exercises

The exercises involve the derivation of some results stated without proof in the text. **A** denotes a square symmetric matrix of order n with real elements. **C** is any matrix with real elements.

1. Show directly that $(\mathbf{AC})' = \mathbf{C}'\mathbf{A}'$.
2. All bases for a given finite dimensional vector space have the same number of vectors.
3. Let V_n be the space of all vectors in E_n (Euclidean n space). The dimension of V_n is n.
4. There does not exist a set of $n + 1$ mutually orthogonal vectors in V_n.
5. The rank of **A** is the order of the largest nonvanishing determinant in **A**.
6. The number of linearly independent rows of **C** is the same as the number of linearly independent columns of **C**.
7. $\mathbf{A}^{-1}$ is unique. If **B** is a right inverse of **A** and **E** is a right inverse, then $\mathbf{B} = \mathbf{E}$. If **B** and **E** are both left inverses, $\mathbf{B} = \mathbf{E}$. We have given in the text a procedure for obtaining one matrix that is both a right and left inverse of **A** when **A** is nonsingular.
8. We defined latent roots as solutions of $|\mathbf{A} - \lambda\mathbf{I}| = 0$. Show that any scalar λ for which there exists a vector $\mathbf{x} \neq \mathbf{0}$ such that $\mathbf{Ax} = \lambda\mathbf{x}$ is a root of $|\mathbf{A} - \lambda\mathbf{I}| = 0$.

9. If λ is a latent root of **A** with multiplicity m, show that we can find a set of m mutually orthogonal latent vectors (but no more) corresponding to λ.

10. **A** is positive semidefinite if, and only if, **A** has at least one zero latent root and no negative latent roots.

11. $\text{tr}(\mathbf{BC}) = \text{tr}(\mathbf{CB})$.

12. If there exists an orthogonal matrix **P** such that $\mathbf{P'AP}$ is diagonal, then $\mathbf{P'AP} = \mathbf{\Lambda}$.

13. A necessary and sufficient condition for **A** to be positive definite is that every principal minor of $|\mathbf{A}|$ be positive.

14. If **A** is idempotent and $\mathbf{I} = \mathbf{A} + \mathbf{B}$, then **B** is idempotent and $\mathbf{AB} = \mathbf{BA} = \mathbf{0}$.

15. If the elements of **C** are real, $\mathbf{C'C} = \mathbf{0}$ if, and only if, $\mathbf{C} = \mathbf{0}$. Show by a counter-example that this is not true when **C** has complex elements.

Bibliography

ADDELMAN, S. (1962). "Orthogonal Main-Effect Plans for Asymmetrical Factorial Experiments." *Technometrics*, **4**, 21–46.

ADDELMAN, S., and O. KEMPTHORNE (1961). "Some Main Effects Plans and Orthogonal Arrays of Strength Two." *Ann. Math. Statist.*, **32**, 1167–1176.

AIA, M. A., R. L. GOLDSMITH, and R. W. MOONEY (1961). "Precipitating Stoichiometric $CaHPO_4 \cdot 2H_2O$." *Industrial and Engineering Chemistry*, **53**, 55–57.

AITKEN, A. C. (1950). "On the Statistical Independence of Quadratic Forms in Normal Variates." *Biometrika*, **37**, 93–96.

ALLAN, R. E., and J. WISHART (1930). "A Method of Estimating the Yield of a Missing Plot in Field Experimental Work." *J. Agr. Sci.*, **20**, 399–406.

ATIQULLAH, M. (1958). "On Configurations and Non-isomorphism of Some Incomplete Block Designs." *Sankhya*, **20**, 227–248.

ATIQULLAH, M. (1961). "On a Property of Balanced Designs." *Biometrika*, **48**, 215–218.

BANERJEE, K. S. (1964). "A Note on Idempotent Matrices." *Ann. Math. Statist.*, **35**, 880–882.

BATCHELDER, A. R., M. J. ROGERS, and J. P. WALKER (1966). "Effects of Subsoil Management Practices on Growth of Flue-Cured Tobacco." *Agron. Journal*, **58**, 345–347.

BAUMERT, L., S. W. GOLOMB, and M. HALL, Jr. (1962). "The Discovery of an Hademard Matrix of Order 92." *Bull. Amer. Math. Soc.*, **68**, 237–238.

BENNETT, C. A., and N. J. FRANKLIN (1954). *Statistical Analysis in Chemistry and the Chemical Industry*. John Wiley and Sons, New York.

BHATTACHARYA, K. N. (1943). "A Note on Two-Fold Triple Systems. *Sankhya*, **6**, 313–314.

BHATTACHARYA, K. N. (1944*a*). "A New Balanced Incomplete Block Design." *Science and Culture*, **9**, 508.

BHATTACHARYA, K. N. (1944*b*). "On a New Symmetrical Balanced Incomplete Block Design." *Bull. Calcutta Math. Soc.*, **36**, 91–96.

BHATTACHARYA, K. N. (1946). "A New Solution in Symmetrical Balanced Incomplete Block Designs ($v = b = 31, r = k = 10, \lambda = 3$)." *Sankhya*, **7**, 423–424.

BINET, F. E., R. T. LESLIE, S. WEINER, and R. L. ANDERSON (1955). "Analysis of Confounded Factorial Experiments in Single Replications." North Carolina Agric. Exp. Station. *Techn. Bull. No. 113.*

BOND, D. A. (1966). "Yield and Components of Yield in Diallel Crosses Between Inbred Lines of Winter Beans (vicia faba)." *J. Agri. Sci.*, **67**, 325–336.

BOSE, R. C. (1938). "On the Application of the Properties of Galois Fields to the Problem of the Construction of Hyper-Graeco-Latin Squares." *Sankhya*, **3**, 323–338.

BOSE, R. C. (1939). "On the Construction of Balanced Incomplete Block Designs." *Annals of Eugenics*, **9**, 353–399.

BOSE, R. C. (1942*a*). "On Some New Series of Balanced Incomplete Block Designs." *Bull. Calcutta Math. Soc.*, **34**, 17–31.

BOSE, R. C. (1942*b*). "A Note on Two Series of Balanced Incomplete Block Designs." *Bull. Calcutta Math. Soc.*, **34**, 129–130.

BOSE, R. C. (1942*c*). "A Note on the Resolvability of Balanced Incomplete Block Designs." *Sankhya*, **6**, 105–110.

BOSE, R. C. (1944). "The Fundamental Theorem of Linear Estimation." *Proc. 31st Indian Sci. Congress*, 2–3 (abstract).

BOSE, R. C. (1947*a*). "Mathematical Theory of the Symmetrical Factorial Design." *Sankhya*, **8**, 107–166.

BOSE, R. C. (1947*b*). "On a Resolvable Series of Balanced Incomplete Block Designs." *Sankhya*, **8**, 249–256.

BOSE, R. C. (1949). "A Note on Fisher's Inequality for Balanced Incomplete Block Designs." *Ann. Math. Statist.*, **20**, 619–620.

BOSE, R. C. (1963). "Combinatorial Properties of Partially Balanced Designs and Association Schemes." *Sankhya*, Ser. A., **25**, 109–136.

BOSE, R. C., and K. A. BUSH (1952). "Orthogonal Arrays of Strength Two and Three." *Ann. Math. Statist.*, **23**, 508–524.

BOSE, R. C. and W. H. CLATWORTHY (1955). "Some Classes of Partially Balanced Designs." *Ann. Math. Statist.*, **26**, 212–232.

BOSE, R. C., W. H. CLATWORTHY, and S. S. SHRIKHANDE (1954). "Tables of Partially Balanced Designs with Two Associate Classes." North Carolina Agric. Exp. Station. *Tech. Bull. No. 107.*

BOSE, R. C., and W. S. CONNOR (1952). "Combinatorial Properties of Group Divisible Incomplete Block Designs." *Ann. Math. Statist.*, **23**, 367–383.

BOSE, R. C., and K. KISHEN (1940). "On the Problem of Confounding in the General Symmetric Factorial Design." *Sankhya*, **5**, 21–36.

BOSE, R. C., and D. M. MESNER (1959). "On Linear Associative Algebras Corresponding to Association Schemes of Partially Balanced Designs." *Ann. Math. Statist.*, **30**, 21–38.

BOSE, R. C., and K. R. NAIR (1939). "Partially Balanced Incomplete Block Designs." *Sankhya*, **4**, 337–372.

BOSE, R. C., and T. SHIMAMOTO (1952). "Classification and Analysis of Designs with Two Associate Classes." *J. Amer. Statist. Ass.*, **47**, 151–184.

BOSE, R. C., S. S. SHRIKHANDE, and K. N. BHATTACHARYA (1953). "On the Construction of Group Divisible Incomplete Block Designs." *Ann. Math. Statist.*, **24**, 167–195.

BOSE, R. C., S. S. SHRIKHANDE, and E. T. PARKER (1960). "Some Further Results on the Construction of Mutually Orthogonal Latin Squares and the Falsity of Euler's Conjecture." *Canad. J. Math.*, **12**, 189–203.

BOX, G. E. P. (1953). "Non-normality and Tests on Variances." *Biometrika*, **40**, 318–335.

BOX, G. E. P. (1957). "Evolutionary Operation: A Method for Increasing Industrial Productivity." *Applied Statist.*, **6**, 3–23.

BOX, G. E. P., and D. W. BEHNKEN (1960). "Simplex Sum Designs: A Class of Second Order Rotatable Designs Derivable from Those of First Order." *Ann. Math. Statist.*, **31**, 838–864.

BOX, G. E. P., and J. S. HUNTER (1957). "Multifactor Experimental Designs for Exploring Response Surfaces." *Ann. Math. Statist.*, **28**, 195–242.

BOX, G. E. P., and J. S. HUNTER (1959). "Condensed Calculations for Evolutionary Operation Programs." *Technometrics*, **1**, 77–95.

BOX, G. E. P., and J. S. HUNTER (1961). "The 2^{k-p} Fractional Factorial Designs." *Technometrics*, **3**, Part I, 311–352; Part II, 449–458.

BOX, G. E. P., and K. J. WILSON (1951). "On the Experimental Attainment of Optimum Conditions." *J. Roy. Statist. Soc.*, Ser. B., **13**, 1–45.

BOX, G. E. P., and P. V. YOULE (1955). "The Exploration and Exploitation of Response Surfaces: An Example of the Link Between the Fixed Surface and the Basic Mechanism of the System." *Biometrics*, **11**, 287–323.

BROWNLEE, K. A., B. K. KELLY, and P. K. LORAINE (1948). "Fractional Replication Arrangements for Factorial Experiments with Factors at Two Levels." *Biometrika*, **35**, 268–276.

BRUCK, R. H., and H. J. RYSER (1949). "The Non-existence of Certain Finite Projective Planes." *Canad. J. Math.*, **1**, 88–93.

BULMER, M. G. (1957). "Approximate Confidence Limits for Components of Variance." *Biometrika*, **44**, 159–167.

BURTON, R. C., and W. S. CONNOR (1957). "On the Identity Relationship for Fractional Replicates in the 2^n Series." *Ann. Math. Statist.*, **28**, 762–767.

CALVIN, L. C. (1954). "Doubly Balanced Incomplete Block Designs for Experiments in Which the Effects Are Correlated." *Biometrics*, **10**, 61–88.

CHANDA, et al. (1952). "The Use of Chromium Sesquioxide to Measure the Digestibility of Carotene by Goats and Cows." *J. Agr. Sci.*, **42**, 179–185.

CHANG, L. C. (1959). "The Uniqueness and Non-uniqueness of the Triangular Association Schemes." *Science Record*, **3**, new series, 604–613.

CHANG, L. C. (1960). "Association Schemes of Partially Balanced Designs with Parameters $v = 28$, $n_1 = 12$, $n_2 = 15$, $p_{11}^2 = 4$." *Science Record*, **4**, new series, 12–18.

CHANG, L. C., and W. R. LIU (1964). "Incomplete Block Designs with Square Parameters for which $k \leq 10$ and $r \leq 10$." *Scientia Sinica*, **13**, 1493–1495.

CHANG, L. C., C. W. LIU, and W. R. LIU (1965). "Incomplete Block Designs with Triangular Parameters for $k \leq 10$ and $r \leq 10$." *Scientia Sinica*, **14**, 329–338.

CHOWLA, S., and H. J. RYSER (1950). "Combinatorial Problems." *Canad. J. Math.*, **2**, 93–99.

CLATWORTHY, W. H. (1956). "Contributions on Partially Balanced Incomplete Block Designs with Two Associate Classes." National Bureau of Standards. *Applied Math. Series, No.* 47.

CLATWORTHY, W. H. (1967). "Some New Families of Partially Balanced Designs of the Latin Square Type and Related Designs." *Technometrics*, **9**, 229–244.

COCHRAN, W. G. (1934). "The Distribution of Quadratic Forms in a Normal System, with Applications to the Analysis of Covariance." *Proc. Camb. Philos. Soc.*, **30**, 178–191.

COCHRAN, W. G., K. M. AUTREY, and C. Y. CANNON (1941). "A Double Change-Over Design for Dairy Cattle Feeding Experiments." *J. Dairy Sci.*, **24**, 937–951.

COCHRAN, W. G., and G. M. COX (1957). *Experimental Designs.* (Second ed.) John Wiley and Sons, New York.

CONNOR, W. S. (1952*a*). "On the Structure of Balanced Incomplete Block Designs." *Ann. Math. Statist.*, **23**, 57–71.

CONNOR, W. S. (1952*b*). "Some Relations Among the Blocks of Symmetrical Group-Divisible Designs." *Ann. Math. Statist.*, **23**, 602–609.

CONNOR, W. S. (1958). "The Uniqueness of the Triangular Association Scheme." *Ann. Math. Statist.*, **29**, 262–266.)

CONNOR, W. S., and W. H. CLATWORTHY (1954). "Some Theorems for Partially Balanced Designs." *Ann. Math. Statist.*, **25**, 100–112.

CORNFIELD, J., and J. W. TUKEY (1956). "Average Values of Mean Squares in Factorials." *Ann. Math. Statist.*, **27**, 907–949.

CRAIG, A. T. (1943). "Note on the Independence of Certain Quadratic Forms." *Ann. Math. Statist.*, **14**, 195–197.

CURNOW, R. N. (1963). "Sampling the Diallel Cross." *Biometrics*, **19**, 287–306.

DANIEL, C. (1956). "Fractional Replication in Industrial Research." *Proc. Third Berkeley Symposium on Mathematical Statistics and Probability*, **5**, 87–98.

DANIEL, C. (1959). "Use of Half-Normal Plots in Interpreting Factorial Two Level Experiments." *Technometrics*, **1**, 311–342.

DANIEL, C. (1962). "Sequences of Fractional Replicates in the 2^{p-q} Series." *J. Amer. Statist. Assoc.*, **57**, 403–429.

DAVID, H. A. (1963). "The Structure of Cyclic Paired-Comparison Designs." *J. Austral. Math. Soc.*, **3**, 117–127.

DAVID, H. A., and F. W. WOLOCK (1965). "Cyclic Designs." *Ann. Math. Statist.*, **36**, 1526–1534.

DAVIES, O. L. (1954). *Design and Analysis of Industrial Experiments.* Oliver and Boyd, London.

DEBAUN, R. (1959). "Response Surface Designs for Three Factors at Three Levels." *Technometrics*, **1**, 1–8.

DRAPER, N. R. (1960*a*). "Third Order Rotatable Designs in Three Dimensions.' *Ann. Math. Statist.*, **31**, 865–874.

DRAPER, N. R. (1960*b*). "A Third Order Rotatable Design in Four Dimensions.' *Ann. Math. Statist.*, **31**, 875–877.

DRAPER, N. R. (1961). "Third Order Rotatable Designs in Three Dimensions: Some Specific Designs." *Ann. Math. Statist.*, **32**, 910–913.

DRAPER, N. R. (1962). "Third Order Rotatable Designs in Three Factors: Analysis." *Technometrics*, **4**, 219–234.

DRAPER, N. R., and H. SMITH (1966). *Applied Regression Theory.* John Wiley and Sons, New York.

DUNCAN, D. B. (1955). "Multiple Range and Multiple *F*-Tests." *Biometrics*, **11**, 1–42.

FEDERER, W. T. (1955). *Experimental Design: Theory and Application.* The Macmillan Company, New York.

FINNEY, D. J. (1945). "The Fractional Replication of Factorial Arrangements." *Annals of Eugenics*, **12**, 291–301.

FINNEY, D. J. (1946). "Recent Developments in the Design of Field Experiments. I. Split-Plot Confounding." *J. Agri. Sci.*, **36**, 56–62.

FISHER, R. A. (1925). *Statistical Methods for Research Workers.* Oliver and Boyd, Edinburgh.

FISHER, R. A. (1940). "An Examination of the Different Possible Solutions of a Problem in Incomplete Blocks." *Annals of Eugenics*, **10**, 52–75.

FISHER, R. A. (1942). "The Theory of Confounding in Factorial Experiments in Relation to the Theory of Groups." *Annals of Eugenics*, **11**, 341–353.

FISHER, R. A. (1947). *The Design of Experiments.* (Fourth ed.) Oliver and Boyd, Edinburgh.

FISHER, R. A., and F. YATES (1953). *Statistical Tables for Biological, Agricultural, and Medical Research.* (Fourth ed.) Oliver and Boyd, Edinburgh.

FOX, M. (1956). "Charts of the Power of the *F* Test." *Ann. Math. Statist.*, **27**, 484–497.

FREEMAN, G. H. (1957). "Some Further Methods of Constructing Regular Group Divisible Incomplete Block Designs," *Ann. Math. Statist.*, **28**, 479–487.

FYFE, J. L., and N. GILBERT (1963). "Partial Diallel Crosses." *Biometrics*, **19**, 278–286.

GARDINER, D. A., A. H. E. GRANDAGE, and R. J. HADER (1959). "Third Order Rotatable Designs for Exploring Response Surfaces." *Ann. Math. Statist.*, **30**, 1082–1096.

GOOD, I. J. (1958). "The Interaction Algorithm and Practical Fourier Analysis." *J. Roy. Statist. Soc.*, Ser. B., **20**, 361–372.

GOOD, I. J. (1960). Addendum to "The Interaction Algorithm and Practical Fourier Analysis," *J. Roy. Statist. Soc.*, Ser. B., **22**, 372–375.

GRAYBILL, F. A. (1961). *An Introduction to Linear Statistical Models.* Vol. I. McGraw-Hill, New York.

GRAYBILL, F. A., and R. B. DEAL (1959). "Combining Unbiased Estimators." *Biometrics*, **15**, 543–550.

GRAYBILL, F. A., and G. MARSAGLIA. (1957). "Idempotent Matrices and Quadratic Forms in the General Linear Hypothesis." *Ann. Math. Statist.*, **28**, 678–686.

GRAYBILL, F. A., and V. SESHADRI (1960). "On the Unbiasedness of Yates' Method of Estimation Using Interblock Information." *Ann. Math. Statist.*, **31**, 786–787.

GRAYBILL, F. A., and D. L. WEEKS (1959). "Combining Interblock and Intrablock Information in Balanced Incomplete Blocks." *Ann. Math. Statist.*, **30**, 799–805.

GRIFFING, B. (1956). "Concept of General and Specific Combining Ability in Relation to Diallel Crossing Systems." *Austral. J. Biol. Sci.*, **9**, 463–493.

GROSS, H. D., E. R. Parvist, and G. A. Ahlgren (1953). "The Response of Alfalfa Varieties to Different Soil Fertility Levels." *Agron. J.*, **45**, 118–120.

GRUNDY, P. M., and M. J. R. HEALY (1950). "Restricted Randomization and Quasi-Latin Squares." *J. Roy. Statist. Soc.*, Ser. B, **12**, 286–291.

HALL, M., and W. S. CONNOR (1953). "An Embedding Theorem for Incomplete Block Designs." *Canad. J. Math.*, **6**, 35–41.

HALL, M., and H. J. RYSER (1951). "Cyclic Incidence Matrices." *Canad. J. Math.*, **3**, 495–502.

HAMMERSLEY, J. M. (1949). "The Unbiased Estimate and Standard Error of the Interclass Variance." *Metron*, **15**, 173–188.

HANANI, H. (1961). "The Existence and Construction of Balanced Incomplete Block Designs." *Ann. Math. Statist.*, **32**, 361–386.

HANANI, H. (1965). "A Balanced Incomplete Block Design." *Ann. Math. Statist.*, **36**, 711.

HARTLEY, H. O. (1959). "Smallest Composite Designs for Quadratic Response Surfaces." *Biometrics*, **15**, 611–624.

HEALY, M. J. R. (1951). "Latin Rectangle Designs for 2^n Factorial Experiments on 32 Plots." *J. Agri. Sci.*, **41**, 315–316.

HILL, W. J., W. G. HUNTER (1966). "A Review of Response Surface Methodology: A Literature Survey." *Technometrics*, **8**, 571–590.

HOFFMAN, A. J. (1960). "On the Uniqueness of the Triangular Association Scheme." *Ann. Math. Statist.*, **31**, 492–497.

HOTELLING, H. (1944*a*). "Some Improvements in Weighing and Other Experimental Techniques." *Ann. Math. Statist.*, **15**, 297–306.

HOTELLING, H. (1944*b*). "On a Matric Theorem of A. T. Craig." *Ann. Math. Statist.*, **15**, 427–429.

HUNTER, J. S. (1960). "Some Applications of Statistics to Experimentation." *Chemical Engineering Progress Symposium*, Series 31, **56**, 10–26.

JAMES, E., and T. A. BANCROFT (1951). "The Use of Half Plants in a Balanced Incomplete Block Design in Investigating the Effect of Calcium, Phosphorus, and Potassium, at Two Levels Each on the Production of Hard Seed in Crimson Clover, Trifolium Incarnatum." *J. Agron.*, **43**, 96–98.

JAMES, G. S. (1952). "Notes on a Theorem of Cochran. "*Proc. Camb. Philos. Soc.*, **48**, 443–446.

JOHN, J. A. (1966). "Cyclic Incomplete Block Designs." *J. Roy. Statist. Soc.*, Ser. B., **28**, 345–360.

JOHN, P. W. M. (1961*a*). "Three-Quarter Replicates of 2^4 and 2^5 Designs." *Biometrics*, **17**, 319–321.

JOHN, P. W. M. (1961*b*). "An Application of a Balanced Incomplete Block Design." *Technometrics*, **3**, 51–54.

JOHN, P. W. M. (1962). "Three-quarter Replicates of 2^n Designs." *Biometrics*, **18**, 172–184.

JOHN, P. W. M. (1963). "Extended Complete Block Designs." *Austral. J. Statist.*, **5**, 147–152.

JOHN, P. W. M. (1964*a*). "Pseudo-Inverses in the Analysis of Variance." *Ann. Math. Statist.*, **35**, 895–896.

JOHN, P. W. M. (1964*b*). "Blocking of $3(2^{n-k})$ Designs." *Technometrics*, **6**, 371–376.

JOHN, P. W. M. (1964*c*). "Balanced Designs with Unequal Numbers of Replicates." *Ann. Math. Statist.*, **35**, 897–899.

JOHN, P. W. M. (1966*a*). "On Identity Relationships for 2^{n-r} Designs Having Words of Equal Length." *Ann. Math. Statist.*, **37**, 1842–1843.

JOHN, P. W. M. (1966*b*). "Augmenting 2^{n-1} Designs." *Technometrics*, **8**, 469–480.

JOHN, P. W. M. (1966*c*). "An Extension of the Triangular Association Scheme to Three Associate Classes." *J. Roy. Statist. Soc.*, Ser. B., **28**, 361–365.

JOHN, P. W. M. (1969). "Some Non-Orthogonal Fractions of 2^n Designs." *J. Roy. Statist. Soc.*, Ser. B., **31**, 270–275.

JOHN, P. W. M. (1970). "A Three-Quarter Replicate of the 4^3 Design." *Austral. J. of Statist.*, **12**, 73–77.

JOHNSON, J. D. (1967). "Adding Partially Balanced Incomplete Block Designs." Doctoral dissertation, University of California, Davis.

KAPADIA, C. H., and D. L. WEEKS (1964). "On the Analysis of Group Divisible Designs." *J. Amer. Statist. Ass.*, **59**, 1217–1219.

KEMPTHORNE, O. (1952). *The Design and Analysis of Experiments.* John Wiley and Sons, New York.

KEMPTHORNE, O. (1953). "A Class of Experimental Designs Using Blocks of Two Plots." *Ann. Math. Statist.*, **24**, 76–84.

KEMPTHORNE, O. (1955). "The Randomization Theory of Experimental Inference." *J. Amer. Statist. Ass.*, **50**, 946–967.

KEMPTHORNE, O. (1956*a*). "The Efficiency Factor of an Incomplete Block Design." *Ann. Math. Statist.*, **27**, 846–849.

KEMPTHORNE, O. (1956*b*). "The Theory of the Diallel Cross." *Genetics*, **41**, 451–459.

KEMPTHORNE, O., and R. N. CURNOW (1961). "The Partial Diallel Cross." *Biometrics*, **17**, 229–250.

KEMPTHORNE, O., and W. T. FEDERER (1948). "The General Theory of Prime Power Lattice Designs." *Biometrics*, **4**, Part I, 54–79; Part II, 109–121.

KEULS, M. (1952). "The Use of the 'Studentized Range' in Connection with an Analysis of Variance." *Euphytica*, **1**, 112–122.

KISHEN, K. (1945). "On the Design of Experiments for Weighing and Making Other Types of Measurements." *Ann. Math. Statist.*, **16**, 294–300.

KRUSKAL, W. H., and W. A. WALLIS (1952). "Use of Ranks in One Criterion Analysis of Variance." *J. Amer. Statist. Ass.*, **47**, 583–621.

LANCASTER, H. O. (1954). "Traces and Cumulants of Quadratic Forms in Normal Variables." *J. Roy. Statist. Soc.*, Ser. B., **16**, 247–254.

LUCAS, H. L. (1951). "Bias in Estimation of Error in Change-Over Trials with Dairy Cattle." *J. Agri. Sci.*, **41**, 146–148.

LUCAS, H. L. (1957). "Extra-Period Latin Square Change-Over Designs." *J. Dairy Sci.*, **40**, 225–239.

MADOW, W. G. (1940). "The Distribution of Quadratic Forms in Non-Central Normal Random Variables." *Ann. Math. Statist.*, **11**, 100–103.

MARGOLIN, B. H. (1967). "Systematic Methods of Analyzing 2^m3^n Factorial Experiments with Applications." *Technometrics*, **9**, 245–260.

MARGOLIN, B. H. (1969*a*). "Results on Factorial Designs of Resolution IV for the 2^n and 2^n3^m Series." *Technometrics*, **11**, 431–444.

MARGOLIN, B. H. (1969*b*). "Resolution IV Fractional Factorial Designs." *J. Roy. Statist. Soc.*, Ser. B., **31**, 514-523.

MARGOLIN, B. H. (1969*c*). "Orthogonal Main Effect Plans Permitting Estimation of All Two-Factor Interactions for the 2^n3^m Factorial Series of Designs." *Technometrics*,, **11**, 747–762.

MESNER, D. M. (1956). "An Investigation of Certain Combinatorial Properties of Partially Balanced Incomplete Block Designs and Association Schemes, with a Detailed Study of Latin Square and Related Types." Doctoral dissertation, Michigan State University.

MESNER, D. M. (1965). "A Note on the Parameters of *PBIB* Association Schemes." *Ann. Math. Statist.*, **36**, 331–336.

MESNER, D. M. (1967). "A New Family of Partially Balanced Incomplete Block Designs with Some Latin Square Design Properties." *Ann. Math. Statist.*, **38**, 571–582.

MIKHAIL, W. F. (1960). "An Inequality for Balanced Incomplete Block Designs." *Ann. Math. Statist.*, **31**, 520–522.

MILLER, R. G., Jr. (1966). *Simultaneous Statistical Inference*. McGraw-Hill, New York.

MOOD, A. M. (1946). "On Hotelling's Weighing Problem." *Ann. Math. Statist.*, **17**, 432–446.

MURTY, V. N. (1961). "An Inequality for Balanced Incomplete Block Designs." *Ann. Math. Statist.*, **32**, 908–909.

NAIR, K. R. (1952). "Analysis of Partially Balanced Incomplete Block Designs Illustrated on the Simple Square and Rectangular Lattices." *Biometrics*, **8**, 122–155.

NAIR, K. R., and C. R. RAO (1942). "A Note on Partially Balanced Incomplete Block Designs." *Science and Culture*, **7**, 568–569.

NANDI, H. K. (1945). "On the Relation Between Certain Types of Tactical Configurations." *Bull. Calcutta Math. Soc.*, **37**, 92–94.

NATIONAL BUREAU OF STANDARDS (1957). "Fractional Factorial Experiment Designs for Factors at Two Levels." *Applied Mathematics Series, No. 48.*

NATIONAL BUREAU OF STANDARDS (1959). "Fractional Factorial Experiment Designs for Factors at Three Levels." *Applied Mathematics Series, No. 54.*

NATIONAL BUREAU OF STANDARDS (1961). "Fractional Factorial Designs for Experiments with Factors at Two or Three Levels." *Applied Mathematics Series, No. 58.*

NELDER, J. A. (1963). "Identification of Contrasts in Fractional Replication of 2^n Experiments." *Applied Statistics*, **12**, 38–43.

NEWMAN, D. (1939). "The Distribution of the Range in Samples from a Normal Population, Expressed in Terms of an Independent Estimate of the Standard Deviation." *Biometrika*, **31**, 20–30.

NEYMAN, J., et al. (1935). "Statistical Problems in Agricultural Experimentation." *J. Roy. Statist. Soc., Suppl.*, **2**, 107–154.

OGAWA, J. (1949). "On the Independence of Bilinear and Quadratic Forms of a Random Sample from a Normal Population." *Ann. Inst. Statist. Math. Tokyo*, **1**, 83–108.

OGAWA, J. (1959). "A Necessary Condition for Existence of Regular and Symmetrical Experimental Designs of Triangular Type, with Partially Balanced Incomplete Blocks." *Ann. Math. Statist.*, **30**, 1063–1071.

PALEY, R. E. A. C. (1933). "On Orthogonal Matrices." *J. Math. Phys.*, **12**, 311–320.

PAPATHANASIOU, G. A., K. J. LESSMAN, and W. E. NYQUIST (1966). "Evaluation of Eleven Introductions of Crambe, *Crambe Abyssinica* Hochst." *Agron. J.*, **58**, 587–589.

PATTERSON, H. D. (1950). "The Analysis of Change-Over Trials." *J. Agri. Sci.*, **40**, 375–380.

PATTERSON, H. D. (1952). "The Construction of Balanced Designs for Experiments Involving Sequences of Treatments." *Biometrika*, **39**, 32–48.

PEARSON, E. S., and H. O. HARTLEY (1951). "Charts of the Power Function of the Analysis of Variance Tests, Derived from the Non-Central F Distribution." *Biometrika*, **38**, 112–130.

PEARSON, E. S., and H. O. HARTLEY (1954). *Biometrika Tables for Statisticians*, vol. 1, 176–177.

PEART, J. N. (1968). "Lactation Studies with Blackface Ewes and Their Lambs." *J. Agri. Sci.*, **70**, 87–94.

PETERSEN, E. M. (1952). "Controlling Tobacco Sucker Growth with Maleic Hydrazide." *Agron. J.*, **44**, 332–334.

PLACKETT, R. L. (1950). "Some Theorems in Least Squares." *Biometrika*, **37**, 149–157.

PLACKETT, R. L. (1960). *Regression Analysis*. Oxford University Press, London.

PLACKETT, R. L., and J. P. BURMAN (1946). "The Design of Optimum Multifactorial Experiments." *Biometrika*, **33**, 305–325.

QUENOUILLE, M. H. (1953). *The Design and Analysis of Experiment*. Charles Griffin and Company, London.

QUENOUILLE, M. H. (1955). "Checks on the Calculation of Main Effects and Interactions in a 2^n Factorial Experiment." *Annals of Eugenics*, **19**, 151–152.

RAESE, J. T., and A. M. DECKER (1966). "Yields, Stand Persistence and Carbohydrate Reserves of Perennial Grasses as Influenced by Spring Harvest Stage, Stubble Height and Nitrogen Fertilization." *Agron. J.*, **58**, 322–326.

RAGHAVARAO, D. (1959). "Some Optimum Weighing Designs." *Ann. Math. Statist.*, **30**, 295–303.

RAGHAVARAO, D. (1960*a*). "A Generalization of Group Divisible Designs." *Ann. Math. Statist.*, **31**, 756–765.

RAGHAVARAO, D. (1960*b*). "On the Block Structure of Certain *PBIB* Designs with Two Associate Classes Having Triangular and L_2 Association Schemes." *Ann. Math. Statist.*, **31**, 787–791.

RAGHAVARAO, D., and K. CHANDRASEKHARARAO (1964). "Cubic Designs." *Ann. Math. Statist.*, **35**, 389–397.

RAO, C. R. (1946). "On Hypercubes of Strength d and a System of Confounding in Factorial Experiments." *Bull. Calcutta Math. Soc.*, **38**, 67–78.

RAO, C. R. (1947*a*). "General Methods of Analysis for Incomplete Block Designs." *J. Amer. Statist. Ass.*, **42**, 541–561.

RAO, C. R. (1947*b*). "Factorial Experiments Derivable from Combinatorial Arrangements of Arrays." *J. Roy. Statist. Soc., Suppl.*, **9**, 128–139.

RAO, C. R. (1961). "A Study of *BIB* Designs with Replications 11 to 15." *Sankhya*, **23**, 117–127.

RAO, C. R. (1962). "A Note on Generalized Inverse of a Matrix with Applications to Problems in Mathematical Statistics." *J. Roy. Statist. Soc.*, Ser. B., **24**, 152–158.

RAO, V. R. (1958). "A Note on Balanced Designs." *Ann. Math. Statist.*, **29**, 290–294.

RAYNER, A. A. (1967). "The Square Summing Check on the Main Effects and Interactions in a 2^n Experiment as Calculated by Yates's Algorithm." *Biometrics*, **23**, 571–573.

ROLLINSON, D. H. L., K. W. HARKER, J. I. TAYLOR, and F. B. LEECH (1956). "Studies on the Habits of Zebu Cattle IV. Errors Associated with Recording Technique." *J. Agron. Sci.*, **46**, 1–5.

ROY, J. (1958). "On the Efficiency Factor of Block Designs." *Sankhya*, **19**, 181–188.

ROY, J., and R. G. LAHA (1957*a*). "On Partially Balanced Linked Block Designs." *Ann. Math. Statist.*, **28**, 488–493.

ROY, J., and R. G. LAHA (1957*b*). "Classification and Analysis of Linked Block Designs." *Sankhya*, **17**, 115–132.

ROY, P. M. (1952). "A Note on the Resolvability of Balanced Incomplete Block Designs." *Bull. Calcutta Statist. Ass.*, **4**, 130–132.

ROY, P. M. (1953). "Hierarchical Group Divisible Incomplete Block Designs with m Associate Classes." *Science and Culture*, **19**, 210–211.

SATTERTHWAITE, F. (1946). "An Approximate Distribution of Estimates of Variance Components." *Biometrics Bulletin*, **2**, 110–114.

SCHEFFÉ, H. (1953). "A Method for Judging All Contrasts in the Analysis of Variance." *Biometrika*, **40**, 87–104.

SCHEFFÉ, H. (1956*a*). "A 'Mixed Model' for the Analysis of Variance." *Ann. Math. Statist.*, **27**, 23–36.

SCHEFFÉ, H. (1956*b*). "Alternative Models for the Analysis of Variance." *Ann. Math. Statist.*, **27**, 251–271.

SCHEFFÉ, H. (1959). *The Analysis of Variance*. John Wiley and Sons, New York.

SCHNEIDER, A. M., and A. L. STOCKETT (1963). "An Experiment to Select Optimum Operating Conditions on the Basis of Arbitrary Preference Ratings." *Chemical Engineering Progress Symposium Series, No. 42*, Vol. 59.

SCHÜTZENBERGER, M. P. (1949). "A Non-Existence Theorem for an Infinite Family of Symmetric Block Designs." *Annals of Eugenics*, **14**, 286–287.

SESHADRI, V. (1963*a*). "Constructing Uniformly Better Estimators." *J. Amer. Statist. Ass.*, **58**, 172–175.

SESHADRI, V. (1963*b*). "Combining Unbiased Estimators." *Biometrics*, **19**, 163–170.

SHRIKHANDE, S. S. (1950). "The Impossibility of Certain Symmetrical Balanced Incomplete Block Designs." *Ann. Math. Statist.*, **21**, 106–111.

SHRIKHANDE, S. S. (1952). "On the Dual of Some Balanced Incomplete Block Designs." *Biometrics*, **8**, 66–72.

SHRIKHANDE, S. S. (1959*a*). "On a Characterization of the Triangular Association Scheme." *Ann. Math. Statist.*, **30**, 39–47.

SHRIKHANDE, S. S. (1959*b*). "The Uniqueness of the L_2 Association Scheme." *Ann. Math. Statist.*, **30**, 781–798.

SINGH, N. K., and K. N. SINGH (1964). "The Non-Existence of Some Partially Balanced Incomplete Block Designs with Three Associate Classes." *Sankhya*, **24**, 239–250.

SMITH, C. A. B., and H. O. HARTLEY (1948). "The Construction of Youden Squares." *J. Roy. Statist. Soc.*, Ser. B., **10**, 262–263.

SPRAGUE, G. F., and L. A. TATUM (1942). "General vs. Specific Combining Ability in Single Crosses of Corn." *J. Amer. Soc. Agron.*, **34**, 923–932.

SPROTT, D. A. (1954). "A Note on Balanced Incomplete Block Designs." *Canad. Jour. Math.*, **6**, 341–346.

SPROTT, D. A. (1956*a*). "A Note on Combined Interblock and Intrablock Estimation in Incomplete Block Designs." *Ann. Math. Statist.*, **27**, 633–641. [Erratum vol. 27, 269 (1957)].

SPROTT, D. A. (1956*b*). "Some Series of Balanced Incomplete Block Designs." *Sankhya*, **17**, 185–192.

SPROTT, D. A. (1962). "A Listing of *BIB* Designs from $r = 16$ to 20." *Sankhya*, **24**, 203–204.

STEEL, R. G. D. (1960). "A Rank-Sum Test for Comparing All Pairs of Treatments." *Technometrics*, **2**, 197–207.

STEINER, J. (1853). "Eine Kombinatorische Aufgabe." *Crelle's J. Reine und Angew. Math.*, **45**, 181–182.

STEVENS, W. L. (1939). "The Completely Orthogonalized Latin Square." *Annals of Eugenics*, **9**, 82–93.

TANDON, R. K. (1949). "The Response of Flax to Rates and Formulations of 2,4-Dichlorophenoxyacetic Acid." *Agron. J.*, **41**, 213–218.

TANG, P. C. (1938). "The Power Function of the Analysis of Variance Tests with Tables and Illustrations of Their Use." *Statist. Research Memoirs*, **2**, 126–149.

TARRY, G. (1901). "Le problème de 36 officieurs." *Compte Rendu de l'Assoc. Française pour l'Avancement de Science Naturel*, **2**, 170–203.

TUKEY, J. W. (1949). "One Degree of Freedom for Non-Additivity." *Biometrics*, **5**, 232–242.

TUKEY, J. W. (1953). "The Problem of Multiple Comparisons." Dittoed manuscript, 396 pp., Princeton University.

VAJDA, S. (1967*a*). "Patterns and Configurations in Finite Spaces." *Griffin's Statistical Monographs and Courses, No. 22.* Charles Griffin and Company, London.

VAJDA, S. (1967*b*). "The Mathematics of Experimental Design: Incomplete Block Designs and Latin Squares." *Griffin's Statistical Monographs and Courses, No. 23.* Charles Griffin and Company, London.

VANCE, F. P. (1962). "Optimization Study of Lube Oil Treatment by Process 'X'." *Proc. of Symp. on Application of Statistics and Computers to Fuel and Lubricant Research Problems. Office of the Chief of Ordnance, U.S. Army*, March 13–15, 1962.

VARTAK, M. N. (1959). "The Non-Existence of Certain *PBIB* Designs." *Ann. Math. Statist.*, **30**, 1051–1062.

WEBB, S. R. (1968). "Non-Orthogonal Designs of Even Resolution." *Technometrics*, **10**, 291–300.

WELCH, B. L. (1947). "The Generalisation of 'Student's' Problems When Several Different Population Variances are Involved." *Biometrika*, **34**, 28–35.

WELCH, B. L. (1956). "On Linear Combinations of Several Variances." *J. Amer. Statist. Ass.*, **51**, 132–148.

WIDDOWSON, F. V., A. PENNY, and R. J. B. WILLIAMS (1966). "An Experiment Measuring Effects of N, P, and K Fertilizers on Yield and N, P, and K Contents of Grazed Grass." *J. Agri. Sci.*, **67**, 121–128.

WILK, M. B., and O. KEMPTHORNE (1955). "Fixed, Mixed, and Random Models." *J. Amer. Statist. Ass.*, **50**, 1144–1167.

WILK, M. B., and O. KEMPTHORNE (1956). "Some Aspects of the Analysis of Factorial Experiments in a Completely Randomized Design." *Ann. Math. Statist.*, **27**, 950–985.

WILK, M. B., and O. KEMPTHORNE (1957). "Non-Additivities in a Latin Square Design." *J. Amer. Statist. Ass.*, **52**, 218–236.

WILLIAMS, E. J. (1949). "Experimental Designs Balanced for the Estimation of Residual Effects of Treatments." *Austral. J. Scientific Res.*, Ser. A., **2**, 149–168.

WILLIAMS, E. J. (1950). "Experimental Designs Balanced for Pairs of Residual Effects." *Austral J. Scientific Res.*, Ser. A., **3**, 351–363.

WOODMAN, R. M., and D. A. JOHNSON (1946). "The Nutrition of the Carrot. III Grown in Gravel Soil." *J. Agri. Sci.*, **36**, 10–17.

YANG, C. H. (1966). "Some Designs for Maximal $(+1, -1)$ Determinant of Order $n = 2$ (mod 4)." *Mathematics of Computation*, **20**, 147–148.

YANG, C. H. (1968). "On Designs of Maximal $(+1, -1)$ Matrices of Order $n = 2$ (mod 4)." *Mathematics of Computation*, **22**, 174–180.

YATES, F. (1933). "The Analysis of Replicated Experiments when the Field Results are Incomplete. *Emp. Jour. Exp. Agr.*, **1**, 129–142.

YATES, F. (1936). Incomplete Randomized Blocks. *Annals of Eugenics*, **7**, 121–140.

YATES, F. (1937). "The Design and Analysis of Factorial Experiments." Imperial Bureau of Soil Science, Harpenden, England.

Yates, F. (1940). "The Recovery of Interblock Information in Balanced Incomplete Block Designs." *Annals of Eugenics*, **10**, 317–325.

Yates, F. (1948). "Discussion on a paper by F. Anscombe." *J. Roy. Statist. Soc.*, Ser. A., **111**, 204–205.

Youden, W. J. (1951). "Linked Blocks, A New Class of Incomplete Block Designs." (Abstract.) *Biometrics*, **7**, 124.

Youden, W. J., and J. S. Hunter (1955). "Partially Replicated Latin Squares." *Biometrics*, **11**, 399–405.

Zelen, M. (1954). "A Note on Partially Balanced Designs." *Ann. Math. Statist.*, **25**, 599–602.

Index